Wolfgang Merzkirch (Editor)

Fluid Mechanics of Flow Metering

Wolfgang Merzkirch (Editor)

Fluid Mechanics of Flow Metering

With 150 Figures

Springer

Editor
Wolfgang Merzkirch

Authors

Wolfgang Merzkirch
Universität Duisburg-Essen
FB Maschinenwesen
Strömungslehre
Schützenbahn 70
45117 Essen, Germany

Ernst von Lavante
Universität Duisburg-Essen
FB Maschinentechnik
Strömungsmaschinen
Schützenbahn 70
45117 Essen, Germany

Klaus Gersten
Franz Peters
Venkatesa Vasanta Ram
Universität Bochum
Fakultät für Maschinenbau
Institut Thermo- und Fluiddynamik
Universitätstr. 150
44780 Bochum, Germany

Volker Hans
Universität Duisburg-Essen
FB 12 Maschinenwesen
Mess- und Regelungstechnik
Schützenbahn 70
45117 Essen, Germany

In cooperation with:
Kathrin Kalkühler, Hans-Dieter Papenfuß, Carsten Ruppel, Franka Schneider, Carsten Wildemann, Wei Xiong

ISBN 978-3-642-06076-2 e-ISBN 978-3-540-26725-6

Springer is a part of Springer Science+Business Media

springeronline.com

Cover-Design: Design & Production, Heidelberg
Printed on acid-free paper 61/3020 Rw 5 4 3 2 1 0

Preface

A flow meter is a measuring device for determining the flow rate of a fluid, normally in a pipe, e.g., the flow of natural gas, or oil, or waste water. With these instruments, an immense amount of money is controlled, and thus, flow metering has an enormous economical significance. In an editorial of the journal "Flow Measurement and Instrumentation" (Vol. 1, no. 1, 1989) it was stated that the costs that are controlled by flow meters are estimated as being worldwide of the order of 10,000 billion US dollars (10,000 milliard EURO) per year. One percent uncertainty in the measurement would still represent a considerable value.

The majority of the existing literature on flow metering is authored by electrical and electronic engineers. The emphasis of this literature is on the processing of the signals generated by the fluid flow and on the methods of calibration, by which a relationship between the signals and the desired quantity, the volumetric flow rate in the pipe, is established. For a long time, the field of flow metering was disregarded by fluid mechanicists, although it poses challenging problems of fluid dynamics. In view of this situation, the authors of this book started a special research programme at Universität Essen and Universität Bochum in 1995 that was aimed at providing a more substantial understanding of "fluid-mechanical fundamentals of flow metering", hence the official title of the programme for which financial support was granted by Deutsche Forschungsgemeinschaft ("Forschergruppe Strömungsmechanische Grundlagen der Durchflussmessung", DFG Az. ME 484/29).

This book summarises the results of this research programme. Two problem areas of particular fluid-mechanical interest are emphasised: the generation of the signals by the fluid flow in the pipe, and the measurement of the volumetric flow rate in pipe flow that is not fully developed, a situation that often occurs in practice and where the state of flow deviates from that during calibration of the meter. The discussion focuses on flows at high Reynolds numbers as they are relevant for many industrial processes, e.g., the transport of natural gas. The book starts with a survey of the theory of turbulent pipe flow; deviations from the fully developed state of flow are characterized; the physics of the signal generation are analyzed for a number of known and newly proposed methods of flow metering; and ways of accounting for flow disturbances, i.e., deviations from the fully developed state, either by flow conditioning or correcting the reading of a meter, are presented.

The authors are grateful for the support received from Deutsche Forschungsgemeinschaft, and they would like to thank the many students and doctoral students who have worked with them in this programme. They express their particular thanks and appreciation to Dr. Carsten Ruppel who brought all the manuscripts into the form required by the publisher.

Essen,
April 2004 *W. Merzkirch*

Contents

1

Fully Developed Turbulent Pipe Flow

Klaus Gersten

Institut für Thermo- und Fluiddynamik, Ruhr-Universität Bochum, 44780 Bochum, Germany

Incompressible pipe flows are fully developed if the velocities are independent of the axial coordinate. The theory of turbulent pipe flows at high Reynolds numbers leads to analytical expressions for the velocity profile and the friction factor, which contain free constants. One of the latter is the well-known Kármán constant. The free constants can be determined from the best available experimental data. Final formulae for the velocity distribution and the friction factor are derived. Further effects on these laws are also considered, such as wall roughness and low Reynolds number effects.

1.1 Basic Equations

Fully developed turbulent pipe flows are the basis for most flow meter concepts and flow meters are usually calibrated in fully developed flows. The fully developed turbulent flow of a fluid with constant physical properties (density ρ^*, viscosity μ^*) in a circular pipe with the radius $R^* = d^*/2$ is considered. Here and in the following text, dimensional quantities are marked by an asterisk. To describe this flow, cylindrical coordinates x^*, r^*, θ with corresponding velocities u^*, v^*, w^* are used, see Fig. 1.1.

A flow is designated as fully developed if the distribution of the (time averaged) mean velocity $\overline{u^*}(r^*)$ is independent of x^*. Since the flow is axisymmetric the continuity equation reads

$$\frac{\partial \overline{u^*}}{\partial x^*} + \frac{1}{r^*}\frac{\partial \left(r^* \overline{v^*}\right)}{\partial r^*} = 0 \tag{1.1}$$

and leads to $\overline{v^*} = 0$ (as consequence of the impermeability of the wall). Hence, the two momentum equations reduce to ($\overline{\tau^*}$ and $\overline{v^{*\prime 2}}$ are also independent of x):

$$\frac{1}{r^*}\frac{\mathrm{d}\left(r^* \overline{\tau^*}\right)}{\mathrm{d}r^*} - \frac{\partial \overline{p^*}}{\partial x^*} = 0\ , \tag{1.2}$$

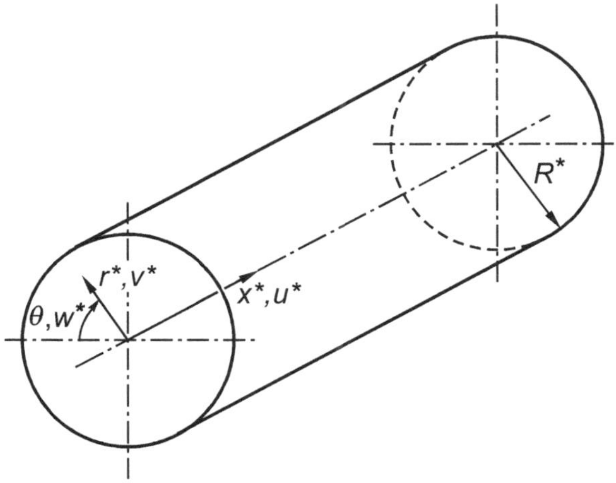

Fig. 1.1. Geometry and coordinate system.

$$\rho^* \left[\frac{\mathrm{d}(\overline{v^{*\prime 2}})}{\mathrm{d}r^*} + \frac{\overline{v^{*\prime 2}} - \overline{w^{*\prime 2}}}{r^*} \right] + \frac{\partial \overline{p^*}}{\partial r^*} = 0 \,, \tag{1.3}$$

where overbars denote the time averaged values while primes denote the turbulent fluctuations. The time averaged pressure $\overline{p^*}(x^*, r^*)$ is not independent of r^* as in the laminar case.

Integrating (1.3) yields

$$\overline{p^*}(x^*, r^*) = \overline{p^*_{\mathrm{w}}} - \rho^* \overline{v^{*\prime 2}} + \rho^* \int\limits_{r^*}^{R^*} \frac{\overline{v^{*\prime 2}} - \overline{w^{*\prime 2}}}{r^*} \mathrm{d}r^* \,. \tag{1.4}$$

The difference between the local pressure in the flow field $\overline{p^*}$ and the wall pressure $\overline{p^*_{\mathrm{w}}}$ is due to the turbulent fluctuations and rather small.

Since the velocity field and, hence, $\overline{v^{*\prime 2}}$ and $\overline{w^{*\prime 2}}$ are independent of x^*, it follows from (1.4)

$$\frac{\partial \overline{p^*}}{\partial x^*} = \frac{\mathrm{d}\overline{p^*_{\mathrm{w}}}}{\mathrm{d}x^*} \tag{1.5}$$

and from (1.2)

$$\frac{1}{r^*} \frac{\mathrm{d}}{\mathrm{d}r^*}(r^* \overline{\tau^*}) - \frac{\mathrm{d}\overline{p^*_{\mathrm{w}}}}{\mathrm{d}x^*} = 0 \,. \tag{1.6}$$

The consequence of this equation is that the axial pressure gradient $\mathrm{d}\overline{p^*_{\mathrm{w}}}/\mathrm{d}x^*$ is constant. Integrating (1.6) leads to

$$\overline{\tau^*} = \frac{\mathrm{d}\overline{p^*_{\mathrm{w}}}}{\mathrm{d}x^*} \frac{r^*}{2} \tag{1.7}$$

and particularly for $r^* = R^*$ to

$$\overline{\tau_{\mathrm{w}}^*} = \frac{\mathrm{d}\overline{p_{\mathrm{w}}^*}}{\mathrm{d}x^*}\frac{R^*}{2} . \tag{1.8}$$

Hence, the wall shear stress $\overline{\tau_{\mathrm{w}}^*}$ is proportional to the pressure gradient and can easily be determined experimentally by measuring the pressure gradient.

It follows from (1.7) and (1.8)

$$\overline{\tau^*} = \frac{r^*}{R^*}\overline{\tau_{\mathrm{w}}^*} , \tag{1.9}$$

i.e., the local shear stress $\overline{\tau^*}$ is proportional to the radius r^*.

If the (total) shear stress $\overline{\tau^*}$ is split into a viscous and a turbulent part

$$\overline{\tau^*} = \overline{\tau_{\mathrm{v}}^*} + \tau_{\mathrm{t}}^* = \rho^*\nu^*\frac{\mathrm{d}\overline{u^*}}{\mathrm{d}r^*} - \rho^*\overline{u^{*\prime}v^{*\prime}} \tag{1.10}$$

($\nu^* = \mu^*/\rho^*$ = kinematic viscosity) the force balance in axial direction, (1.9), leads to

$$\rho^*\nu^*\frac{\mathrm{d}\overline{u^*}}{\mathrm{d}r^*} + \tau_{\mathrm{t}}^* = \frac{r^*}{R^*}\overline{\tau_{\mathrm{w}}^*} . \tag{1.11}$$

Note 1.1 *(Sign of Shear Stress)* The sign of the shear stress depends on the coordinate system chosen. The shear stress in the fully developed circular pipe flow is negative if cylindrical coordinates are used, cf. Fig. 1.1. In this case, the velocity component $v^{*\prime}$ is positive in radially outward direction.

Equation (1.11) will be applied to solve the following problem: the pressure gradient $\mathrm{d}\overline{p_{\mathrm{w}}^*}/\mathrm{d}x^*$ (and hence the wall shear stress $\overline{\tau_{\mathrm{w}}^*}$ according to (1.8)) and the quantities R^*, ρ^* and ν^* are prescribed and the distribution of the velocity gradient

$$\frac{\mathrm{d}\overline{u^*}}{\mathrm{d}r^*} = f\left(r^*, R^*, \nu^*, \frac{\overline{\tau_{\mathrm{w}}^*}}{\rho^*}\right) \tag{1.12}$$

has to be found. Integration leads finally to the velocity distribution $\overline{u^*}(r^*)$ and to the volume flux.

Using the dimensionless quantities

$$r = \frac{r^*}{R^*} , \quad u^+ = \frac{\overline{u^*}}{u_\tau^*} , \quad \tau_{\mathrm{t}}^+ = \frac{\tau_{\mathrm{t}}^*}{\overline{\tau_{\mathrm{w}}^*}} , \quad u_\tau^* = \sqrt{\frac{-\overline{\tau_{\mathrm{w}}^*}}{\rho^*}} , \quad Re_\tau = \frac{R^* u_\tau^*}{\nu^*} \tag{1.13}$$

reduces (1.11) to

$$-\frac{1}{Re_\tau}\frac{\mathrm{d}u^+}{\mathrm{d}r} + \tau_{\mathrm{t}}^+ = r . \tag{1.14}$$

Now the function $\mathrm{d}u^+/\mathrm{d}r = F(r, Re_\tau)$ which also follows from (1.12) by dimensional analysis must be found. This function is here of particular interest for high Reynolds numbers Re_τ.

It should be mentioned that (1.14) is not sufficient to solve the problem. In addition to (1.14) a turbulence model is required that generates a relation between $\tau_{\mathrm{t}}^+(r)$ and $\mathrm{d}u^+/\mathrm{d}r$ to close the system of equations.

1.2 Two-Layer Structure for $\mathrm{Re}_\tau \to \infty$

For high Reynolds numbers Re_τ equation (1.14) reduces to

$$\tau_{\mathrm{t}}^{+} = r \qquad \text{(core region)} . \tag{1.15}$$

This equation is valid in the core region where the turbulent shear stress τ_{t}^* is much larger than the viscous shear stress $\overline{\tau_{\mathrm{v}}^*}$. This equation, however, does not satisfy the condition $\tau_{\mathrm{t}}^{+} = 0$ at the wall which follows from the no-slip condition. Hence, (1.15) is not valid near the wall where $\overline{\tau_{\mathrm{v}}^*}$ is not negligible any more. Hence, the flow field has a two-layer structure. It consists of the core region where (1.15) is valid and viscosity effects are negligible, and the wall layer where the viscous shear stress has the same order of magnitude as the turbulent shear stress.

The two regions clearly have thicknesses of different order of magnitude. Whereas the thickness of the core region is of the order of magnitude of R^*, a *wall-layer thickness* δ_{v}^* may be determined from the two characteristic quantities ν^* and u_τ^*

$$\delta_{\mathrm{v}}^* = \frac{\nu^*}{u_\tau^*} = \frac{R^*}{Re_\tau} . \tag{1.16}$$

This tends to zero for high Reynolds numbers Re_τ. The wall layer thickness is therefore small compared to R^* for high values Re_τ. Consequently, the flow quantities in the wall layer are *independent* of R^* in the leading order.

In the following the velocity gradient will be found separately for the wall layer and for the core layer. The two solutions must be matched properly, i.e., they have to agree in an *overlap layer*.

1.3 Wall Layer

In order to describe the wall layer it is suitable to introduce a dimensionless wall distance

$$y^{+} = \frac{y^*}{\delta_{\mathrm{v}}^*} = \frac{y^* u_\tau^*}{\nu^*} = Re_\tau(1 - r) . \tag{1.17}$$

With this coordinate (1.14) changes into

$$\frac{\mathrm{d}u^{+}}{\mathrm{d}y^{+}} + \tau_{\mathrm{t}}^{+} = 1 - \frac{y^{+}}{Re_\tau} . \tag{1.18}$$

Since the second term on the right side vanishes for high Reynolds numbers, the wall layer is a layer of constant total shear stress (equal to the wall shear stress), where the effect of the pressure gradient can be neglected. It is also independent of R^*, as mentioned earlier. Hence, the velocity $u^{+}(y^{+})$ is determined when $\overline{\tau_{\mathrm{w}}^*}/\rho^*$ and ν^* are given. It is therefore a universal function, called the *universal law of the wall* $u^{+}(y^{+})$. This universal law is valid not only for turbulent pipe flow but also for all turbulent flows with finite wall shear stress, cf. Schlichting and Gersten (2000).

Some properties of this law are known a priori. The velocity gradient $\mathrm{d}u^+/\mathrm{d}y^+$ must become independent of the viscosity for large wall distance (in the overlap layer). This yields:

$$\lim_{y^+\to\infty} \frac{\mathrm{d}u^+}{\mathrm{d}y^+} = \frac{1}{\kappa y^+} \qquad \text{(overlap layer)}\ , \tag{1.19}$$

where the constant κ is called *Kármán constant* after v. Kármán (1930).

Integrating (1.19) leads to the well-known *universal logarithmic law*

$$\lim_{y^+\to\infty} u^+(y^+) = \frac{1}{\kappa} \ln\, y^+ + C^+ \qquad \text{(overlap layer)}\ . \tag{1.20}$$

As will be shown in Sect. 1.7, C^+ generally depends on the wall roughness.

The no-slip condition and the continuity equation yield the asymptote close to the wall:

$$\lim_{y^+\to 0} \frac{\mathrm{d}u^+}{\mathrm{d}y^+} = 1 - A{y^+}^3 + \ldots \tag{1.21}$$

$$\lim_{y^+\to 0} u^+ = y^+ - \frac{A}{4}{y^+}^4 + \ldots\ . \tag{1.22}$$

The following analytical description for the universal law of the wall satisfies the conditions (1.19) to (1.22); cf. Gersten and Herwig (1992), p. 378:

$$\frac{\mathrm{d}u^+}{\mathrm{d}y^+} = \frac{1}{1+(A+B){y^+}^3} + \frac{B{y^+}^3}{1+\kappa B{y^+}^4} \tag{1.23}$$

$$u^+ = \frac{1}{\Lambda}\left[\frac{1}{3}\ln\frac{\Lambda y^+ + 1}{\sqrt{(\Lambda y^+)^2 - \Lambda y^+ + 1}} + \frac{1}{\sqrt{3}}\left(\arctan\frac{2\Lambda y^+ - 1}{\sqrt{3}} + \frac{\pi}{6}\right)\right] + \frac{1}{4\kappa}\ln\left(1+\kappa B{y^+}^4\right) \tag{1.24}$$

where

$$\Lambda = (A+B)^{\frac{1}{3}}\ ;\quad C^+ = \frac{2\pi}{3\sqrt{3}\Lambda} + \frac{1}{4\kappa}\ln(\kappa B)\ . \tag{1.25}$$

The universal constants κ, C^+ and A can be determined from experiments, cf. Sect. 1.6.2. When κ, C^+ and A are known, the constants Λ and B follow from (1.25).

1.4 Core Region

The distribution $\mathrm{d}u^+/\mathrm{d}r$ in the core region vanishes at the pipe axis ($r = 0$) and must agree with the equivalent wall layer distribution in the overlap layer. It is analogous to (1.19)

$$\lim_{r\to 1} \frac{\mathrm{d}u^+}{\mathrm{d}r} = -\frac{1}{\kappa(1-r)} \qquad \text{(overlap layer)} . \tag{1.26}$$

If $\mathrm{d}u^+/\mathrm{d}r$ is known, this can be integrated to obtain the velocity distribution in the form of a *velocity defect law* as follows

$$u^+(r) - u_c^+ = \int_0^r \frac{\mathrm{d}u^+}{\mathrm{d}r}\mathrm{d}r . \tag{1.27}$$

In this equation u_c^+ is the velocity u^+ on the axis or center (maximum velocity).

To determine the velocity gradient, a turbulence modeling is necessary. Boussinesq (1872) proposed that, in analogy to Newton's law of friction, see (1.10), the following ansatz should be used for τ_t^*:

$$\tau_t^* = \rho^* \nu_t^* \frac{\mathrm{d}\overline{u^*}}{\mathrm{d}r^*} \tag{1.28}$$

where ν_t^* is called the *(kinematic) eddy viscosity*.

Using dimensionless quantities according to (1.13) yields

$$\tau_t^+ = -\nu_t^+ \frac{\mathrm{d}u^+}{\mathrm{d}r} \tag{1.29}$$

with the dimensionless eddy viscosity

$$\nu_t^+ = \frac{\nu_t^*}{u_\tau^* R^*} . \tag{1.30}$$

Combining (1.15) and (1.29) leads to the formula for the velocity gradient in the core region

$$\frac{\mathrm{d}u^+}{\mathrm{d}r} = -\frac{r}{\nu_t^+} . \tag{1.31}$$

Comparing (1.26) with (1.31) yields

$$\lim_{r\to 1} \nu_t^+ = \kappa(1-r) \qquad \text{(overlap layer)} . \tag{1.32}$$

The eddy-viscosity distribution $\nu_t^+(r)$ in the core region is a symmetrical function and must satisfy the condition (1.32).

The following function has these properties and will be used as eddy-viscosity distribution:

$$\frac{\nu_t^+(r)}{\kappa} = \frac{\left(1-r^2\right)\left(1+ar^2\right)\left(1+br^2\right)}{2(1+a)(1+b)} \tag{1.33}$$

or

$$\frac{\kappa}{\nu_t^+(r)} = \frac{2}{1-r^2} + \frac{\alpha}{1+ar^2} + \frac{\beta}{1+br^2} . \tag{1.34}$$

These formulae have, beside the Kármán constant κ, the two free parameters a and b, which can be determined by comparison of the resulting velocity distribution with experimental data. The abbreviations α and β depend on a and b as follows:

$$\alpha = \frac{2a^2(1+b)}{a-b} \quad , \quad \beta = \frac{2b^2(1+a)}{b-a} \, . \tag{1.35}$$

Integration of (1.31) can be carried out easily because of (1.34) which leads to the velocity defect distribution

$$u^+(r) - u_c^+ = \frac{1}{\kappa}\left[\ln\left(1-r^2\right) - \frac{\alpha}{2a}\ln\left(1+ar^2\right) - \frac{\beta}{2b}\ln\left(1+br^2\right)\right] \, . \tag{1.36}$$

This distribution shows the logarithmic law in the overlap layer:

$$\lim_{r\to 1} u^+(r) = u_c^+ + \frac{1}{\kappa}\ln(1-r) - \overline{C} \tag{1.37}$$

where

$$\overline{C} = \frac{1}{\kappa}\left[\frac{\alpha}{2a}\ln(1+a) + \frac{\beta}{2b}\ln(1+b) - \ln 2\right] \, . \tag{1.38}$$

In the overlap layer the two solutions (core region and wall layer) must be identical. The matching condition

$$\lim_{r\to 1} u^+(r) = \lim_{y^+\to\infty} u^+\left(y^+\right) \tag{1.39}$$

combines (1.20) and (1.37) and leads because of (1.17) to a dependence of the maximum velocity on the Reynolds number Re_τ:

$$u_c^+ = \frac{1}{\kappa}\ln Re_\tau + C^+ + \overline{C} \, . \tag{1.40}$$

The velocity averaged over the cross section of the pipe can be obtained by integration:

$$u_m^+ = \frac{u_m^*}{u_\tau^*} = \frac{2}{u_\tau^* R^{*\,2}}\int_0^{R^*} \overline{u^*} r^* \mathrm{d}r^* = 2\int_0^1 u^+ r \mathrm{d}r = u_c^+ + \overline{\overline{C}} \tag{1.41}$$

with

$$\begin{aligned}\overline{\overline{C}} &= -2\int_0^1 \left[u_c^+ - u^+(r)\right] r\mathrm{d}r \\ &= -\frac{1}{\kappa}\left[\frac{\alpha}{2a^2}(1+a)\ln(1+a) + \frac{\beta}{2b^2}(1+b)\ln(1+b)\right] \, .\end{aligned} \tag{1.42}$$

The average velocity or *bulk velocity* shows the following dependence on Re_τ when (1.40) and (1.41) are combined

$$\boxed{u_m^+ = \frac{1}{\kappa}\ln Re_\tau + C^+ + \overline{C} + \overline{\overline{C}} \, .} \tag{1.43}$$

This is the *friction law* for turbulent pipe flow at high Reynolds numbers Re_τ in case that the pressure gradient (wall shear stress) is given and the volumen flux (averaged velocity) is to be determined.

It is

$$\kappa = 0.421\ , \qquad C^+ + \overline{C} + \overline{\overline{C}} = 2.55 \tag{1.44}$$

according to the latest results of the Superpipe Experiments at Princeton University, see McKeon et al. (2003b). The velocity distribution according to (1.36) is shown in Fig. 1.2.

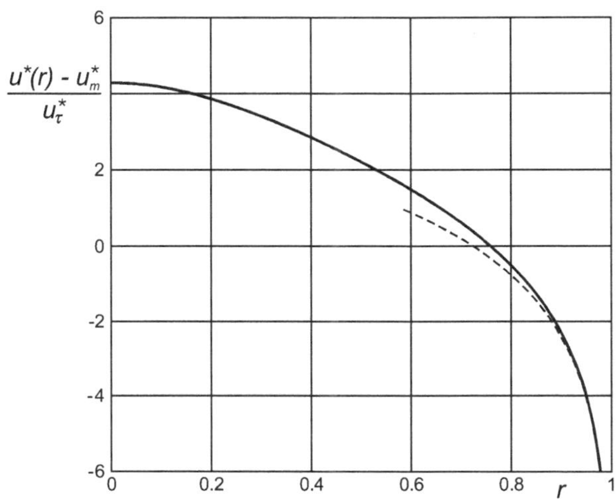

Fig. 1.2. Velocity defect distribution for fully developed pipe flow according to (1.36), (1.40) and (1.43). (- - - -) Asymptote for $r \to 1$ according to (1.37) and (1.41).

1.5 Friction Law for Given Volume Flux

When the volume flux is given the friction law is taken to be the dependence of the pipe *friction factor*

$$\lambda = -\frac{8\overline{\tau_{\mathrm{w}}^*}}{\rho^* u_{\mathrm{m}}^{*\,2}} = \frac{8}{u_{\mathrm{m}}^{+\,2}} = -\frac{d^* \mathrm{d}\overline{p_{\mathrm{w}}^*}/\mathrm{d}x^*}{\rho^* u_{\mathrm{m}}^{*\,2}/2} \tag{1.45}$$

on the Reynolds number based on the pipe diameter d^*

$$Re = \frac{u_{\mathrm{m}}^* d^*}{\nu^*} = 2Re_\tau \sqrt{\frac{8}{\lambda}}\ . \tag{1.46}$$

Inserting λ and Re into (1.43) leads to the *implicit* form of the friction law for given Reynolds number Re

$$\boxed{\frac{1}{\sqrt{\lambda}} = C_1 \log\left(Re\sqrt{\lambda}\right) - C_2} \tag{1.47}$$

where

$$C_1 = \frac{\ln 10}{\sqrt{8}\kappa}, \qquad C_2 = \frac{1}{\sqrt{32}\kappa}\left[\ln 32 - 2\kappa\left(C^+ + \overline{C} + \overline{\overline{C}}\right)\right] . \tag{1.48}$$

It is

$$C_1 = 1.934 , \qquad C_2 = 0.554 \tag{1.49}$$

by using the constants in (1.44). It is worth mentioning that these constants are valid in the range $3 \times 10^5 \leq Re \leq 10^7$, see McKeon et al. (2003b). For $Re < 3 \times 10^5$, lower Reynolds number effects must be taken into account. This will be explained in Sect. 1.8. The data points in the Superpipe Experiments beyond $Re = 10^7$ have obviously been affected by roughness as will be discussed in Sect. 1.7.

The *explicit* form of (1.47) is

$$\lambda = 8\left[\frac{\kappa}{\ln Re} G(\Lambda; D)\right]^2 \tag{1.50}$$

with the G-function after Gersten and Herwig (1992), p. 782. This newly introduced function $G(\Lambda; D)$ is defined by

$$\frac{\Lambda}{G} + 2\ln\left(\frac{\Lambda}{G}\right) - D = \Lambda \tag{1.51}$$

and is tabulated in the mentioned reference. It satisfies the asymptotic condition

$$\lim_{\Lambda \to \infty} G(\Lambda; D) = 1 . \tag{1.52}$$

In the application here, it was set

$$\Lambda = 2\ln Re , \qquad D = 2\left[\ln \kappa + \kappa\left(C^+ + \overline{C} + \overline{\overline{C}}\right)\right] . \tag{1.53}$$

It is $D = 0.417$ by using (1.44).

1.6 Velocity Distribution

1.6.1 General Formula

After the velocity distributions have been determined for the two regions in the flow field, i.e., the $u^+_{\mathrm{WL}}(y^+)$ according to (1.24) for the wall layer and $u^+_{\mathrm{CR}}(r)$ according to (1.36) for the core region, a *composite solution* can be constructed, cf. Van Dyke (1975). The so-called *additive composition* is the sum of these two solutions which is corrected by subtracting the part they have in common, i.e., the solution in the overlap layer according to (1.20):

$$u^+ (r, Re_\tau) = u^+_{\mathrm{CR}}(r) + u^+_{\mathrm{WL}} \left(y^+\right) - \frac{1}{\kappa} \ln y^+ - C^+ \tag{1.54}$$

where y^+ and r are connected by (1.17).

When (1.24), (1.36) and (1.40) are used, the composite velocity distribution has the form:

$$\begin{aligned} u^+(r) = {} & \frac{1}{\kappa} \left[\ln(1+r) - \frac{\alpha}{2a} \ln\left(1 + ar^2\right) - \frac{\beta}{2b} \ln\left(1 + br^2\right) \right] + \overline{C} \\ & + \frac{1}{\Lambda} \left[\frac{1}{3} \ln \frac{\Lambda y^+ + 1}{\sqrt{(\Lambda y^+)^2 - \Lambda y^+ + 1}} + \frac{1}{\sqrt{3}} \left(\arctan \frac{2\Lambda y^+ - 1}{\sqrt{3}} + \frac{\pi}{6} \right) \right] \\ & + \frac{1}{4\kappa} \ln\left(1 + \kappa B y^{+4}\right) \end{aligned} \tag{1.55}$$

where

$$y^+ = \frac{Re\sqrt{\lambda}}{2\sqrt{8}}(1 - r) \; . \tag{1.56}$$

The average velocity according to (1.41) has been determined in (1.42) by taking into account only the velocity distribution in the core region. When the integration is carried out over the composite velocity distribution, the relevant constant $\overline{\overline{C}}$ turns out to be

$$\overline{\overline{C}} = \overline{\overline{C}}_{\mathrm{CR}} + \Delta\overline{\overline{C}} \; . \tag{1.57}$$

Here $\overline{\overline{C}}_{\mathrm{CR}}$ is the constant in (1.42), when only the velocity distribution in the core region is taken into account. The effect of the velocity distribution in the wall layer is given by the correction term $\Delta\overline{\overline{C}}$, for which it follows

$$\Delta\overline{\overline{C}} = -\frac{2}{Re_\tau} \int\limits_0^\infty \left[\frac{1}{\kappa} \ln y^+ + C^+ - u^+_{\mathrm{WL}} \left(y^+\right) \right] \mathrm{d}y^+ \; . \tag{1.58}$$

Using (1.24) for $u^+_{\mathrm{WL}} \left(y^+\right)$ with the constants $\kappa = 0.421$; $C^+ = 5.60$; $A = 6 \times 10^{-4}$ (the values of C^+ and A are explained later) leads to

$$\Delta\overline{\overline{C}} = -\frac{102}{Re_\tau} \; , \tag{1.59}$$

see also Gersten and Herwig (1992), p. 630.

For Reynolds numbers higher than $Re = 10^6$ ($Re_\tau > 2378$) the correction $\Delta\overline{\overline{C}}$ results in a reduction of u^+_{m} according to (1.43) which is smaller than 0.2 percent. Only for Reynolds numbers less than $Re = 3 \times 10^5$ the correction $\Delta\overline{\overline{C}}$ should be taken into account, see also Sect. 1.8.

1.6.2 Determination of the Model Constants

The velocity distribution (1.55) contains five free constants κ, C^+, a, b, and A. These can be determined by comparison with experiments as follows:

1. From the friction law, (1.43) or (1.47), κ and $C^+ + \overline{C} + \overline{\overline{C}}$ can be found.

2. The velocity defect distribution (1.36) shows a quadratic behavior near the axis

$$\lim_{\eta\to 0}\left[u^+(r) - u_{\mathrm{c}}^+\right] = -\frac{2+\alpha+\beta}{2\kappa}r^2 = -\frac{(1+a)(1+b)}{\kappa}r^2 \,. \tag{1.60}$$

Note 1.2 *(Alternative for (1.60))* Instead of using (1.60) one could take the location where $u^+ = u_{\mathrm{m}}^+$ as one condition to determine the model constants. Experiments at high Reynolds numbers show for this location $r \approx 0.75$, see Fig. 1.2.

3. The combination of (1.35), (1.41) and (1.42) results in

$$u_{\mathrm{c}}^+ - u_{\mathrm{m}}^+ = -\overline{\overline{C}} = \frac{(1+a)(1+b)}{\kappa(a-b)}\ln\frac{1+a}{1+b}\,. \tag{1.61}$$

When κ has been determined via the friction law, the constants a and b can be calculated from (1.60) and (1.61). Then $\overline{C}$ follows from (1.38) and C^+ from the combination $C^+ + \overline{C} + \overline{\overline{C}}$, found via the friction law.

4. The fifth constant A is given by

$$A = -\frac{1}{6}\lim_{y^+\to 0}\frac{\mathrm{d}^4 u^+}{\mathrm{d}y^{+4}} \tag{1.62}$$

because of (1.22). When (1.24) is applied for the law of the wall, A is connected with the velocity u^+ at $y^+ = 15$ by

$$A = \frac{u^+\,(y^+ = 15) - 4.4}{1.07}10^{-4}\,. \tag{1.63}$$

The newest results of the Superpipe Experiments at Princeton University yield

$$\begin{aligned} &\kappa = 0.421 \quad , C^+ + \overline{C} + \overline{\overline{C}} = 2.55\,, \\ &\overline{\overline{C}} = -4.28\,,\, C^+ = 5.60 \qquad\qquad , \end{aligned} \tag{1.64}$$

where C^+ was determined by using (1.20). From these constants it follows

$$\overline{C} = 1.23\,, \tag{1.65}$$

which is in very good agreement with the Superpipe Experiments by using (1.37), although McKeon et al. (2004) found the slightly different value $\overline{C} = 1.20 \pm 0.1$.

By applying the values κ, $\overline{C}$ and $\overline{\overline{C}}$, (1.38) and (1.42) lead to

$$a = -0.2714 \qquad b = 5.567 \tag{1.66}$$

and via (1.35) to

$$\alpha = -0.1656 \qquad \beta = 7.735\,. \tag{1.67}$$

These constants have been applied to calculate the universal velocity distribution for high Reynolds numbers according to the combination of (1.36) and (1.41). The result is shown in Fig. 1.2.

The constant A can be found in the literature as a number in the interval $5 \leq A \times 10^4 \leq 7$, see Gersten and Herwig (1992), p. 377. Numerous measurements have brought out $u^+\,(y^+ = 15) = 10.6$ cf. Reichardt (1951) and Kestin and Richardson (1963). At higher Reynolds number, this value is only slightly higher. McKeon et al. (2003a) found $u^+\,(y^+ = 15) = 10.7$. Therefore applying (1.63) leads to the rough estimation

$$A = 6 \times 10^{-4}\,. \tag{1.68}$$

Note 1.3 *(The Tragicomedy of Fluid Mechanics)* Fluid mechanics is supposed to be a great science. Outstanding progress in theoretical, numerical and experimental fluid mechanics has been made in the last decades.
Ironically enough, however, for the simplest technical flow one can think of, namely the incompressible, fully developed turbulent flow in a straight circular pipe, a reliable friction law is not available yet. Millions of such flows have been realized under extremely precise conditions in the numerious national Offices of Weights and Measures or Bureau of Standards but, unfortunately, only the volume flow has been measured, not simultaneously the pressure gradient. The opposite case, measuring the pressure gradient, but not the volume flow, is also available, see AGARD-AR-345 (1998), Case PCH02. As a result only a few experimental data for the friction law are available. The uncertainty, however, is unacceptable. In the recent collection of test data in the AGARD-Report just mentioned (selected for the validation of Large-Eddy-Simulations!) the most important Kármán constant κ varies in a range between $\kappa = 0.39$ and $\kappa = 0.436$. It is hoped that the constants, chosen in this chapter, will bring us very close to the truth.

Note 1.4 *(Power-Law Velocity Distributions)* Sometimes the velocity distributions of fully developed turbulent pipe flows have been modeled by power laws. For example, Nikuradse (1932) tried power law representations and found a dependence of the exponents on the Reynolds number. But he abandoned this mode of correlation as inferior to the log-law representation described in this chapter.
Recently, Barenblatt (1993) reviewed the power-law model and rationalized the form of the power law

$$u^+ = B(Re)\left[y^+\right]^{\alpha(Re)} \tag{1.69}$$

by using scaling arguments, see also Barenblatt et al. (1997a, b). On the basis of the data of Nikuradse (1932) he proposed expressions for $B(Re)$ and $\alpha(Re)$ as functions of the Reynolds number. Unfortunately, the power-law model for the velocity distribution is seriously in error near the pipe axis and close to the pipe wall. These errors might have been anticipated from the prediction by (1.69) of an unbounded velocity gradient at the wall and a finite velocity gradient at the centerline. Afzal (2001) has shown that power law and log-law velocity profiles are equivalent when the power law velocity distribution is restricted to the overlap layer, see also Barenblatt and Chorin (1998). The drawback of the power-law model is again that the matching condition in the overlap layer leads to a function of the Reynolds number rather than to the universal Kármán constant. For practical applications, the log-law model is obviously much more attractive than the power-law model, see also Churchill (2001).

1.7 Influence of Wall Roughness

1.7.1 Hydraulically Defined Roughness

So far in the analysis of this chapter it has been assumed that the inner pipe wall is smooth. It turns out that everything is valid for rough walls as well, with one exception: the integration constant C^+ according to (1.20) now depends on the wall roughness.

From dimensional analysis it follows that $C^+(k^+)$ is a function of the dimensionless *roughness parameter* (sometimes called *roughness Reynolds number*)

$$k^+ = \frac{k^* u_\tau^*}{\nu^*} \, . \tag{1.70}$$

Here k^* is the *roughness height* or simply *roughness*. It can be interpreted geometrically in connection with the well-known experiments by Nikuradse (1933). For his investigations on artificially roughened pipes, Nikuradse coated the inside of the pipes with carefully graded sand grains, glued in place. He used the mesh size of the grading screens as roughness k^*. Therefore, k^* is also called *sandgrain roughness* or *Nikuradse roughness.*

The roughness Reynolds number k^+ is the ratio of the Nikuradse roughness k^* and the wall-layer thickness δ_v^* defined in (1.16). It is plausible that the viscosity effect must disappear completely when $k^* \gg \delta_\mathrm{v}^*$ or $k^+ \to \infty$. In this case, the friction factor must become independent of the viscosity. This so-called *fully rough pipe flow* is governed by *Kármán's law*

$$u_\mathrm{m}^+ = \sqrt{\frac{8}{\lambda}} = \frac{1}{\kappa} \ln \frac{3.71 d^*}{k^*} \, . \tag{1.71}$$

Inversion of this law leads to

$$\frac{k^*}{d^*} = 3.71 \exp\left[-\sqrt{8}\kappa/\sqrt{\lambda}\right] \, , \tag{1.72}$$

which defines the roughness k^*. For a given condition of the internal wall, each pipe has a roughness k^*. However, k^* is not defined in a geometrical but rather in a hydraulical sense. It can be determined via (1.72) by measuring the friction factor λ at a high enough Reynolds number such that fully rough pipe flow is guaranteed. In principle, λ must be measured at several different Reynolds numbers to ensure that it becomes independent of the Reynolds number.

Combining (1.43) and (1.71) leads to

$$\lim_{k^+\to\infty} C^+\left(k^+\right) = \frac{2.004}{\kappa} - \overline{C} - \overline{\overline{C}} - \frac{1}{\kappa}\ln k^+ \,. \tag{1.73}$$

When the constants of the Superpipe Experiments (1.64) are used, the limits of the function $C^+\left(k^+\right)$ are:

$$\lim_{k^+\to 0} C^+(k^+) = 5.60 \qquad \text{hydraulically smooth} \tag{1.74}$$

$$\lim_{k^+\to\infty} C^+(k^+) = 7.81 - \frac{1}{\kappa}\ln k^+ \qquad \text{fully rough} \tag{1.75}$$

It should be mentioned that k^* according to (1.72) depends on the Kármán constant κ being used. Hence, any roughness k^* determined with $\kappa \neq 0.407$ is not directly comparable with the Nikuradse roughness ($\kappa = 0.407$) as it is usually found in the literature. This, however, is not so detrimental because the friction factor is not very sensitive with respect to the roughness.

Now the following question arises: What is the correct formula for the function $C^+\left(k^+\right)$ in the transition regime between smooth flow and fully rough flow? Such a formula must satisfy the two limiting cases for smooth flow ($k^+ \to 0$) and fully rough flow ($k^+ \to \infty$).

Roughness may be classified in various ways. For instance, roughness elements may be discrete or distributed and their geometry may be deterministic (regular) or stochastic (random). Experiments have shown that for many roughness geometries a single parameter is inadequate to specify the roughness effect on the flow. Parameters such as the roughness height, shape, density and manner of distribution may be important in determining the net effect of the roughness on the pipe flow, see Waigh and Kind (1998). For all possible roughness geometries, the so-called *equivalent sandgrain roughness* can be determined by using (1.72) in fully rough flow. But various roughness geometries show in general a different behavior in the transition regime between smooth and fully rough flow.

1.7.2 Natural Roughness

Colebrook (1938/1939), in collaboration with White [see (Colebrook and White, 1937)] found that the parameter k^+ is sufficient to describe most forms of *naturally rough commercial pipes* in the transition regime.

Therefore, in the following only naturally rough commercial pipes with the technical roughnesss will be considered. According to Colebrook the function $C^+\left(k^+_{\text{tech}}\right)$ is a simple combination of the two limiting cases mentioned above

$$\begin{aligned} C^+(k^+_{\text{tech}}) &= 7.81 - \frac{1}{\kappa}\ln\left(2.54 + k^+_{\text{tech}}\right) \\ &= 5.60 - \frac{1}{\kappa}\ln\left(1 + \frac{k^+_{\text{tech}}}{2.54}\right) \,. \end{aligned} \tag{1.76}$$

Inserting this formula for $C^+\left(k^+_{\text{tech}}\right)$ in (1.47) and using the constants (1.64) leads to the following friction law for rough commercial pipes

$$\frac{1}{\sqrt{\lambda}} = -1.934\log\left[\frac{k^*/d^*}{3.71} + \frac{1.934}{Re\sqrt{\lambda}}\right] . \tag{1.77}$$

In the limiting case $k^*/d^* \to 0$ or $k^+ \to 0$ at finite Reynolds number, (1.77) reduces to (1.47). In the other limiting case $Re \to \infty$ at finite k^*/d^* or $k^+ \to \infty$ (1.77) is reduced to (1.71).

The roughness function (1.76) after Colebrook-White leads for small values k^+_{tech} to

$$\lim_{k^+_{\text{tech}}\to 0} C^+(k^+_{\text{tech}}) = 5.60 - k^+_{\text{tech}}/(2.54\kappa) . \tag{1.78}$$

However, Bradshaw (2000) has shown that the correct form is

$$\lim_{k^+_{\text{tech}}\to 0} C^+\left(k^+_{\text{tech}}\right) = 5.60 - \alpha\, k^{+2}_{\text{tech}} . \tag{1.79}$$

The coefficient α for commercial pipes is not known yet. But it can be estimated from the Superpipe Experiment, cf. Zagarola and Smits (1998) and McKeon et al. (2003b). This particular pipe ($d^* = 0.129\,\text{m}$) had the roughness $k^*_{\text{tech}} = 0.45\,\mu\text{m}$ and the flow was hydraulically smooth for $Re \leq 10^7$ or for $k^+_{\text{tech}} \leq 1$. The correct smoothness condition can be derived from (1.43) and (1.79)

$$k^{+2}_{\text{tech}} \leq \frac{\sqrt{2}}{\alpha}\frac{1}{\sqrt{\lambda}}\frac{\Delta\lambda}{\lambda} \approx \frac{14}{\sqrt{\lambda}}\frac{\Delta\lambda}{\lambda} \tag{1.80}$$

when $\alpha \approx 0.1$ was chosen so that (1.80) is valid for the Superpipe Experiments ($Re = 10^7, 1/\sqrt{\lambda} = 11, \Delta\lambda/\lambda = 0.00005/0.0082 = 0.006, k^+ \leq 1$). This formula shows clearly that the smoothness criterion for surfaces with sandgrain roughness, $k^+ \leq 5$, cannot be applied for natural roughness of commercial pipes. The behavior of these two roughness types have a quite different behavior in the transition regime between hydraulically smooth and fully rough flows.

The roughness effects can therefore be divided into the following three regimes

$$\left.\begin{array}{lll} \text{hydraulically smooth} & 0 \leq k^+_{\text{tech}} \leq \sqrt{\left(14/\sqrt{\lambda}\right)\left(\Delta\lambda/\lambda\right)} & \\ \text{transition region} & \sqrt{\left(14/\sqrt{\lambda}\right)\left(\Delta\lambda/\lambda\right)} < k^+_{\text{tech}} < 420\sqrt{\lambda}/\left(\Delta\lambda/\lambda\right) & \\ \text{fully rough} & 420\sqrt{\lambda}/\left(\Delta\lambda/\lambda\right) \leq k^+_{\text{tech}} & \end{array}\right\} \tag{1.81}$$

Here $\Delta\lambda/\lambda$ is the relative uncertainty of the friction factor.

1.7.3 Velocity Distribution

The velocity defect distribution for rough walls is the same as for smooth walls, see (1.36). But the maximum velocity u^+ according to (1.40) is reduced by roughness because of (1.76). Therefore, the rough-wall velocity profile when plotted in wall-layer coordinates (u^+, y^+) is *shifted downward* by a constant amount

$$\begin{aligned}\Delta u^+ &= u^+_{\text{smooth}} - u^+_{\text{rough}} = C^+(0) - C^+(k^+) \\ &= \left[\sqrt{\frac{8}{\lambda_{\text{smooth}}}} - \sqrt{\frac{8}{\lambda_{\text{rough}}}}\right]_{Re_\tau=\text{const}} = \frac{1}{\kappa}\ln\left(1 + \frac{k^+_{\text{tech}}}{2.54}\right)\end{aligned} \tag{1.82}$$

The function $\Delta u^+\left(k^+_{\text{tech}}\right)$ is called *roughness function.*

If the surface is rough, it is difficult to determine the origin of the coordinate system $y = 0$. It is common practice to choose this origin so that the overlap law (1.20) is satisfied, cf. Grigson (1984), Acharya et al. (1986).

1.7.4 Geometric Roughness Parameters

To describe the geometric profile of a rough surface, the following geometric roughness parameters are in use:

a) Arithmetic mean roughness

$$R^*_{\text{a}} = \frac{1}{\ell^*}\int\limits_0^{\ell^*} |y^*(x) - y^*_{\text{m}}|\,\mathrm{d}x^* \tag{1.83}$$

where $y^*\left(x^*\right)$ describes the roughness profile and y^*_{m} is the mean of $y^*\left(x^*\right)$:

$$y^*_{\text{m}} = \frac{1}{\ell^*}\int\limits_0^{\ell^*} y^*\left(x^*\right)\mathrm{d}x^*\ . \tag{1.84}$$

b) Root-mean-square roughness

$$R^*_{\text{q}} = \sqrt{\frac{1}{\ell^*}\int\limits_0^{\ell^*} \left[y^*\left(x^*\right) - y^*_{\text{m}}\right]^2 \mathrm{d}x^*} \tag{1.85}$$

This value is also written as R^*_{rms} .

c) Maximum profile valley depth

$$R^*_{\text{t}} = R^*_{\text{max}} = y^*_{\text{max}} - y^*_{\text{min}} \tag{1.86}$$

for a given reference length ℓ^*.

d) Average maximum profile valley depth (also: ten point height of irregularities)

$$R_z^* = \frac{1}{5}\sum_{i=1}^{5} (y_{max}^* - y_{min}^*)_i \tag{1.87}$$

In this case the reference length ℓ^* is divided into five lengths $\ell_i^* = \ell^*/5$ and for each length ℓ_i^* the difference between the peak y_{max}^* and the valley y_{min}^* is determined.

Table 1.1. Examples of the ratio of the geometric roughness parameters R_a^*, R_q^*, R_z^*, R_t^* for surfaces of commercial steel pipes. Effects of painting, sanding, polishing and coating.

Surface	Reference	R_a^* / μm	R_q^*/R_a^*	R_z^*/R_a^*	R_t^*/R_a^*
steel	Gersten et al. (2000)	8.5	1.3	6.8	8.0
steel painted unsanded or sanded	Schultz (2002)	0.4 – 2.7	1.8	14	16
steel painted sanded polished	Schultz (2002)	0.18	1.7	11	12
steel coated	Gersten et al. (2000)	3.9	1.3	6.3	7.9

Table 1.1 shows examples of geometric roughness parameters and their ratios. The geometric roughness parameters can be determined by using mechanical profilometers. Laser interferometry can also be applied which leads to about 12% higher roughness parameters.

1.7.5 Relation Between Hydraulic and Geometric Roughness Parameters

There is general agreement that a relation between the hydraulic roughness parameter k_{tech}^* and the geometric roughness parameters in (1.83) to (1.87) exists. But only very few investigations have led to quantitative results for this relation. The results of investigations of the natural roughness of commercial pipes are summarized in Table 1.2. From these data it follows that the natural roughness of commercial pipes is approximately

$$k_{tech}^* = 3.5 R_a^* \, . \tag{1.88}$$

It is worth mentioning that in DIN EN ISO 5167-1 (January 2004) the relation $k_{tech}^* = \pi R_a^*$ is used. The correlation of k_{tech}^* with other geometric roughness parameters can be estimated by the results in Table 1.1.

Table 1.2. Relationship between roughness k^*_{tech} of commercial pipes and geometric roughness R^*_{a} according to various authors.

Reference	$k^*_{\text{tech}}/R^*_{\text{a}}$
Speidel (1962)	3.1
Uhl (1965)	3.3
Acharya et al. (1986)	4.2
Revell (1998)	3.1
Zagarola and Smits (1998)	3.9

1.8 Low Reynolds Number Pipe Flow

In the previous Sections pipe flows at high Reynolds numbers have been treated. The empirical constants according to the Superpipe Experiment in (1.44) and (1.64) turned out to be valid only for about $Re \geq 3 \times 10^5$. In this Section, turbulent pipe flows for $Re < 3 \times 10^5$ will therefore be dicussed.

Pipe flows for Reynolds numbers below about $Re = 2000$ are laminar. For pipe Reynolds numbers between 2000 and about 4000 the friction factor can have large uncertainties and is essentially indeterminate. This region is, therefore, called *critical zone*, cf. Benedict (1980). Detailed experimental investigations, for instance by Rotta (1956), have shown that the flow in this zone has an *intermittent character*. This means that the flow is occasionally laminar and occasionally turbulent. Hence, the velocity distribution alternates between a corresponding fully developed laminar and a corresponding fully developed turbulent flow. However, for $Re > 4000$ the pipe flow becomes steady and reasonably determinate. It will be therefore discussed next, how the analysis presented so far can be extended to lower Reynolds numbers in the range $4000 < Re < 3 \times 10^5$.

As already mentioned the analysis in the previous Sections referred to pipe flows at very high Reynolds numbers. These solutions can be extended systematically to lower Reynolds numbers. For that purpose the solutions are set as asymptotic series where $\epsilon = 1/Re_\tau$ is the perturbation parameter. The leading terms of these series are the solutions presented so far. The terms that follow in the asymptotic series represent the higher order terms which describe the deviations from the known asymptotic solutions at lower Reynolds numbers. Details of the procedure for such singular perturbation problem are given in Gersten and Herwig (1992).

The extension of the friction-factor formula (1.43) to lower Reynolds number is as follows:

$$u^+_{\text{m}} = \frac{1}{\kappa} \ln Re_\tau + \widetilde{C}_1 + \frac{1}{Re_\tau}\left(\frac{1}{\kappa_2} \ln Re_\tau + \widetilde{C}_2\right) + \ldots \tag{1.89}$$

where

$$\widetilde{C}_1 = C^+ + \overline{C} + \overline{\overline{C}}\,. \tag{1.90}$$

The higher order term contains two more constants $1/\kappa_2$ and $\widetilde{C}_2$. The correction term (1.59) is included in $\widetilde{C}_2$. The new constants can again be determined by comparison with experimental data. Afzal and Yajnik (1973) have made such an adjustment and found:

$$1/\kappa = 2.5 \quad ; \quad \widetilde{C}_1 = 1.28 \quad ; \quad 1/\kappa_2 = 0 \quad ; \quad \widetilde{C}_2 = 75 \tag{1.91}$$

see also Afzal (1976). Since nowadays experimental data on pipe flows over a wider range of Reynolds numbers are available, the numbers in (1.91) must obviously be revised.

When (1.89) is written in the form

$$u_{\mathrm{m}}^{+} = \left(\frac{1}{\kappa} + \frac{1}{Re_\tau}\frac{1}{\kappa_2}\right)\ln Re_\tau + \left(\widetilde{C}_1 + \frac{1}{Re_\tau}\widetilde{C}_2\right) \tag{1.92}$$

it is clear that the function $u_{\mathrm{m}}^{+}(Re_\tau)$ plotted in a semi-logarithmic diagram is in general not any more a straight line. In order to determine the two new constants in the higher-order term, reliable data in the lower Reynolds number range must be applied. In AGARD-AR-345 (1998) a selection of test cases of turbulent pipe flows have been collected for the validation of computer codes.

For the following two test cases sufficient information is available:

1. Durst et al. (1995):

$$Re = 7442 \quad , \quad Re_\tau = 250 \quad , \quad 1/\sqrt{\lambda} = 5.26 \tag{1.93}$$

2. den Toonder (1995):

$$Re = 24580 \quad , \quad Re_\tau = 691 \quad , \quad 1/\sqrt{\lambda} = 6.29 \tag{1.94}$$

   It is worth mentioning that this data point satisfies exactly the well-known friction law by Blasius (1913).

Taking these data and those of the Superpipe Experiments, cf. McKeon et al. (2004), the following constants can be determined:

$$1/\kappa = 2.375 \quad ; \quad \widetilde{C}_1 = 2.55 \quad ; \quad 1/\kappa_2 = 0 \quad ; \quad \widetilde{C}_2 = -201\,. \tag{1.95}$$

This leads to the friction law for the so-called transmission factor $1/\sqrt{\lambda}$ valid for *all* turbulent pipe flows above $Re = 4000$ (hydraulically smooth):

$$\frac{1}{\sqrt{\lambda}} = 0.8398 \ln Re_\tau + 0.9016 - \frac{71}{Re_\tau} \tag{1.96}$$

or

$$\boxed{\frac{1}{\sqrt{\lambda}} = 1.934 \log\left(Re\sqrt{\lambda}\right) - 0.554 - \frac{402}{Re\sqrt{\lambda}}} \tag{1.97}$$

The comparison of the formula (1.96) with the experimental data is shown in Fig. 1.3. It should be mentioned that Zagarola and Smits (1998) found empirically the higher-order term in (1.97) as $-233/\left(Re\sqrt{\lambda}\right)^{0.9}$, which can be considered as a confirmation of the analysis presented here. It is worth noting that Wosnik et al. (2000) worked out a slightly different analysis, but ended up with a similar higher-order term proportional to $(\ln Re_\tau)^{-0.441}$.

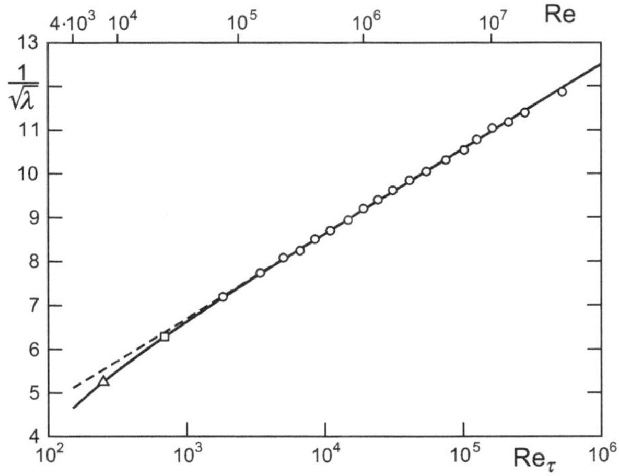

Fig. 1.3. Friction law for turbulent pipe flows for $Re > 4000$ (smooth surface). (- - - -) Asymptote for $Re \to \infty$ according to (1.43), $u_\mathrm{m}^+ = \sqrt{8/\lambda}$; (——) Formula (1.96); (△) Experiment by Durst et al. (1995); (□) Experiment by den Toonder (1995); (○) Experiment by McKeon et al. (2004).

References

Acharya M, Bornstein J, Escudier MP (1986) Turbulent boundary layers on rough surfaces. Experiments in Fluids 4:33–47

Afzal N (1976) Millikan's argument at moderately large Reynolds number. Phys. Fluids 19:600–602

Afzal N (2001): Power law and log law velocity profiles in fully developed turbulent pipe flow: equivalent relations at large Reynolds numbers. Acta Mechanica, Vol. 151:171–183

Afzal N, Yajnik K (1973) Analysis of turbulent pipe and channel flows at moderately large Reynolds number. J. Fluid Mech. 61:23–31

AGARD (1998) A selection of test cases for the validation of large-eddy simulations of turbulent flows. AGARD Advisory Report 345

Barenblatt GI (1993) Scaling laws for fully developed turbulent shear flows. Part I: Basic hypothesis and analysis. J. Fluid Mech. 248:513–520

Barenblatt GI, Chorin AJ (1998) Scaling of the intermediate region in wall-bounded turbulence: The power law. Phys. Fluids, Vol. 10:1043–1044

Barenblatt GI, Chorin AJ, Prostokishin VM (1997a) Scaling laws for fully developed turbulent flow in pipes. Appl. Mech. Rev., Vol. 50:413–429

Barenblatt GI, Chorin AJ, Prostokishin VM (1997b): Scaling laws for fully developed turbulent flow in pipes: Discussion of experimental data. Proc. Natl. Acad. Sci. USA, Vol. 94, 773

Benedict RP (1980) Fundamentals of Pipe Flow. John Wiley Sons, New York

Blasius H (1913) Das Ähnlichkeitsgesetz bei Reibungsvorgängen in Flüssigkeiten. Forschg. Arb. Ing.-Wesen, Heft 131, Berlin

Boussinesq J (1872) Essai sur la theorie des eaux courantes. Memoires Acad. des Sciences, Vol. 13, No. 1, Paris

Bradshaw P (2000) A note on "critical roughness height" and "transitional roughness". Phys. Fluids 12:1611–1614

Churchill SW (2001) Turbulent flow and convection: the prediction of turbulent flow and convection in a round tube. In: Advances in Heat Transfer, Academic Press, San Diego, Vol. 34:255–361

Colebrook CF (1938/1939) Turbulent flow in pipes with particular references to the transition region between the smooth and the rough pipe laws. J. Inst. Civil Eng., London, 11:133–156 and 12:393–422

Colebrook CF, White CM (1937) Experiments with fluid friction in roughened pipes. Proc. Royal Soc., London, Series A 161:367–381

den Toonder JMJ (1995) Drag reduction by polymer additives in a turbulent pipe flow: Laboratory and numerical results. Ph. D. thesis, Delft University of Technology

Durst F, Jovanovic J, Sender J (1995) LDA measurements in the near-wall region of a turbulent pipe flow. J. Fluid Mech. 295:305–335

Gersten K, Herwig H (1992) Strömungsmechanik - Grundlagen der Impuls-, Wärme- und Stoffübertragung aus asymptotischer Sicht. Vieweg-Verlag, Braunschweig, Wiesbaden

Gersten K, Papenfuss HD, Kurschat T, Genillon P, Fernández Pérez F, Revell N (2000) New transmission-factor formula proposed for gas pipelines. Oil & Gas Journal, February 14

Grigson CWB (1984) Nikuradse's experiment. AIAA Journal 22:999–1001

Kestin J, Richardson PD (1963) Heat transfer across turbulent incompressible boundary layers. Int. J. Heat Mass Transfer 6:147–189. Also: Forsch. Ing.-Wesen 29:93–104

McKeon BJ, Li J, Jiang W, Morrison JF, Smits A (2003a) Pitot probe corrections in fully-developed turbulent pipe flow. Meas. Sci. Technol. 14:1449–1458

McKeon BJ, Morrison JF, Jiang W, Li J, Smits AJ (2003b) Revised log-law constants for fully-developed turbulent pipe flow. In: AJ Smits (Ed.): IUTAM Symposium on Reynolds Number Scaling in Turbulent Flow. Kluwer Academic Publishers, Dordrecht

McKeon BJ, Li J, Jiang W, Morrison JF, Smits AJ (2004): Further observations on the mean velocity in fully-developed pipe flow. J. Fluid Mech. 501:135–147

Nikuradse J (1932) Gesetzmäßigkeiten der turbulenten Strömung in glatten Rohren. VDI-Forsch. Arb. Ing.-Wesen

Nikuradse J (1933) Strömungsgesetze in rauhen Rohren. VDI-Forsch.-Heft 361; VDI-Verlag, Berlin

Revell N (1998) Internal Report of BG Technology, PR 091

Reichardt H (1951) Vollständige Darstellung der turbulenten Geschwindigkeitsverteilung in glatten Leitungen. Z. angew. Math. Mech. 31:203–219

Rotta JC (1956) Experimenteller Beitrag zur Entstehung turbulenter Strömung im Rohr. Ing. Arch. 24:258–281

Schlichting H, Gersten K (2000) Boundary-Layer Theory. Springer-Verlag, Berlin, Heidelberg; 8th Edition, Corrected Printing 2003

Schultz MP (2002) The relationship between frictional resistance and roughness for surfaces smoothed by sanding. J. Fluids Eng. 124:492–499

Speidel L (1962) Determination of the necessary surface quality and possible losses due to roughness in steam turbines. Elektrizitätswirtschaft 61:799–804

Uhl AE (1965) Steady flow in gas pipelines. American Gas Association. Technical Report No. 10

Van Dyke M (1975) Perturbation Methods in Fluid Mechanics. The Parabolic Press, Stanford, California von Kármán Th (1930) Mechanische Ähnlichkeit und Turbulenz. Nachr. Ges. Wiss. Göttingen. Math. Phys. Klasse: 58-76 und Verhandlg. des III. Intern. Kongresses für Techn. Mechanik, Stockholm, Teil I:85–93

Waigh DR, Kind RJ (1998) Improved aerodynamic characterization of regular three-dimensional roughness. AIAA J. 36: 1117-1119

Wosnik M, Castillo L, George WK (2000) A theory for turbulent pipe and channel flows. J. Fluid Mech. 421:115-145

Zagarola MW, Smits AJ (1998) Mean-flow scaling of turbulent pipe flow. J. Fluid Mech. 373:33–79

2

Decay of Disturbances in Turbulent Pipe Flow

Klaus Gersten, Heinz-Dieter Papenfuss

Institut für Thermo- und Fluiddynamik, Ruhr-Universität Bochum, 44780 Bochum, Germany

The basic equations for turbulent flows in circular pipes at high Reynolds numbers are derived. At a certain cross section a three-dimensional velocity field is prescribed that differs only slightly from the fully developed state. The axial decay of the disturbances is investigated by solving the basic equations using an asymptotic expansion of the solutions. The resulting system of linear ordinary differential equations leads to eigenvalue problems. Eigenvalues and eigenfunctions are presented. The general solution can be expanded into series of eigensolutions. The decaying flow consists of various flow structures that fall into two categories: the U-type (ring vortex, source-sink pair, source-sink quadrupole, etc.) and the W-type (single, double and multiple streamwise vortices).

Each of these flow structures has its own decay behavior. Swirl (single streamwise vortex) has the lowest decay followed by the source-sink pair. The strength of the individual flow structure is described by means of characteristic parameters that are the coefficients in the series expansions of the solution. Comparisons of the theory with experimental results from the literature show very good agreement.

The different dependence of the decay rate on the Reynolds number for flows with and without swirl can be clarified. The theory is valid for smooth as well as for rough pipe walls.

2.1 Introduction

Flows in straight pipes are often not fully developed. Deviations from the fully developed state may occur due to bends, valves or other pipe fittings. Even swirl can be found downstream of a pair of bends installed in perpendicular planes. The decay of the deviations from the fully developed condition is of great practical importance. For example, these disturbances can cause significant errors in the measurement of the mass flow rate. It is therefore important to know which length of a straight pipe is required for a given upstream flow disturbance to reduce to a specified acceptable deviation from the fully developed flow condition. It is also worth knowing the additional pressure drop due to the disturbances.

Experimental as well as theoretical investigations on such disturbed turbulent pipe flows have been carried out to a great extent. The simplest case is the axisymmetric flow without swirl as it occurs, for instance, in the final stage of a turbulent pipe inlet flow. The experimental studies by Klein (1981), Barbin and Jones (1963) and Laws et al. (1987) and the theoretical work by Cebeci and Chang (1978) and Voigt (1995) are worth mentioning.

More recent investigations on axisymmetric pipe flow with swirl have been carried out by Kitoh (1991), Parchen (1993), Reader-Harris (1994), Steenbergen (1995) and Steenbergen and Voskamp (1998). The general case of three-dimensional disturbances in the pipe has been studied only in a few experimental investigations, for instance by Norman et al. (1989), Sudo et al. (1998), Kalkühler (1998) and Mickan (1999). Theoretical studies with general results on the decay of three-dimensional disturbances in a pipe are not available apart from the work by Gersten and Klika (1998) for the three-dimensional laminar case.

In order to quantify the intensity of various flow disturbances in the pipe certain characteristic parameters (sometimes denoted as "flow indices") are used. The axisymmetric flow without swirl can be characterized by the *blockage factor*, which is essentially the ratio between the maximum velocity and the average velocity; see Klein (1981). The swirl intensity can be determined by the *swirl number*, which is a dimensionless radial moment of the circumferencial momentum flux. Corresponding characteristic parameters can be defined for three-dimensional disturbances; see Sudo et al. (1998) and Mickan (1999).

Many experiments have shown that these characteristic parameters decay downstream in form of an exponential function: $S(x) = S_0 \exp(-\alpha x/D)$, where S stands for the swirl number or any other characteristic parameter describing the flow disturbances. Experimental as well as theoretical investigations showed a dependence of α on the Reynolds number. With respect to swirl, α turned out to be proportional to the friction factor λ of the fully developed pipe flow according to Norman et al. (1989), Reader-Harris (1994), Steenbergen (1995) and Steenbergen and Voskamp (1998). For the axisymmetric disturbance without swirl, however, the coefficient α was found to be proportional to the square-root of the friction factor λ according to Herwig and Voigt (1995) and Voigt (1995). The same behavior was found for the three-dimensional disturbance downstream of an elbow (essentially two counter-rotating longitudinal vortices) according to Norman et al. (1989).

The purpose of the present investigation is to treat general three-dimensional disturbances in turbulent pipe flow in a similar way as it was done by Gersten and Klika (1998) for the laminar case. Of particular interest is the correct description of the decaying of the various disturbances as function of the Reynolds number.

2.2 Basic Equations

2.2.1 Two-Layer Structure

The turbulent incompressible steady flow in a straight circular pipe is considered. At high Reynolds numbers the turbulent pipe flow consists of two regions: 1) the core region where the viscous effects are negligible compared to the turbulent momentum transfer and 2) the viscous wall layer. The latter is characterized by a universal untwisted velocity distribution depending on the local shear stress τ_w^* and the viscosity ν^*; see Schlichting and Gersten (2000). This is in complete analogy to the two-layer structure of turbulent three-dimensional boundary layers as described in the textbook by Schlichting and Gersten (2000) on page 638. The basic equations for the core region are therefore free of viscous effects. The boundary conditions for these equations are not the conditions at the pipe wall, but rather the matching conditions with the wall-layer solution, i.e., the well-known logarithmic velocity distribution in the overlap layer of the core region and wall region.

2.2.2 Core Region

The basic equations in the core region in vector form are as follows; see Schlichting and Gersten (2000), page 73.
Continuity equation

$$\operatorname{div} \vec{w}^* = 0 \tag{2.1}$$

momentum equation

$$\operatorname{grad}\left(\frac{1}{2}\,\vec{w}^{*2}\right) - \vec{w}^* \times \vec{\omega}^* = -\frac{1}{\rho^*}\operatorname{grad} p^* + \operatorname{Div} \boldsymbol{\tau}_t^* \tag{2.2}$$

where

$$\vec{\omega}^* = \operatorname{curl} \vec{w}^* \tag{2.3}$$

is the vorticity and $\boldsymbol{\tau}_t^*$ the turbulent stress tensor. Dimensional quantities are marked by an asterisk. A cylindrical coordinate system is used, see Fig. 2.1. The following

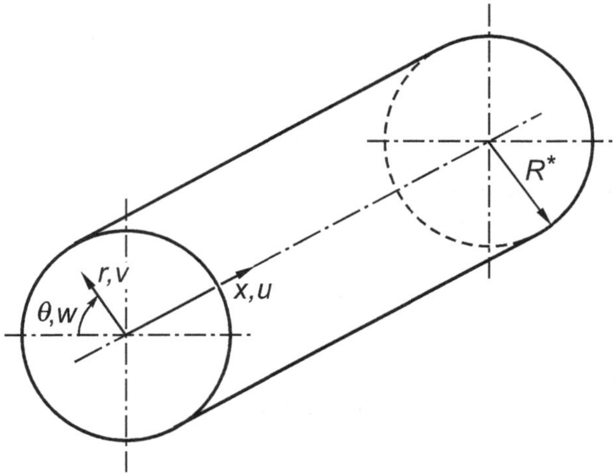

Fig. 2.1. Geometry and coordinate system.

dimensionless coordinates are introduced:

$$x = \frac{x^*}{L^*} \qquad r = \frac{r^*}{R^*} \tag{2.4}$$

where R^* is the pipe radius and L^* is a length in axial direction characteristic for the decaying of the flow disturbances considered. The ratio

$$\epsilon_1 = \frac{R^*}{L^*} \tag{2.5}$$

is assumed to be a small number. The axial velocity component is based on the mean velocity or *bulk velocity*

$$u^*_{\mathrm{m}} = \frac{1}{\pi\, R^{*2}} \int_0^{2\pi} \int_0^{R^*} u^*(x^*, r^*, \theta) r^* \mathrm{d}r^* \mathrm{d}\theta \; . \tag{2.6}$$

To avoid a degeneration of the continuity equation, the velocity components v^* and w^* are based on $\epsilon_1 u^*_{\mathrm{m}}$. In other words, these velocity components are small compared with u^*_{m}.

An eddy-viscosity concept is applied to the turbulent stress tensor. Therefore, the basic equations for the fully turbulent core region are the same as those for laminar flow, when the viscosity μ^* is replaced by the eddy viscosity μ^*_{t}. Hence, the following dimensionless quantities are used:

$$\begin{gathered} u = \frac{u^*}{u^*_{\mathrm{m}}} \qquad v = \frac{v^*}{\epsilon_1 u^*_{\mathrm{m}}} \qquad w = \frac{w^*}{\epsilon_1 u^*_{\mathrm{m}}} \qquad p = \frac{p^* - p^*_0}{\rho^* u^{*2}_{\mathrm{m}}} \; , \\ \nu^+_{\mathrm{t}} = \frac{\mu^*_{\mathrm{t}}}{\rho^* R^* u^*_\tau} = \frac{\mu^*_{\mathrm{t}}}{\rho^* R^* u^*_{\tau\infty}} \frac{1}{\gamma} \; . \end{gathered} \tag{2.7}$$

The eddy viscosity is based on the local skin-friction velocity u^*_τ defined by the (negative) local resultant wall shear stress

$$\tau^*_{\mathrm{w}}(x^*, \theta) = -\rho^* u^{*2}_\tau (x^*, \theta) \; . \tag{2.8}$$

The skin-friction velocity $u^*_\tau(x^*, \theta)$ tends to the value $u^*_{\tau\infty} = u^*_\tau(x^* \to \infty; \theta)$ of the fully developed flow far downstream. The function

$$\gamma(x^*, \theta) = \frac{u^*_\tau(x^*, \theta)}{u^*_{\tau\infty}} \tag{2.9}$$

tends to 1 for $x^* \to \infty$.

From the asymptotic theory for the fully developed turbulent pipe flow it is well-known that the value

$$\epsilon_2 = \frac{u^*_{\tau\infty}}{u^*_{\mathrm{m}}} = \sqrt{\frac{\lambda}{8}} = \mathrm{O}\left(\frac{1}{\ln Re}\right) \tag{2.10}$$

is a small number, too, when the Reynolds number

$$Re = \frac{\rho^* u_{\mathrm{m}}^* 2R^*}{\mu^*} \tag{2.11}$$

becomes large. The friction factor λ is a measure of the friction resistance in the fully developed pipe flow.

In principle, the flow problem under consideration is a two-parameter problem with the two small perturbation parameters ϵ_1 and ϵ_2.

By using the dimensionless quantities, the basic equations for the core region in the cylindrical coordinate system are as follows:

$$\frac{\partial u}{\partial x} + \frac{1}{r}\frac{\partial (rv)}{\partial r} + \frac{1}{r}\frac{\partial w}{\partial \theta} = 0 \tag{2.12}$$

$$\begin{aligned} u\frac{\partial u}{\partial x} + v\frac{\partial u}{\partial r} + \frac{w}{r}\frac{\partial u}{\partial \theta} &= -\frac{\partial p}{\partial x} + 2\epsilon_1\epsilon_2\frac{\partial}{\partial x}\left(\gamma\nu_{\mathrm{t}}^+\frac{\partial u}{\partial x}\right) \\ &+\frac{\epsilon_2}{\epsilon_1}\left\{\frac{1}{r}\frac{\partial}{\partial r}\left[r\gamma\nu_{\mathrm{t}}^+\left(\frac{\partial u}{\partial r} + \epsilon_1^2\frac{\partial v}{\partial x}\right)\right] + \frac{1}{r}\frac{\partial}{\partial \theta}\left[\gamma\nu_{\mathrm{t}}^+\left(\frac{1}{r}\frac{\partial u}{\partial \theta} + \epsilon_1^2\frac{\partial w}{\partial x}\right)\right]\right\} \end{aligned} \tag{2.13}$$

$$\begin{aligned} u\frac{\partial v}{\partial x} + v\frac{\partial v}{\partial r} + \frac{w}{r}\frac{\partial v}{\partial \theta} - \frac{w^2}{r} &= -\frac{1}{\epsilon_1^2}\frac{\partial p}{\partial r} + \frac{\epsilon_2}{\epsilon_1}\left\{\frac{\partial}{\partial x}\left[\gamma\nu_{\mathrm{t}}^+\left(\frac{\partial u}{\partial r} + \epsilon_1^2\frac{\partial v}{\partial x}\right)\right]\right. \\ &\left.+\frac{2}{r}\frac{\partial}{\partial r}\left(r\gamma\nu_{\mathrm{t}}^+\frac{\partial v}{\partial r}\right) + \frac{1}{r}\frac{\partial}{\partial \theta}\left[\gamma\nu_{\mathrm{t}}^+\left(r\frac{\partial}{\partial r}\left(\frac{w}{r}\right) + \frac{1}{r}\frac{\partial v}{\partial \theta}\right)\right] - \frac{2\gamma\nu_{\mathrm{t}}^+}{r}\left(\frac{1}{r}\frac{\partial w}{\partial \theta} + \frac{v}{r}\right)\right\} \end{aligned} \tag{2.14}$$

$$\begin{aligned} u\frac{\partial w}{\partial x} + v\frac{\partial w}{\partial r} + \frac{w}{r}\frac{\partial w}{\partial \theta} + \frac{vw}{r} &= -\frac{1}{\epsilon_1^2}\frac{1}{r}\frac{\partial p}{\partial \theta} + \frac{\epsilon_2}{\epsilon_1}\left\{\frac{\partial}{\partial x}\left[\gamma\nu_{\mathrm{t}}^+\left(\frac{1}{r}\frac{\partial u}{\partial \theta} + \epsilon_1^2\frac{\partial w}{\partial x}\right)\right]\right. \\ &\left.+\frac{1}{r^2}\frac{\partial}{\partial r}\left[r^2\gamma\nu_{\mathrm{t}}^+\left(r\frac{\partial}{\partial r}\left(\frac{w}{r}\right) + \frac{1}{r}\frac{\partial v}{\partial \theta}\right)\right] + \frac{2}{r}\frac{\partial}{\partial \theta}\left[\gamma\nu_{\mathrm{t}}^+\left(\frac{1}{r}\frac{\partial w}{\partial \theta} + \frac{v}{r}\right)\right]\right\}. \end{aligned} \tag{2.15}$$

Here the solution of the system of equations is of interest for the double limit $\epsilon_1 \to 0$ and $\epsilon_2 \to 0$.

The momentum equation in x-direction degenerates except for the case that the two parameters are coupled such that

$$\epsilon_1 = \epsilon_2 = \epsilon\,. \tag{2.16}$$

This coupling of the viscous effects (ϵ_2) and the geometry (ϵ_1) is typical for the so-called *slender-channel theory*; see Schlichting and Gersten (2000), page 552.

The dimensionless axial coordinate is now built with the square root of the friction factor λ of the fully developed pipe flow

$$x = \epsilon \frac{x^*}{R^*} = \sqrt{\frac{\lambda}{8}} \frac{x^*}{R^*} \,. \tag{2.17}$$

Hence, the dimensionless velocity components v and w are also connected with λ

$$v = \frac{1}{\epsilon} \frac{v^*}{u_{\mathrm{m}}^*} = \sqrt{\frac{8}{\lambda}} \frac{v^*}{u_{\mathrm{m}}^*} \,, \qquad w = \frac{1}{\epsilon} \frac{w^*}{u_{\mathrm{m}}^*} = \sqrt{\frac{8}{\lambda}} \frac{w^*}{u_{\mathrm{m}}^*} \,. \tag{2.18}$$

When terms of the order $O\left(\epsilon^2\right)$ are neglected, (2.12) and (2.13) yield the following equations for the leading order solutions:

$$\frac{\partial u}{\partial x} + \frac{1}{r}\frac{\partial (rv)}{\partial r} + \frac{1}{r}\frac{\partial w}{\partial \theta} = 0 \tag{2.19}$$

$$u\frac{\partial u}{\partial x} + v\frac{\partial u}{\partial r} + \frac{w}{r}\frac{\partial u}{\partial \theta} = -\frac{\partial p}{\partial x} + \frac{1}{r}\frac{\partial}{\partial r}\left(r\gamma\nu_{\mathrm{t}}^+ \frac{\partial u}{\partial r}\right) + \frac{1}{r}\frac{\partial}{\partial \theta}\left(\frac{\gamma\nu_{\mathrm{t}}^+}{r}\frac{\partial u}{\partial \theta}\right) \,. \tag{2.20}$$

The equations (2.14) and (2.15) are reduced accordingly whereby the terms $\epsilon^2 \partial v/\partial x$ in (2.14) and $\epsilon^2 \partial w/\partial x$ in (2.15) are neglected.

Eliminating the pressure in (2.14) and (2.15) leads to the vorticity transport equation for the vorticity component

$$\omega_x = \frac{\omega_x^* R^*}{\epsilon u_{\mathrm{m}}^*} = \frac{1}{r}\left[\frac{\partial (rw)}{\partial r} - \frac{\partial v}{\partial \theta}\right] \,. \tag{2.21}$$

This transport equation has the form:

$$\begin{aligned}
& u\frac{\partial \omega_x}{\partial x} + v\frac{\partial \omega_x}{\partial r} + \frac{w}{r}\frac{\partial \omega_x}{\partial \theta} - \omega_x\frac{\partial u}{\partial x} + \frac{\partial u}{\partial r}\frac{\partial w}{\partial x} - \frac{1}{r}\frac{\partial u}{\partial \theta}\frac{\partial v}{\partial x} = \\
& \frac{1}{r}\frac{\partial}{\partial r}\left\{\frac{1}{r}\frac{\partial}{\partial r}\left[\gamma\nu_{\mathrm{t}}^+ r^2\left(r\frac{\partial}{\partial r}\left(\frac{w}{r}\right) + \frac{1}{r}\frac{\partial v}{\partial \theta}\right)\right] + 2\frac{\partial}{\partial \theta}\left[\gamma\nu_{\mathrm{t}}^+\left(\frac{1}{r}\frac{\partial w}{\partial \theta} + \frac{v}{r}\right)\right]\right\} \\
& -\frac{1}{r}\frac{\partial}{\partial \theta}\left\{\frac{2}{r}\frac{\partial}{\partial r}\left(\gamma\nu_{\mathrm{t}}^+ r\frac{\partial v}{\partial r}\right) + \frac{1}{r}\frac{\partial}{\partial \theta}\left[\gamma\nu_{\mathrm{t}}^+\left(r\frac{\partial}{\partial r}\left(\frac{w}{r}\right) + \frac{1}{r}\frac{\partial v}{\partial \theta}\right)\right]\right. \\
& \left. -\frac{2\gamma\nu_{\mathrm{t}}^+}{r}\left(\frac{1}{r}\frac{\partial w}{\partial \theta} + \frac{v}{r}\right)\right\} + \frac{1}{r}\frac{\partial}{\partial x}\left[\frac{\partial u}{\partial \theta}\frac{\partial (\gamma\nu_{\mathrm{t}}^+)}{\partial r} - \frac{\partial u}{\partial r}\frac{\partial (\gamma\nu_{\mathrm{t}}^+)}{\partial \theta}\right] \,.
\end{aligned} \tag{2.22}$$

As it will turn out later the three equations (2.19), (2.20) and (2.22) are sufficient to determine the three velocity components u, v and w.

2.2.3 Boundary Conditions

As mentioned in Sect. 2.2.1, the boundary conditions for the core region solution follow from matching this solution with the solution in the viscous wall layer. In the overlap layer the well-known logarithmic velocity distribution occurs.

When the resultant velocity in the wall layer is denoted as V^* with the components $u^* = V^* \cos\beta_{\mathrm{w}}$ and $w^* = V^* \sin\beta_{\mathrm{w}}$, the following matching condition results

$$\lim_{r^* \to R^*} \frac{\partial V^*}{\partial r^*} = -\frac{u_\tau^*}{\kappa(R^* - r^*)} \tag{2.23}$$

where u_τ^* was defined in (2.8) and κ is the Kármán constant. Using dimensionless quantities

$$V = \frac{V^*}{u_{\mathrm{m}}^*} \;, \qquad u_\tau = \frac{u_\tau^*}{u_{\mathrm{m}}^*} = \epsilon\gamma \tag{2.24}$$

the boundary conditions for the functions u, v and w follow from (2.23):

$$\lim_{r \to 1} \frac{\partial u}{\partial r} = -\frac{\epsilon\gamma \cos\beta_{\mathrm{w}}}{\kappa(1-r)} \tag{2.25}$$

$$\lim_{r \to 1} v = 0 \tag{2.26}$$

$$\lim_{r \to 1} \frac{\partial w}{\partial r} = -\frac{\epsilon\gamma \sin\beta_{\mathrm{w}}}{\kappa(1-r)\epsilon} = -\frac{\gamma \sin\beta_{\mathrm{w}}}{\kappa(1-r)} \tag{2.27}$$

where $\beta_{\mathrm{w}}(x,\theta)$ is the constant angle between the resultant velocity and the axial velocity component in the wall layer.

Due to the constant resultant shear stress in the wall layer it follows

$$\lim_{r^* \to R^*} \nu_{\mathrm{t}}^* \frac{\partial V^*}{\partial r^*} = -u_\tau^{*2} \tag{2.28}$$

and according to (2.7) and (2.23)

$$\lim_{r \to 1} \nu_{\mathrm{t}}^+ = \kappa(1-r) \,. \tag{2.29}$$

2.3 Fully Developed Pipe Flow

2.3.1 Velocity Distribution in the Core Region

In the fully developed flow (index ∞) $u_\infty(r)$ is independent of x and θ. Because of $w = 0$, $\beta_{\mathrm{w}} = 0$ and $\gamma = 1$ it follows from (2.19) $v = 0$ and from (2.20)

$$\frac{\mathrm{d}p_\infty}{\mathrm{d}x} = \frac{1}{r}\frac{\mathrm{d}}{\mathrm{d}r}\left(r\nu_{\mathrm{t}}^+ \frac{\mathrm{d}u_\infty}{\mathrm{d}r}\right) \,. \tag{2.30}$$

Multiplying by $r\mathrm{d}r$ and integrating over the cross section lead to

$$\frac{1}{2}\frac{\mathrm{d}p_\infty}{\mathrm{d}x} = \lim_{r \to 1}\left(r\nu_{\mathrm{t}}^+ \frac{\mathrm{d}u_\infty}{\mathrm{d}r}\right) = -\epsilon \,, \tag{2.31}$$

where (2.25) and (2.29) are taken into account.

The limiting solution for the pipe flow at infinite Reynolds number is the homogeneous flow with the constant velocity u_{m}^{*}. The solution for large but finite Reynolds numbers is considered a small perturbation of this limiting solution, where ϵ is the perturbation parameter, cf. Schlichting and Gersten (2000). Hence $u_{\infty}(r)$ has the following asymptotic expansion:

$$u_{\infty}(r) = 1 + \epsilon \widetilde{u}_{\infty}(r) + ... \tag{2.32}$$

Combining (2.30), (2.31) and (2.32) ends up with the differential equation for $\widetilde{u}_{\infty}(r)$

$$\frac{\mathrm{d}}{\mathrm{d}r}\left(r\nu_{\mathrm{t}}^{+}\frac{\mathrm{d}\widetilde{u}_{\infty}}{\mathrm{d}r}\right) = -2r\ . \tag{2.33}$$

Integration yields

$$\frac{\mathrm{d}\widetilde{u}_{\infty}}{\mathrm{d}r} = -\frac{r}{\nu_{\mathrm{t}}^{+}(r)}\ . \tag{2.34}$$

If the function $\nu_{\mathrm{t}}^{+}(r)$ is known, (2.34) can be integrated leading to the so-called *defect law*

$$\widetilde{u}_{\infty}(r) - \widetilde{u}_{\infty\mathrm{c}} = \int_{0}^{r} \frac{\mathrm{d}\widetilde{u}_{\infty}}{\mathrm{d}r}\mathrm{d}r = F(r) + \frac{1}{\kappa}\ln(1-r)\ , \tag{2.35}$$

where the function

$$F(r) = \int_{0}^{r}\left[\frac{\mathrm{d}\widetilde{u}_{\infty}}{\mathrm{d}r} + \frac{1}{\kappa(1-r)}\right]\mathrm{d}r \tag{2.36}$$

is not singular for $r \to 1$. It is

$$F(0) = 0\ , \qquad F(1) = \lim_{r\to 1}\int_{0}^{r}\left[\frac{\mathrm{d}\widetilde{u}_{\infty}}{\mathrm{d}r} + \frac{1}{\kappa(1-r)}\right]\mathrm{d}r = -\overline{C}\ . \tag{2.37}$$

It follows from (2.6), (2.7) and (2.32)

$$\lim_{r\to 1}\int_{0}^{r}\widetilde{u}_{\infty}(r)r\mathrm{d}r = 0 \tag{2.38}$$

which determines $\widetilde{u}_{\infty\mathrm{c}} = \widetilde{u}_{\infty}(0)$ in (2.35):

$$\widetilde{u}_{\infty\mathrm{c}} = \frac{3}{2\kappa} - 2\int_{0}^{1}F(r)r\mathrm{d}r = -\overline{\overline{C}}\ . \tag{2.39}$$

The final form of the velocity distribution results from (2.35) as

$$\widetilde{u}_{\infty}(r) = F(r) + \frac{1}{\kappa}\ln(1-r) - \overline{\overline{C}} \tag{2.40}$$

which yields in the overlap layer

$$\lim_{r\to 1}\widetilde{u}_{\infty}(r) = \frac{1}{\kappa}\ln(1-r) - \overline{C} - \overline{\overline{C}}\ . \tag{2.41}$$

2.3.2 Friction Law

The velocity u_∞^* based on the skin friction velocity $u_{\tau\infty}^*$ is connected with $\widetilde{u}_\infty(r)$ by

$$u_\infty^+ = \frac{u_\infty^*}{u_{\tau\infty}^*} = \frac{u_\mathrm{m}^*}{u_{\tau\infty}^*} + \widetilde{u}_\infty \,. \tag{2.42}$$

Matching of u_∞^+ in the overlap layer leads to

$$\lim_{r\to 1} u_\infty^+(r) = \frac{u_\mathrm{m}^*}{u_{\tau\infty}^*} + \lim_{r\to 1} \widetilde{u}_\infty(r) = \lim_{y^+\to\infty} \left(\frac{1}{\kappa}\ln y^+ + C^+\right) \tag{2.43}$$

where

$$y^+ = \frac{(R^* - r^*)u_{\tau\infty}^*}{\nu^*} = (1-r)\frac{R^* u_{\tau\infty}^*}{\nu^*} \tag{2.44}$$

is the wall-layer coordinate and C^+ a constant; see Schlichting and Gersten (2000). Using (2.41), (2.43) and (2.44) yields finally the *friction law*

$$\frac{u_\mathrm{m}^*}{u_{\tau\infty}^*} = \frac{1}{\kappa}\ln\frac{R^* u_{\tau\infty}^*}{\nu^*} + C^+ + \overline{C} + \overline{\overline{C}} \,. \tag{2.45}$$

The friction law is frequently considered to be the dependence of the *friction factor*

$$\lambda = \frac{8u_{\tau\infty}^{*2}}{u_\mathrm{m}^{*2}} \tag{2.46}$$

on the Reynolds number

$$Re = \frac{u_\mathrm{m}^* D^*}{\nu^*} = \frac{u_\mathrm{m}^* 2R^*}{\nu^*} \,. \tag{2.47}$$

From so-called Superpipe Experiments Zagarola and Smits (1998) derived a new friction law which is particularly valid for high Reynolds numbers

$$\frac{1}{\sqrt{\lambda}} = 1.869\log\left(Re\sqrt{\lambda}\right) - 0.241 \,. \tag{2.48}$$

This new friction law is identical with (2.45), when the constants are taken as follows:

$$\kappa = \frac{\ln 10}{1.869\sqrt{8}} = 0.4356$$
$$C^+ = 6.15\ \text{(smooth)}\,, \qquad \overline{C} = 1.51\,, \qquad \overline{\overline{C}} = -4.36\,. \tag{2.49}$$

All four constants have independently been determined from the Superpipe Experiments.

Note 2.1 *(Revised Constants Derived from Superpipe Experiments)* When the calculations had been finished and brought out the results presented in this chapter new and more accurate values of the Kármán constant κ and the additional constants became known, see

McKeon et al. (2003). These improvements of the constants were possible due to corrections of the static pressure measurement error and the velocity gradient Pitot displacement error. The new constants are discussed in Chap. 1 of this book. They are:

$$\kappa = 0.421; \quad C^+ = 5.60; \quad \overline{C} = 1.23; \quad \overline{\overline{C}} = -4.28$$

The calculation have not been repeated by using these new constants, because the authors are convinced, that the results would change only slightly as will be demonstrated later on.

2.3.3 Eddy-Viscosity Distribution

To be in accordance with the friction law, the eddy-viscosity distribution $\nu_t^+(r)$ must satisfy the conditions (2.29), (2.37) and (2.39) with the constants κ, $\overline{C}$ and $\overline{\overline{C}}$ specified in (2.49). The following distribution satisfies all three conditions:

$$\frac{\kappa}{\nu_t^+(r)} = \frac{2}{1-r^2} + \frac{\alpha}{1+ar^2} + \frac{\beta}{1+br^2} \ . \tag{2.50}$$

or

$$\frac{\nu_t^+(r)}{\kappa} = \frac{(1-r^2)(1+ar^2)(1+br^2)}{2(1+a)(1+b)} \tag{2.51}$$

with

$$\alpha = \frac{2a^2(1+b)}{a-b} \qquad \beta = \frac{2b^2(1+a)}{b-a} \ . \tag{2.52}$$

The constants are

$$\begin{aligned} \alpha &= -0.1411 \ , \quad a = -0.2517 \\ \beta &= 9.0919 \ , \quad b = 6.3171 \ . \end{aligned} \tag{2.53}$$

In Fig. 2.2 the eddy-viscosity distribution $\nu_t^+(r)$ is shown. The same distribution, but with the new constants according to McKeon et al. (2003) is also shown for comparison.

Integration of (2.34) can be carried out easily because of (2.51), which leads to the velocity defect distribution:

$$\widetilde{u}_\infty(r) = \frac{1}{\kappa}\left[\ln(1-r^2) - \frac{\alpha}{2a}\ln(1+ar^2) - \frac{\beta}{2b}\ln(1+br^2)\right] - \overline{\overline{C}} \ . \tag{2.54}$$

Its representation in Fig. 2.3 shows also the asymptote for $r \to 1$ according to (2.41). Again the distribution calculated by using the new constants after McKeon et al. (2003) is shown. The deviation is hardly visible.

In principle, the eddy-viscosity distribution should be determined by using a turbulence model. However, one of the most popular two-equation turbulence models, the k-ϵ-model, produces a solution that does not satisfy Zagarola's friction law, see Gersten and Herwig (1992), p. 534. It seems that even by adjusting the model constants it is not possible to obtain a solution in accordance with Zagarola's friction

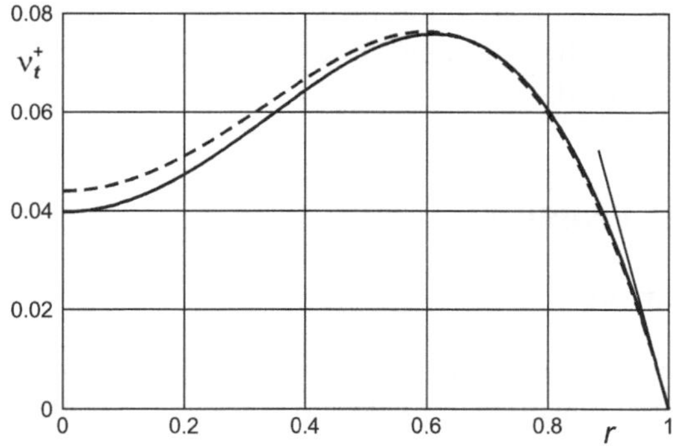

Fig. 2.2. Eddy-viscosity distribution of fully developed pipe flow according to (2.51) and (2.53). Asymptote for $r \to 1$ according to (2.29). (——) constants after Zagarola and Smits (1998). (- - - -) constants after McKeon et al. (2003).

law. The k-ϵ-model is obviously not able to produce the typical distribution of the eddy viscosity known from experiments. The distribution of the eddy viscosity is described quite well by (2.50) which is characterized by a minimum on the pipe axis and a maximum at about $r = 0.6$. In this simple case of fully developed pipe flow, other two-equation turbulence models and even Reynolds-stress models lead more or less to the same form of a differential equation for $\nu_t^+(r)$. It can therefore be concluded that (2.50) is presently the best available description of the eddy viscosity in fully developed pipe flow.

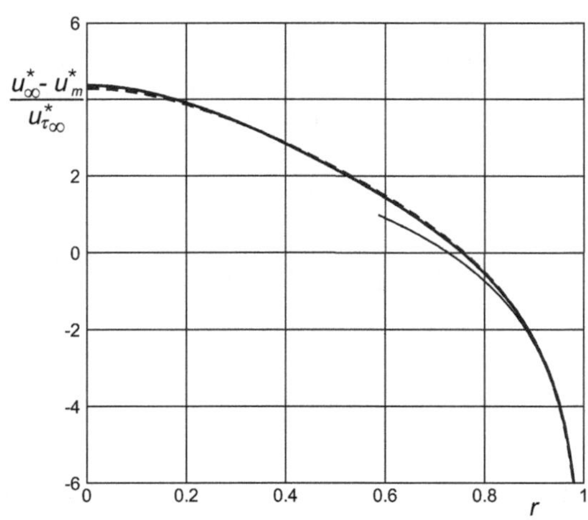

Fig. 2.3. Velocity defect distribution of fully developed pipe flow according to (2.54). Asymptote for $r \to 1$ according to (2.41). (——) constants after Zagarola and Smits (1998); (- - - -) constants after McKeon et al. (2003).

In the following this distribution for $\nu_{\mathrm{t}}^{+}(r)$ will also be applied to pipe flows which deviate from the fully developed state. This is justified as long as only small deviations are considered.

2.4 Eigenvalue Problem

2.4.1 Basic Equations

The general solution of the system of equations (2.19), (2.20) and (2.22) is assumed to be a small perturbation of the fully developed flow. It has the form

$$u(x,r,\theta) = u_\infty(r) + U(r)e^{-\Lambda x}\left[A\cos k\theta + B\sin k\theta\right] \tag{2.55}$$

$$v(x,r,\theta) = V(r)e^{-\Lambda x}\left[A\cos k\theta + B\sin k\theta\right] \tag{2.56}$$

$$w(x,r,\theta) = W(r)e^{-\Lambda x}\left[A\sin k\theta - B\cos k\theta\right] \tag{2.57}$$

$$\frac{\partial p}{\partial x}(x) = \frac{\mathrm{d}p_\infty}{\mathrm{d}x} + \frac{\mathrm{d}P}{\mathrm{d}x}e^{-\Lambda x}A \tag{2.58}$$

$$\gamma(x,\theta) = 1 + c_{\mathrm{U}}e^{-\Lambda x}\left[A\cos k\theta + B\sin k\theta\right]\ . \tag{2.59}$$

Here, A and B are small numbers. Equation (2.59) says that the circumferential component of the wall shear stress has been neglected because of $O\left(A^2\right)$, $O\left(B^2\right)$ and $O\left(AB\right)$, cf. also (2.68). Since U, V, W and $\mathrm{d}P/\mathrm{d}x$ are determined apart from an arbitrary factor, c_{U} is used to fix the solution as will be shown later.

Inserting these functions into (2.19), (2.20) and (2.22) and neglecting terms of $O\left(A^2\right)$, $O\left(B^2\right)$ and $O\left(AB\right)$ result in the following three linear ordinary differential equations:

$$(rV)' + kW - \Lambda rU = 0 \tag{2.60}$$

$$\left[\nu_{\mathrm{t}}^{+}rU'\right]' - \left(\frac{k^2}{r^2}\nu_{\mathrm{t}}^{+} - \Lambda u_\infty\right)rU - ru_\infty' V = -c_{\mathrm{U}}r\frac{\mathrm{d}p_\infty}{\mathrm{d}x} + \frac{\mathrm{d}P}{\mathrm{d}x}r\delta_{0k} \tag{2.61}$$

$$\begin{aligned}&\left\{\frac{1}{r}\left[\nu_{\mathrm{t}}^{+}r^3\left(\frac{W}{r}\right)'\right]' - \frac{k}{r}\left(\nu_{\mathrm{t}}^{+}rV\right)' - 2\frac{\nu_{\mathrm{t}}^{+}}{r}\left(k^2W + kV\right)\right\}'\\ &+\frac{2k}{r}\left(\nu_{\mathrm{t}}^{+}rV'\right)' + k^2\nu_{\mathrm{t}}^{+}\left(\frac{W}{r}\right)' - \frac{\nu_{\mathrm{t}}^{+}}{r^2}\left(k^3V + 2k^2W + 2kV\right)\\ &+\Lambda\left[u_\infty\langle(rW)' + kV\rangle + u_\infty' rW + kU\left(\nu_{\mathrm{t}}^{+}\right)' - c_{\mathrm{U}}ku_\infty'\nu_{\mathrm{t}}^{+}\right] = 0\ .\end{aligned} \tag{2.62}$$

In (2.61) δ_{0k} is the Kronecker delta ($=1$ if $k=0$, $=0$ if $k\neq 0$). The primes refer to differentiation with respect to r.

The boundary conditions are:

$$
\begin{aligned}
r = 0: \quad & k = 0: & & U' = 0\,, \quad V = 0\,, \quad W = 0 \\
& k = 1: & & U = 0\,, \quad V' = 0\,, \quad W' = 0 \\
& k = 2, 3 \ldots: & & U = 0\,, \quad V = 0\,, \quad W = 0
\end{aligned}
\tag{2.63}
$$

$$
\begin{aligned}
r \to 1: \quad & U = c_{\mathrm{U}} \left\{ 1 + \epsilon \left[\frac{1}{\kappa} \ln(1-r) - \overline{C} - \overline{\overline{C}} + \frac{1}{\kappa} \right] \right\} \\
& V = 0 \\
& W = c_{\mathrm{W}} \left\{ 1 + \epsilon \left[\frac{1}{\kappa} \ln(1-r) - \overline{C} - \overline{\overline{C}} \right] \right\} .
\end{aligned}
\tag{2.64}
$$

The boundary conditions at $r = 0$ have been derived by expanding the solutions into power series of r. The boundary conditions for $r \to 1$ follow from matching with the velocities in the viscous wall layer. In the overlap layer it is in analogy to (2.43)

$$
\begin{aligned}
\lim_{r \to 1} \frac{V^*}{u_\tau^*} &= \frac{1}{\kappa} \ln \frac{R^* u_\tau^*}{\nu^*} + \frac{1}{\kappa} \ln(1-r) + C^+ \\
&= \frac{1}{\kappa} \ln \frac{R^* u_{\tau\infty}^*}{\nu^*} + \frac{1}{\kappa} \ln(1-r) + C^+ + \frac{1}{\kappa} \ln \gamma \\
&= \frac{1}{\epsilon} - \overline{C} - \overline{\overline{C}} + \frac{1}{\kappa} \ln(1-r) + \frac{1}{\kappa} \ln \gamma ,
\end{aligned}
\tag{2.65}
$$

which leads to the boundary conditions (2.64) for $U(r)$ and $W(r)$ for $r \to 1$.

For axisymmetric flow ($k = 0$), integration of (2.60) over the cross section yields

$$
\lim_{r \to 1} \int_0^r U(r) r \mathrm{d}r = 0 \,, \tag{2.66}
$$

i.e., the disturbance velocity distribution has no effect on the volume flux.

Integrating (2.61) for $k = 0$ gives the final formula of the disturbance pressure gradient:

$$
\frac{1}{\epsilon} \frac{\mathrm{d}P}{\mathrm{d}x} = -4 c_{\mathrm{U}} + 4 \Lambda \lim_{r \to 1} \int_0^r \widetilde{u}_\infty U r \mathrm{d}r \,. \tag{2.67}
$$

Because of this relation between $\mathrm{d}P/\mathrm{d}x$ and $U(r)$, equations (2.60) to (2.62) can be considered as a system of three ordinary differential equations for the unknown velocity components $U(r)$, $V(r)$ and $W(r)$.

The wall shear stress follows from the velocity gradient at the wall. Since the shear stress in the wall layer is constant and the velocity distribution untwisted, the components of the wall shear stress can also be determined by the velocity gradient in the overlap layer. The formulae for the wall shear stress components are

$$\frac{\tau^*_{\mathrm{w}x}}{\tau^*_{\mathrm{w}\infty}} = \gamma^2 = 1 + 2c_{\mathrm{U}}e^{-\Lambda x}\left[A\cos k\theta + B\sin k\theta\right] \tag{2.68}$$

$$\frac{1}{\epsilon}\frac{\tau^*_{\mathrm{w}\theta}}{\tau^*_{\mathrm{w}\infty}} = \frac{\tan\beta_{\mathrm{w}}}{\epsilon} = c_{\mathrm{W}}e^{-\Lambda x}\left[A\sin k\theta - B\cos k\theta\right] . \tag{2.69}$$

The system of equations (2.61) to (2.62) with the boundary conditions (2.63) and (2.64) is an eigenvalue problem with an infinite number of eigenvalues for each parameter $k = 0, 1, 2 \ldots$. Hence, the general solution of the system is a linear superposition of all eigensolutions.

The eigensolutions depend on ϵ because $u_\infty(r)$ according to (2.32) and the boundary conditions (2.64) are linear functions of ϵ.

2.4.2 Limiting Solution for $\epsilon = 0$

In the limit $\epsilon = 0$ the system (2.60) to (2.64) is simplified considerably. In (2.61) the terms on the right hand side vanish because of (2.31) and (2.67). The same is true for the term proportional to V. The remaining equation contains only the unknown function $U(r)$ and can be used to determine the eigenvalues Λ. To simplify further it can be set $c_{\mathrm{U}} = 1$. When Λ and $U(r)$ are found, $V(r)$, $W(r)$ and the factor c_{W} can be determined via (2.60), (2.62) and the corresponding boundary conditions.

There exists another type of eigensolutions $W(r)$ and $V(r)$, when $U(r)$ vanishes ($c_{\mathrm{U}} = 0$). Then it can be set $c_{\mathrm{W}} = 1$. These eigensolutions describe two-dimensional secondary flows in a pipe.

Therefore, two different types of eigensolutions can be distinguished which are independent of each other: the U-type ($c_{\mathrm{U}} = 1$) and the W-type ($c_{\mathrm{U}} = 0$, $c_{\mathrm{W}} = 1$). It can be shown that in both cases the eigensolutions are sets of orthogonal functions with respect to the weight function $r\,u_\infty$.

2.4.3 Numerical Solutions

As already mentioned the function u_∞ in (2.61) and (2.62) and the boundary conditions (2.64) depend linearly on ϵ. This suggests a splitting of the solution in the form

$$\Lambda = \overline{\Lambda} + \epsilon\widetilde{\Lambda}\,, \quad U = \overline{U} + \epsilon\widetilde{U}\,, \quad V = \overline{V} + \epsilon\widetilde{V}\,, \quad W = \overline{W} + \epsilon\widetilde{W}\,. \tag{2.70}$$

When quadratic terms $O\left(\epsilon^2\right)$ are neglected, the splitting leads for each eigensolution type and for each k to two eigenvalue problems *independent of* ϵ as is expected in an asymptotic analysis. The numerical solution was carried out by iteration. The eigenvalues had to be estimated such that all boundary conditions were satisfied.

Because of the singularities in the differential equations for $r = 0$ and $r \to 1$ and in the boundary conditions for $r \to 1$ the solutions were approximated by series expansions near the ends of the interval. Near the axis power series have been used whereas for $r \to 1$ the series consisted of powers $(1-r)^n$ as well as $(1-r)^n \ln(1-r)$.

In Tables 2.1 and 2.2 the main results of the calculation for the first three eigenvalues are summarized. The velocity components of the first three eigensolutions are shown for $Re = 10^6$ ($\epsilon = 0.0385$) and $Re = \infty$ ($\epsilon = 0$) in Fig. 2.4. It is clear that the eigensolutions will not change much for Reynolds numbers higher than $Re = 10^6$.

Each set of eigensolutions represents a particular flow structure characterized by the secondary flow described by the velocity components $V(r)$ and $W(r)$. When only the first three k-values ($k = 0, 1, 2$) are considered, the following six flow structures can be identified: U-type: ring vortex, source-sink pair, source-sink quadrupole (two sources and two sinks). W-type: single longitudinal vortex, longitudinal vortex pair, longitudinal vortex quadrupole.

All these flow structures have their individual decay law which is described by an exponential function $\exp(-\alpha x^*/D^*)$ with D^* as the pipe diameter. The decay rate α is for all flow structures proportional to $\epsilon = \sqrt{\lambda/8}$:

$$\alpha = 2\Lambda\epsilon = \frac{\Lambda}{\sqrt{2}}\sqrt{\lambda}\,. \tag{2.71}$$

There is only one exception. As can be seen from Table 2.2, the eigenvalue of the W-type for $k = 0$ is one order of magnitude smaller than all the others, i.e., $\Lambda_{\mathrm{W}01} = 4\epsilon$. In this special case the decay rate is identical with the friction factor, $\alpha = \lambda$.

If the length x^*, within which a value is decayed to one tenth of its value is called $L^*_{0.1}$, then it follows

$$\frac{L^*_{0.1}}{D^*} = \frac{2.305}{\alpha}\,. \tag{2.72}$$

This *tenth-value length* is given in Table 2.3 for the various flow structures in descending order. The single longitudinal vortex has the smallest decay rate followed by the source-sink pair.

Any given function can be expanded in terms of the set of eigenfunctions. The velocity distributions in the cross section $x = 0$ have the form

$$u(0,r,\theta) = u_\infty(r) + \sum_{k=0,\ldots}\sum_{i=1,\ldots} U_{\mathrm{U}ki}(r)\left[A_{\mathrm{U}ki}\cos k\theta + B_{\mathrm{U}ki}\sin k\theta\right] \tag{2.73}$$

$$w(0,r,\theta) = \sum_{k=0,\ldots}\sum_{i=1,\ldots} W_{\mathrm{W}ki}(r)\left[A_{\mathrm{W}ki}\sin k\theta - B_{\mathrm{W}ki}\cos k\theta\right] \tag{2.74}$$

where the subscripts U and W indicate the eigensolution type and i the number of the eigensolution in a particular set.

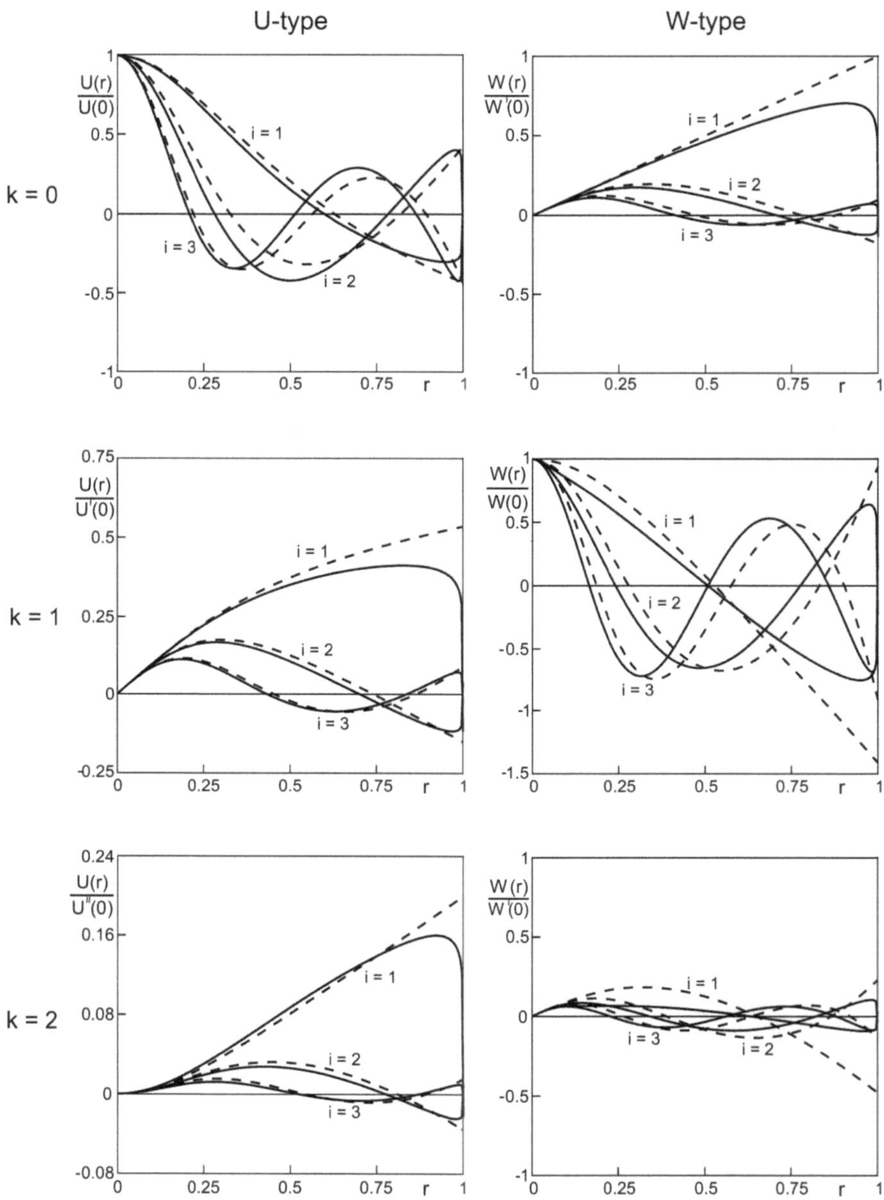

Fig. 2.4. Velocity components of the first three eigensolutions for $k = 0$, 1 and 2; (——) $Re = 10^6$ ($\epsilon = 0.0385$); (- - - -) $Re = \infty$ ($\epsilon = 0$).

Table 2.1. Characteristic results of the eigenvalue problem (U-type). After definition in (2.64): $c_{\mathrm{W}} = (1+\epsilon/\kappa)\overline{W}_{ki} + \epsilon \lim_{r\to 1}\left(\widetilde{W}_{ki} - \overline{W}_{ki}\widetilde{U}_{ki}\right)$

$k=0$				$k=1$				$k=2$			
	$i=1$	2	3		$i=1$	2	3		$i=1$	2	3
$\overline{\Lambda}_{0i}$	0.9555	2.6761	5.2860	$\overline{\Lambda}_{1i}$	0.1980	1.6871	3.8735	$\overline{\Lambda}_{2i}$	0.5299	2.6161	5.2882
$\overline{U}_{0i}(0)$	-2.3400	2.3400	-2.3400	$(\mathrm{d}\overline{U}_{1i}/\mathrm{d}r)_{r=0}$	1.8699	-6.6563	10.928	$(\mathrm{d}^2\overline{U}_{2i}/\mathrm{d}r^2)_{r=0}$	5.0349	-28.125	66.551
$\int_0^1 \overline{U}_{0i}{}^2 r\mathrm{d}r$	0.3521	0.2007	0.1429	$\int_0^1 \overline{U}_{1i}{}^2 r\mathrm{d}r$	0.3297	0.2476	0.1664	$\int_0^1 \overline{U}_{2i}{}^2 r\mathrm{d}r$	0.2213	0.1899	0.1415
				$\overline{W}_{1i}(0)$	0.0444	-0.4349	0.5665	$(\mathrm{d}\overline{W}_{2i}/\mathrm{d}r)_{r=0}$	0.2163	-2.0193	3.9938
				$\overline{W}_{1i}(1)$	0.1342	0.1365	0.0327	$\overline{W}_{2i}(1)$	0.1493	0.1416	0.0250
$\widetilde{\Lambda}_{0i}$	2.1514	5.9031	9.0425	$\widetilde{\Lambda}_{1i}$	3.0332	4.0385	6.0103	$\widetilde{\Lambda}_{2i}$	4.5194	5.2667	7.0681
$\widetilde{U}_{0i}(0)$	-15.624	-11.002	15.531	$\left(\mathrm{d}\widetilde{U}_{1i}/\mathrm{d}r\right)_{r=0}$	7.0182	12.481	-45.927	$\left(\mathrm{d}^2\widetilde{U}_{2i}/\mathrm{d}r^2\right)_{r=0}$	-55.952	1299.5	-7361.8
$2\int_0^1 \overline{U}_{0i}\widetilde{U}_{0i} r\mathrm{d}r$	2.3658	-1.0432	-1.6216	$2\int_0^1 \overline{U}_{1i}\widetilde{U}_{1i} r\mathrm{d}r$	-0.7349	-1.9716	-1.8131	$2\int_0^1 \overline{U}_{2i}\widetilde{U}_{2i} r\mathrm{d}r$	-0.8743	-1.7572	-1.7087
$\int_0^1 \widetilde{u}_\infty \overline{U}_{0i}^2 r\mathrm{d}r$	0.3903	0.1144	0.0647	$\int_0^1 \widetilde{u}_\infty \overline{U}_{1i}^2 r\mathrm{d}r$	-0.3338	0.1137	0.0708	$\int_0^1 \widetilde{u}_\infty \overline{U}_{2i}^2 r\mathrm{d}r$	-0.4250	-0.0079	0.0336
				$\widetilde{W}_{1i}(0)$	1.2726	1.4716	-1.4863	$\left(\mathrm{d}\widetilde{W}_{2i}/\mathrm{d}r\right)_{r=0}$	3.6519	3.2103	-19.955
				$\left(\widetilde{W}_{1i} - \overline{W}_{1i}\widetilde{U}_{1i}\right)_{r\to 1}$	1.1868	0.4817	0.9891	$\left(\widetilde{W}_{2i} - \overline{W}_{2i}\widetilde{U}_{2i}\right)_{r\to 1}$	0.1838	0.6128	1.1215

Table 2.2. Characteristic results of the eigenvalue problem (W-type).

	$k=0$				$k=1$				$k=2$		
	$i=1$	2	3		$i=1$	2	3		$i=1$	2	3
$\overline{\Lambda}_{0i}$	0.0000	1.7149	3.9061	$\overline{\Lambda}_{1i}$	0.5299	2.6300	5.3762	$\overline{\Lambda}_{2i}$	0.9467	3.6500	6.8339
$(\mathrm{d}\overline{W}_{0i}/\mathrm{d}r)_{r=0}$	1.000	-5.6174	10.3291	$\overline{W}_{1i}(0)$	-0.7101	1.0662	-1.0973	$(\mathrm{d}\overline{W}_{2i}/\mathrm{d}r)_{r=0}$	-2.0867	4.3573	-7.2139
$\int_0^1 \overline{W}_{0i}^{\,2} r\mathrm{d}r$	0.2500	0.2409	0.1666	$\int_0^1 \overline{W}_{1i}^{\,2} r\mathrm{d}r$	0.1553	0.1444	0.1171	$\int_0^1 \overline{W}_{2i}^{\,2} r\mathrm{d}r$	0.1105	0.1005	0.0892
$\widetilde{\Lambda}_{0i}$	4.0000	3.2131	3.8305	$\widetilde{\Lambda}_{1i}$	4.7976	3.5246	3.8926	$\widetilde{\Lambda}_{2i}$	5.3000	3.6197	0.1909
$\left(\mathrm{d}\widetilde{W}_{0i}/\mathrm{d}r\right)_{r=0}$	4.3586	-12.4343	-10.5386	$\widetilde{W}_{1i}(0)$	-8.1690	1.6878	2.0782	$\left(\mathrm{d}\widetilde{W}_{2i}/\mathrm{d}r\right)_{r=0}$	-155.344	63.117	-18.553
$2\int_0^1 \overline{W}_{0i}\widetilde{W}_{0i} r\mathrm{d}r$	-0.8029	-1.5840	-1.6165	$2\int_0^1 \overline{W}_{1i}\widetilde{W}_{1i} r\mathrm{d}r$	-0.3832	-0.6247	-1.0009	$2\int_0^1 \overline{W}_{2i}\widetilde{W}_{2i} r\mathrm{d}r$	-0.2788	-0.2080	-0.6166
$\int_0^1 \widetilde{u}_\infty \overline{W}_{0i}^2 r\mathrm{d}r$	-0.4012	0.1317	0.0926	$\int_0^1 \widetilde{u}_\infty \overline{W}_{1i}^2 r\mathrm{d}r$	-0.3618	-0.0926	-0.0272	$\int_0^1 \widetilde{u}_\infty \overline{W}_{2i}^2 r\mathrm{d}r$	-0.3374	-0.1518	-0.0739

Table 2.3. Tenth-value length according to (2.72) for various flow structures in descending order.

	1. flow structure	$\overline{\Lambda}$	$L^*_{0.1}/D^*$ $(Re = 10^6)$
1	single longitudinal vortex	0	194
2	source-sink pair	0.198	151
3	vortex pair	0.530	56
4	source-sink quadrupole	0.530	56
5	vortex quadrupole	0.947	32
6	ring vortex	0.956	31

The coefficients are in a first approximation determined by

$$A_{\mathrm{U}0i} = \frac{\int_0^{2\pi}\int_0^1 u_\infty(r)\left[u(0,r,\theta) - u_\infty(r)\right] U_{\mathrm{U}0i} r \mathrm{d}r \mathrm{d}\theta}{2\pi \int_0^1 u_\infty U_{\mathrm{U}0i}^2 r \mathrm{d}r} \tag{2.75}$$

$$A_{\mathrm{U}ki} = \frac{\int_0^{2\pi}\int_0^1 u_\infty(r)\left[u(0,r,\theta) - u_\infty(r)\right] U_{\mathrm{U}ki}\cos(k\theta) r \mathrm{d}r \mathrm{d}\theta}{\pi \int_0^1 u_\infty U_{\mathrm{U}ki}^2 r \mathrm{d}r} \qquad k = 1, 2 \ldots \tag{2.76}$$

$$B_{\mathrm{U}ki} = \frac{\int_0^{2\pi}\int_0^1 u_\infty(r)\left[u(0,r,\theta) - u_\infty(r)\right] U_{\mathrm{U}ki}\sin(k\theta) r \mathrm{d}r \mathrm{d}\theta}{\pi \int_0^1 u_\infty U_{\mathrm{U}ki}^2 r \mathrm{d}r} \qquad k = 1, 2... \tag{2.77}$$

$$B_{\mathrm{W}0i} = -\frac{\int_0^{2\pi}\int_0^1 u_\infty(r) w(0,r,\theta) W_{\mathrm{W}0i} r \mathrm{d}r \mathrm{d}\theta}{2\pi \int_0^1 u_\infty W_{\mathrm{W}0i}^2 r \mathrm{d}r} \tag{2.78}$$

$$A_{\mathrm{W}ki} = \frac{\int_0^{2\pi}\int_0^1 u_\infty(r) w(0,r,\theta) W_{\mathrm{W}ki}\sin(k\theta) r \mathrm{d}r \mathrm{d}\theta}{\pi \int_0^1 u_\infty W_{\mathrm{W}ki}^2 r \mathrm{d}r} \qquad k = 1, 2... \tag{2.79}$$

$$B_{\mathrm{W}ki} = -\frac{\int_0^{2\pi}\int_0^1 u_\infty(r) w(0,r,\theta) W_{\mathrm{W}ki}\cos(k\theta) r \mathrm{d}r \mathrm{d}\theta}{\pi \int_0^1 u_\infty W_{\mathrm{W}ki}^2 r \mathrm{d}r} \qquad k = 1, 2... \tag{2.80}$$

The integrals in the denominators can be determined by superposition of the integrals given in Tables 2.1 and 2.2.

The general solutions for $u(x,r,\theta)$ and $w(x,r,\theta)$ are obtained when in (2.73) and (2.74) the eigensolutions are multiplied by $\exp(-\Lambda_{\mathrm{U}ki}x)$ and $\exp(-\Lambda_{\mathrm{W}ki}x)$, respectively.

2.5 Comparisons with Experimental Results

2.5.1 Axisymmetric Flow without Swirl

Experimental investigations of the decay of distorted axisymmetric turbulent pipe flows have been carried out by Laws et al. (1987). The disturbed inlet profiles were generated by gauze screens. The test Reynolds number was $Re = 2.52 \times 10^5$ ($\epsilon = 0.0434$). Figure 2.5 shows for a typical example (case 3 from $x^*/D^* = 8$ to $x^*/D^* = 19.5$) the comparison between theory and experiment.

The coefficients of the series

$$u(x,r) = u_\infty(r) + \sum_{i=1}^{3} A_{\mathrm{U}0i} U_{\mathrm{U}0i} \exp\left[-2\epsilon\Lambda_{\mathrm{U}0i}\left(\frac{x^*}{D^*} - 8\right)\right] \tag{2.81}$$

are $A_{\mathrm{U}01} = 0.0648$, $A_{\mathrm{U}02} = 0.0012$, $A_{\mathrm{U}03} = -0.0392$. In cross section $x^*/D^* = 19.5$ the series in (2.81), reduced to only the first term, is in good agreement with the experimental results.

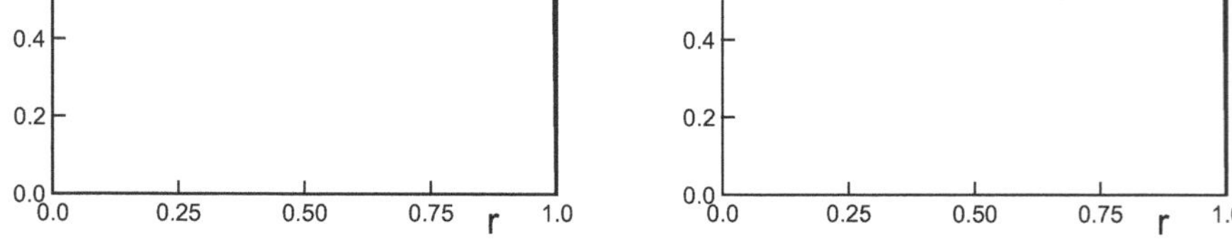

Fig. 2.5. Distorted velocity distributions of an axisymmetric turbulent pipe flow without swirl $Re = 2.52 \times 10^5$ ($\epsilon = 0.0434$); (○) experiments by Laws et al. (1987); (——) theory according to (2.81)

2.5.2 Axisymmetric Flow with Swirl

Analysis

In axisymmetric fow with swirl (W-type, $k = 0$) equation (2.62) can be integrated and reduces to the simple equation

$$\left[\nu_t^+ r^3 \left(\frac{W_{W0i}}{r}\right)'\right]' + \Lambda_{W0i} r^2 u_\infty W_{W0i} = 0 \, . \tag{2.82}$$

Hence, the disturbances in the circumferential and axial direction are not coupled in this case. It can be shown that the eigensolutions are a set of orthogonal functions with respect to the weight function $r u_\infty$, i.e.,

$$\lim_{r \to 1} \int_0^1 r u_\infty W_{W0i} W_{W0j} \mathrm{d}r = \begin{cases} 0 \text{ for } i \neq j \\ c \text{ for } i = j \end{cases} \tag{2.83}$$

The first eigensolution can be given in analytical form:

$$\Lambda_{W01} = 4\epsilon \quad , \quad W_{W01} = r u_\infty \, . \tag{2.84}$$

Hence, the circumferential velocity distribution far downstream is

$$w(x, r) = -B_{W01} r u_\infty(r) \exp\left[-\lambda x^* / D^*\right] \, . \tag{2.85}$$

The decay rate α according to (2.71) is in this case identical with the friction factor λ of the fully developed pipe flow at the given Reynolds number. In the limit $Re \to \infty$ ($u_\infty = 1$) the circumferential velocity $w(x, r)$ is proportional to the radius r (rigid body rotation).

It is worth mentioning that the coefficient B_{W01} according to (2.80) is proportional to the commonly used swirl number, cf. Steenbergen (1995) p. 15,

$$S = 2 \int_0^1 \frac{u^*}{u_m^*} \frac{w^*}{u_m^*} r^2 \mathrm{d}r = -\frac{\epsilon}{2} B_{W01} \tag{2.86}$$

when the Reynolds number is high ($W_{W01} \approx r$) and $u(r)$ is substituted by $u_\infty(r)$. In practice this latter condition is usually satisfied because the decay rate for the axial disturbances is much higher than that for the circumferential disturbances, see Table 2.3 (cases 1 and 6). Under the conditions mentioned only the first eigensolution determines the swirl number. All other eigensolutions have no contribution to S because of the orthogonality.

The radial distribution of the swirl angle far downstream (axial and circumferential disturbances of second and higher eigensolutions already decayed) follows from (2.85)

$$\frac{\tan \beta(x, r)}{\tan \beta_w(x)} = r \tag{2.87}$$

when (2.69) is taken into account.

Comparison with Results of Reader-Harris

Reader-Harris (1994) has analyzed the turbulent pipe flow with swirl by a method similar to the special case (W-type, $k = 0$) presented here. In contrast to the Reynolds number independence of the asymptotic analysis he calculated the eigensolutions individually for each Reynolds number. He found the mean value of the first decay rate for the Reynolds numbers $10^5 \leq Re \leq 10^7$ as $\overline{\alpha}_1 = 1.07\lambda$ in fairly good agreement with $\alpha_1 = \lambda$ in (2.85). An extrapolation of his α-values to $Re \to \infty$ would give $\alpha = \lambda$.

Reader-Harris compared his results of the decay rate α as function of λ and of the swirl angle distribution according to (2.87) with various experiments and found good agreement.

Table 2.4. Model constants and the first three decay rates α according to (2.71) for four different turbulence models.

	turbulence model	Reader-Harris	Prandtl	Zagarola	McKeon
constants	κ	0.40	0.407	0.436	0.421
	C^+ (smooth)	5.5	5.0	6.15	5.6
	$-\left(\overline{C}+\overline{\overline{C}}\right)$	3.77	3.01	2.85	3.05
$Re = 10^6$	λ	0.0115	0.0116	0.0119	0.0118
	α_1	0.0123	0.0116	0.0119	0.0118
	α_2	0.172	0.145	0.144	0.141
	α_3	0.391	0.344	0.303	0.312

The decay rates depend slightly on the turbulence model used, as can be seen in Table 2.4. In this table the first three decay rates are listed for four different turbulence models, the model of Reader-Harris and the model of the present paper applied to the friction law by Prandtl, by Zagarola and by McKeon. The two last models show only very little difference, as mentioned earlier. The type of turbulence model seems to have an increasing effect on higher decay rates.

Reader-Harris investigated also the effect of roughness on the decay of swirl. Since C^+ in (2.45) is the only number affected by the roughness, the whole analysis presented here is the same. Only the value ϵ for a given Reynolds number increases due to roughness. Increasing roughness is equivalent to decreasing the Reynolds number at smooth surface condition. Roughness increases the decay rates in complete agreement with the results by Reader-Harris and with further investigations on roughness effects by Senoo and Nagata (1972) and Morrison and Tung (1999).

2.5.3 Three-Dimensional Flow

Experimental data which are accurate and detailed enough in the whole field of the pipe flow with small disturbances for the purpose of comparison with the theory are not known to the authors.

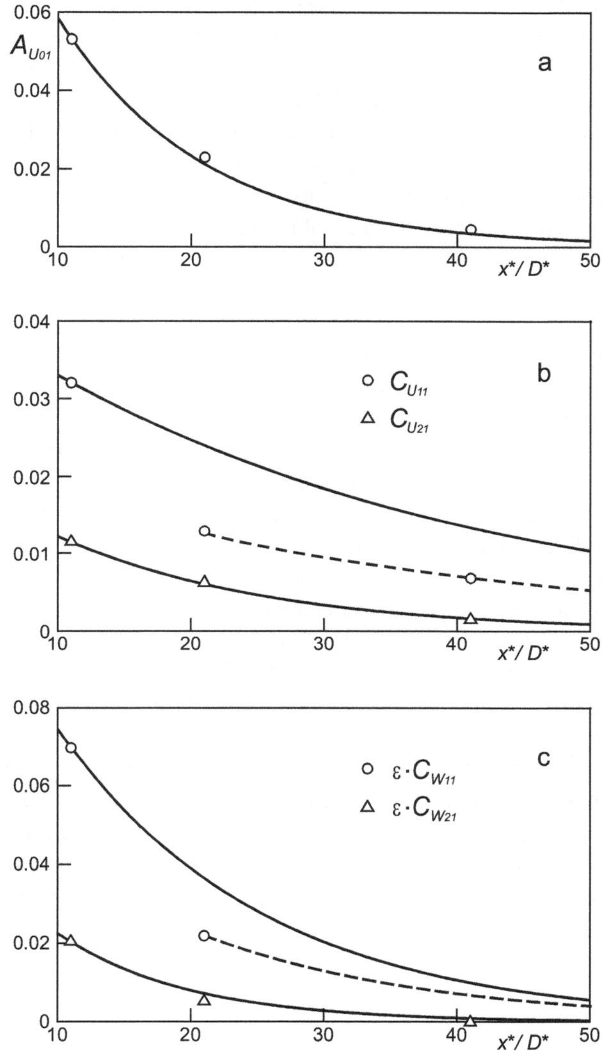

Fig. 2.6. Decay of the coefficients (2.73), (2.74) and (2.88) downstream of a bend. $Re = 2.2 \times 10^5$ ($\epsilon = 0.0439$). (o) (△) experiments by Wildemann (2000); (——) $\sim \exp\left[-\alpha\left(x^*/D^* - 11\right)\right]$ adjusted to experiment at $x^*/D^* = 11$, α according to (2.71), Λ from Tables 2.1 and 2.2. (- - - -) $\sim \exp\left[-\alpha\left(x^*/D^* - 21\right)\right]$ adjusted to experiment at $x^*/D^* = 21$

In technical practice the decay process is quite often already so close to the fully developed state that already the first eigensolutions are sufficient to describe the flow. In this case the coefficients in (2.73) and (2.74) can be determined easily as Gersten et al. (2001) and Wildemann et al. (2002) have shown. The authors measured the circumferential distributions of the wall shear stress components and could determine the coefficients by using (2.68) and (2.69). Figure 2.6 shows as a typical example the decay of the coefficients in the pipe downstream of a bend. For $k = 1$ and $k = 2$ the coefficients A_{k1} and B_{k1} have been combined to

$$C_{k1} = \sqrt{A_{k1}^2 + B_{k1}^2} \tag{2.88}$$

characterizing the intensity of the corresponding flow structure. The coefficient $A_{\mathrm{W}01}$ could not be determined since the swirl in the experiment was very small and showed no decaying tendency.

If the data points lie exactly on an exponential curve, the first eigensolution is sufficient to describe the decay process. This is the case for $A_{\mathrm{U}01}$, $C_{\mathrm{U}21}$ and $C_{\mathrm{W}21}$ where the agreement between theory and experiment is very good. On the other hand, the first eigensolution is not sufficient to describe the decay process for $C_{\mathrm{U}11}$ and this is also true, though to a lesser degree, for $C_{\mathrm{W}11}$. The value $C_{\mathrm{U}11}$ represents the structure of a source-sink pair. This structure has the smallest decay rate of all structures present in this flow according to Table 2.3. The dotted line in Fig. 2.6b, adjusted to the experiments at $x^*/D^* = 21$, shows excellent agreement between theory and experiment. The same may be true for $C_{\mathrm{W}11}$ in Fig. 2.6c, representing the vortex-pair structure.

From this comparison between theory and experiment it can be concluded that the decay process downstream of the bend under consideration is purely exponential at $x^* > 11D^*$ for $A_{\mathrm{U}01}$, $C_{\mathrm{U}21}$, $C_{\mathrm{W}21}$ (ring vortex, source-sink quadrupole, vortex quadrupole) and at about $x^* > 20D^*$ for $C_{\mathrm{U}11}$ and $C_{\mathrm{W}11}$ (source-sink pair, vortex pair). The terms for $k \geq 3$ are in practice negligible as could also be seen from the experimental data.

References

Barbin AR, Jones JB (1963) Turbulent flow in the inlet region of a smooth pipe. J. Basic Eng. 85:29–34

Cebeci T, Chang KC (1978) A general method for calculating momentum and heat transfer in laminar and turbulent duct flows. Numerical Heat Transfer 1:39–68

Gersten K, Herwig H (1992) Strömungsmechanik. Grundlagen der Impuls-, Wärme- und Stoffübertragung aus asymptotischer Sicht. Vieweg-Verlag, Braunschweig/Wiesbaden

Gersten K, Klika M (1998) The decay of three-dimensional deviations from the fully developed state in laminar pipe flow. In: Rath HJ, Egbers Ch. (Eds.): Advances in Fluid Mechanics and Turbomachinery, Springer-Verlag, Berlin/Heidelberg, pp. 17–28

Gersten K, Merzkirch W, Wildemann C (2001) Verfahren und Vorrichtung zur Korrektur fehlerhafter Messwerte von Durchflussmessgeräten infolge gestörter Zuströmung. Patent No. 197 24 116

Herwig H, Voigt M (1995) Turbulent entrance flow in a channel: an asymptotic approach. In: Bois PA et al. (Eds.):Asymptotic Modelling in Fluid Mechanics, Springer-Verlag, Berlin/Heidelberg/New York, pp. 51–58

Kalkühler K (1998) Experimente zur Entwicklung der Geschwindigkeitsprofile und Turbulenzgrößen hinter verschiedenen Gleichrichtern. Fortschritt-Berichte, Reihe 7, Nr. 339, VDI-Verlag, Düsseldorf

Kitoh O (1991) Experimental study of turbulent swirling flow in a straight pipe. J. Fluid Mech. 225:445–479

Klein A (1981) REVIEW: Turbulent developing pipe flow. J. Fluids Eng. 103:243–249

Laws EM, Lim EH, Livesey JL (1987) Momentum balance in highly distorted turbulent pipe flow. Exp. in Fluids 5:36–42

McKeon BJ, Morrison JF, Jiang W, Li J, Smits AJ (2003) Revised log-law constants for fully-developed turbulent pipe flow. In: AJ Smits (Ed.): IUTAM Symposium on Reynolds Number Scaling in Turbulent Flow. Kluwer Academic Publishers, Dordrecht

Mickan B (1999): Systematische Analyse von Installationseffekten sowie der Effizienz von Strömungsgleichrichtern in der Großgasmengenmessung. Doctor thesis, University Essen

Morrison GL, Tung K (1999) The effect of pipe wall roughness upon the flow field downstream of two close coupled 90° out of plane elbows. Proceedings 4th International Symposium on Fluid Flow Measurement, Denver, Colorado, June 27–30, 1999

Norman RS, Mattingly GE, Mc Faddin SE (1989) The decay of swirl in pipes. Proceedings Intern. Gas Research Conference, 312–321

Parchen RR (1993) Decay of swirl in turbulent pipe flows. Ph. D. thesis, Technical University Eindhoven

Reader-Harris MJ (1994) The decay of swirl in a pipe. Int. J. Heat and Fluid Flow 15:212–217

Senoo Y, Nagata T (1972) Swirl flow in long pipes with different roughness. Bulletin of the JSME, 15:1514–1521

Schlichting H, Gersten K (2000) Boundary-Layer Theory. Springer-Verlag, Berlin / Heidelberg

Steenbergen W (1995) Turbulent pipe flow with swirl. Ph. D. thesis, Technical University Eindhoven

Steenbergen W, Voskamp J (1998) The rate of decay of swirl in turbulent pipe flow. Flow Meas. Instrum. 9:67–78

Sudo K, Sumida M, Hibara H (1998) Experimental investigation of turbulent flow in a circular-sectioned 90-degree bend. Exp. in Fluids 25:42–49

Voigt M (1995) Die Entwicklung von Geschwindigkeits- und Temperaturfeldern in laminaren und turbulenten Kanal- und Rohrströmungen aus asymptotischer Sicht. VDI-Fortschritt-Berichte, Reihe 7, Nr. 262, VDI-Verlag, Düsseldorf

Wildemann C (2000) Ein System zur automatischen Korrektur der Messabweichung von Durchflussmessgeräten bei gestörter Anströmung. Doctor thesis, Universität Essen. See also: Fortschritt-Berichte VDI, Reihe 8, Nr. 868

Wildemann C, Merzkirch W, Gersten K (2002) A universal, nonintrusive method for correcting the reading of a flow meter in pipe flow disturbed by installation effects. J. of Fluids Eng. 124:650-656

Zagarola MV, Smits AJ (1998) Mean-flow scaling of turbulent pipe flow. J. Fluid Mech. 373, 33-79

3

Optimal Characteristic Parameters for the Disturbances in Turbulent Pipe Flow

Klaus Gersten, Heinz-Dieter Papenfuss

Institut für Thermo- und Fluiddynamik, Ruhr-Universität Bochum, 44780 Bochum, Germany

A set of optimal characteristic parameters for turbulent pipe flows is proposed to quantify the disturbances, i.e., the deviations from the fully developed flow. These optimal characteristic parameters are based on the eigensolutions of the three-dimensional flow field in the pipe and have the following features: Any desired accuracy can be achieved. The number of parameters is a minimum for a given accuracy. Each parameter is independent of the number of parameters being used. The flow field can be reconstructed from the characteristic parameters. The characteristic parameters can be easily determined by a nonintrusive experiment. As an example, the flow field approximately ten diameters downstream of a single bend at a Reynolds number of $Re \approx 2 \times 10^5$ is analyzed.

3.1 Introduction

Flows in straight pipes are often not fully developed. Deviations from the fully developed state may occur due to bends, valves or other pipe fittings. Even swirl can be found downstream of a pair of bends installed in perpendicular planes. The knowledge of these deviations from the fully developed condition is of great importance. For example, these deviations can cause significant errors in the measurement of the mass flow rate.

In order to quantify the intensity of various flow disturbances in the pipe certain characteristic parameters are used. These parameters are sometimes denoted as "flow indices". The axisymmetric flow without swirl can be characterized by the *blockage factor*, which is essentially the ratio between the maximum velocity and the average velocity, cf. Klein (1981) and Cebeci and Chang (1978). The swirl intensity can be determined by the *swirl number*, which is a dimensionless radial moment of the circumferential momentum flux, see Steenbergen and Voskamp (1998). Corresponding characteristic parameters have been defined for three-dimensional disturbances by Morrison et al. (1992, 1995), Sudo et al. (1998) and Mickan (1999). All these characteristic parameters mentioned have been applied only to special cases and are not sufficient to describe properly and accurately enough the general three-dimensional flow field.

3.2 Objective

In the following a set of characteristic parameters will be proposed to describe the general three-dimensional turbulent pipe flow. The proposal has the following properties:

1. The number of characteristic parameters can be as high as necessary to *guarantee* the desired accuracy.
2. The number of parameters is a *minimum* for a given accuracy in comparison with other possible sets of parameters.
3. Each parameter is *independent* of the total number of parameters used.
4. From a given set of characteristic parameters the flow field can be *reconstructed* ("inversion").
5. The characteristic parameters can be easily determined by *experiment.*

Because of these properties the designation "optimal characteristic parameters" has been chosen.

3.3 Formulae for the Optimal Characteristic Parameters

The decay of three-dimensional deviations from the fully developed state in turbulent pipe flow far downstream at high Reynolds numbers has be investigated by Gersten and Papenfuss in Chap. 2 of this book. The general solution has been expanded into series of eigensolutions. These are for the axial velocity component

$$u(x,r,\theta) = u_\infty(r) + \sum_{k=0,\dots}\sum_{i=1,\dots} U_{\mathrm{U}ki}(r)e^{-\Lambda_{\mathrm{U}ki}x}\left[A_{\mathrm{U}ki}\cos k\theta + B_{\mathrm{U}ki}\sin k\theta\right] \tag{3.1}$$

and for the azimuthal velocity component

$$w(x,r,\theta) = \sum_{k=0,\dots}\sum_{i=1,\dots} W_{\mathrm{W}ki}(r)e^{-\Lambda_{\mathrm{W}ki}x}\left[A_{\mathrm{W}ki}\sin k\theta - B_{\mathrm{W}ki}\cos k\theta\right] . \tag{3.2}$$

A cylindrical coordinate system has been used according to Fig. 3.1. It should be noted, that all velocity components are based on the average velocity u_{m}, in contrast to Chap. 2 of this book, where the radial and the circumferencial velocity components are based on ϵu_{m}. The velocity components are double series. In circumferential direction it is a Fourier series with $k = 0, 1, 2, \dots$ as running parameter. The decay in downstream direction is given by a linear combination of exponential functions. The eigensolutions $U_{\mathrm{U}ki}(r)$ and $W_{\mathrm{W}ki}(r)$ have been documented in detail by Gersten and Papenfuss. They have been normalized in a way that they have the same wall shear

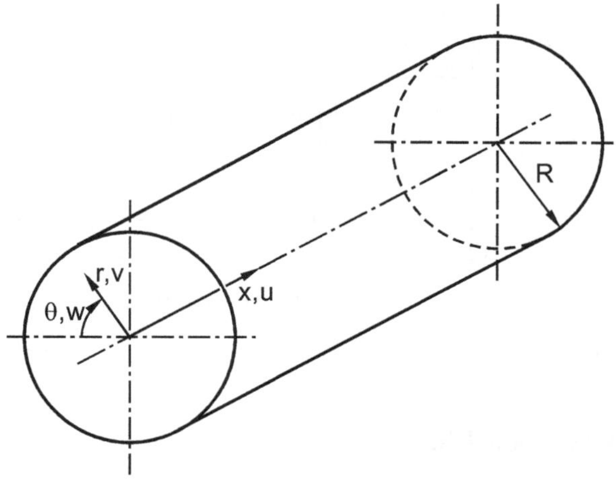

Fig. 3.1. Geometry and coordinate system. The radial coordinate r is normalized by the pipe radius R, the velocity components u, v, w by the average velocity u_{m}.

stress as the fully developed flow at the same Reynolds number. The eigensolutions do not contribute to the flow rate and depend only slightly on the Reynolds number.

The characteristic parameters are given by:

$$A_{\mathrm{U}0i} = \frac{\int_0^{2\pi}\int_0^1 u_\infty(r)\big[u(0,r,\theta) - u_\infty(r)\big] U_{\mathrm{U}0i} r \mathrm{d}r \mathrm{d}\theta}{2\pi \int_0^1 u_\infty U_{\mathrm{U}0i}^2 r \mathrm{d}r} \tag{3.3}$$

$$A_{\mathrm{U}ki} = \frac{\int_0^{2\pi}\int_0^1 u_\infty(r)\big[u(0,r,\theta) - u_\infty(r)\big] U_{\mathrm{U}ki} \cos(k\theta) r \mathrm{d}r \mathrm{d}\theta}{\pi \int_0^1 u_\infty U_{\mathrm{U}ki}^2 r \mathrm{d}r} \tag{3.4}$$

$$B_{\mathrm{U}ki} = \frac{\int_0^{2\pi}\int_0^1 u_\infty(r)\big[u(0,r,\theta) - u_\infty(r)\big] U_{\mathrm{U}ki} \sin(k\theta) r \mathrm{d}r \mathrm{d}\theta}{\pi \int_0^1 u_\infty U_{\mathrm{U}ki}^2 r \mathrm{d}r} \tag{3.5}$$

$$B_{\mathrm{W}0i} = -\frac{\int_0^{2\pi}\int_0^1 u_\infty(r) w(0,r,\theta) W_{\mathrm{W}0i} r \mathrm{d}r \mathrm{d}\theta}{2\pi \int_0^1 u_\infty W_{\mathrm{W}0i}^2 r \mathrm{d}r} \tag{3.6}$$

$$A_{\mathrm{W}ki} = \frac{\int\limits_0^{2\pi}\int\limits_0^1 u_\infty(r) w(0,r,\theta) W_{\mathrm{W}ki} \sin(k\theta) r \mathrm{d}r \mathrm{d}\theta}{\pi \int\limits_0^1 u_\infty W_{\mathrm{W}ki}^2 r \mathrm{d}r} \tag{3.7}$$

$$B_{\mathrm{W}ki} = -\frac{\int\limits_0^{2\pi}\int\limits_0^1 u_\infty(r) w(0,r,\theta) W_{\mathrm{W}ki} \cos(k\theta) r \mathrm{d}r \mathrm{d}\theta}{\pi \int\limits_0^1 u_\infty W_{\mathrm{W}ki}^2 r \mathrm{d}r} \tag{3.8}$$

$(k = 1, 2 \ldots)$

3.4 Fully Developed Flow

The velocity distribution of the fully developed flow is

$$u_\infty(r) = 1 + \epsilon \tilde{u}_\infty(r) \,, \tag{3.9}$$

where $\tilde{u}_\infty(r)$ can be described by

$$\tilde{u}_\infty(r) = \frac{1}{\kappa}\left[\ln\left(1 - r^2\right) - \frac{\alpha}{2a}\ln\left(1 + ar^2\right) - \frac{\beta}{2b}\ln\left(1 + br^2\right)\right] - \overline{\overline{C}} \,. \tag{3.10}$$

The constants α und β depend on a and b. Integration of this velocity distribution leads finally to a friction law which is a correlation between the friction factor

$$\lambda = \frac{\Delta p_\infty}{\frac{\rho}{2} u_{\mathrm{m}}^2} \frac{D}{L} = \frac{8 \tau_{\mathrm{w}\infty}}{\rho\, u_{\mathrm{m}}^2} = 8\epsilon^2 \tag{3.11}$$

and the Reynolds number

$$Re = \frac{u_{\mathrm{m}} D}{\nu} \,. \tag{3.12}$$

The constants a, b, κ and $\overline{\overline{C}}$ in (3.10) have been chosen in such a way that the friction law derived from the Superpipe Experiments by Zagarola and Smits (1998) is reproduced:

$$\frac{1}{\sqrt{\lambda}} = 1.869 \log\left(Re\sqrt{\lambda}\right) - 0.241 \,. \tag{3.13}$$

The constants are:

$$\begin{aligned} a &= -0.2517 \,, & b &= 6.3171 \\ \alpha &= -0.1411 \,, & \beta &= 9.0919 \\ \kappa &= 0.4356 \,, & \overline{\overline{C}} &= -4.36 \,. \end{aligned} \tag{3.14}$$

3.5 Minimal Program

In practice it is sufficient to use three terms of the Fourier series, i.e., $k = 0$, 1 and 2. As already mentioned, the series are linear combinations of exponential functions. Experiments, however, often show a pure exponential decay. This means that all higher eigensolutions are already damped out except the leading one. Far downstream a pure exponential decay is justified. The minimal number of eigensolutions involved is therefore 3 times 2 = 6. Each eigensolution represents a particular flow structure, which has its individual decay law. Table 3.1 gives the first six flow structures. Their decay behavior is determined by the *tenth-value length* $L_{0.1}$, which is the length within which a value is decayed to one tenth of its original value. In Table 3.2 the six eigensolutions are listed. It is

$$\begin{aligned} U_{\mathrm{U}ki}(r) &= \overline{U}_{\mathrm{U}ki}(r) + \epsilon\widetilde{U}_{\mathrm{U}ki}(r) \;, \\ W_{\mathrm{W}ki}(r) &= \overline{W}_{\mathrm{W}ki}(r) + \epsilon\widetilde{W}_{\mathrm{W}ki}(r) \;. \end{aligned} \tag{3.15}$$

The eigensolutions are sets of orthogonal functions, i.e.,

$$\left.\begin{aligned} \int_0^1 u_\infty(r) U_{\mathrm{U}ki} U_{\mathrm{U}kj} r \mathrm{d}r &= 0 \\ \int_0^1 u_\infty(r) W_{\mathrm{W}ki} W_{\mathrm{W}kj} r \mathrm{d}r &= 0 \end{aligned}\right\} \quad \text{for } i \neq j \tag{3.16}$$

with the consequence that the characteristic parameters according to (3.3) to (3.8) are independent of the number of parameters being used.

It is noteworthy that $B_{\mathrm{W}0.1}$ is proportional to the commonly used swirl number S, i.e., $B_{\mathrm{W}0.1} = 2S$, when $u(r)$ is substituted by the velocity distribution of the fully developed flow $u_\infty(r)$. In practice this latter condition is usually satisfied because the decay rate for axial disturbances is much higher than that for circumferential disturbances, see Table 3.1.

Table 3.1. Tenth-value length $L_{0.1}$ for the first six flow structures in descending order. $Re = 10^6$.

	flow structure	$L_{0.1}/D$
1	single longitudinal vortex	194
2	source-sink pair	151
3	vortex pair	56
4	source-sink quadrupole	56
5	vortex quadrupole	32
6	ring vortex	31

Since the flow structures are usually determined by two characteristic parameters (intensity and orientation) except the axisymmetric ones, ten characteristic parameters are needed to describe the first six flow structures listed in Table 3.1. These can be determined using (3.3) to (3.8). However, if the velocities and their first and second derivatives at the pipe axis are known, these ten characteristic parameters are given by:

$$A_{\mathrm{U01}} = \left[u(0) - u_\infty(0)\right] / U_{\mathrm{U01}}(0) \tag{3.17}$$

$$B_{\mathrm{U11}} = u'\left(0, \theta = 90°\right) / U'_{\mathrm{U11}}(0) \tag{3.18}$$

$$A_{\mathrm{U11}} = u'\left(0, \theta = 90°\right) / U'_{\mathrm{U11}}(0) \tag{3.19}$$

$$A_{\mathrm{U21}} = \left[u''\left(0, \theta = 0°\right) - u''_\infty(0) - U''_{\mathrm{U01}}(0) A_{\mathrm{U01}}\right] / U''_{\mathrm{U21}}(0) \tag{3.20}$$

$$B_{\mathrm{U21}} = \left[u''\left(0, \theta = 45°\right) - u''_\infty(0) - U''_{\mathrm{U01}}(0) A_{\mathrm{U01}}\right] / U''_{\mathrm{U21}}(0) \tag{3.21}$$

$$\begin{aligned} B_{\mathrm{W01}} = & -\left\{ \left[w'\left(0, \theta = 45°\right) + w'\left(0, \theta = 90°\right)\right] / \sqrt{2} \right. \\ & \left. + w'\left(0, \theta = 112.5°\right) \right\} / \left(1 + \sqrt{2}\right) W'_{\mathrm{W01}}(0) \end{aligned} \tag{3.22}$$

$$A_{\mathrm{W11}} = w\left(0, \theta = 90°\right) / W_{\mathrm{W11}}(0) \tag{3.23}$$

$$B_{\mathrm{W11}} = -w\left(0, \theta = 0°\right) / W_{\mathrm{W11}}(0) \tag{3.24}$$

$$A_{\mathrm{W21}} = \left[w'\left(0, \theta = 45°\right) - W'_{\mathrm{W01}}(0) B_{\mathrm{W01}}\right] / W'_{\mathrm{W21}}(0) \tag{3.25}$$

$$B_{\mathrm{W21}} = -\left[w'\left(0, \theta = 90°\right) - W'_{\mathrm{W01}}(0) B_{\mathrm{W01}}\right] / W'_{\mathrm{W21}}(0) \tag{3.26}$$

The values and their derivatives of the eigenfunctions at the axis are given in Table 3.2. Furthermore it is

$$u_\infty(0) = 1 + 4.36\epsilon \quad , \quad u''_\infty(0) = -25.14\epsilon \,. \tag{3.27}$$

Table 3.2. Components of the six eigensolutions $U_{\mathrm{U}k0}(r)$ and $W_{\mathrm{W}k0}(r)$ for $k = 0, 1, 2$ according to (3.15) after Gersten and Papenfuss in Chap. 2 of this book.

	$\overline{U}''_{\mathrm{U01}}(0)$ = 28.101	$\widetilde{U}''_{\mathrm{U01}}(0)$ = 303.73	$\overline{U}'_{\mathrm{U11}}(0)$ = 1.8699	$\widetilde{U}'_{\mathrm{U11}}(0)$ = 7.0182	$\overline{U}''_{\mathrm{U21}}(0)$ = 5.0349	$\widetilde{U}''_{\mathrm{U21}}(0)$ = -55.952	$\overline{W}'_{\mathrm{W01}}(0)$ = 1.0000	$\widetilde{W}'_{\mathrm{W01}}(0)$ = 4.3586			$\overline{W}'_{\mathrm{W21}}(0)$ = -2.0867	$\widetilde{W}'_{\mathrm{W21}}(0)$ = -155.34
r	$\overline{U}_{\mathrm{U01}}$	$\widetilde{U}_{\mathrm{U01}}$	$\overline{U}_{\mathrm{U11}}$	$\widetilde{U}_{\mathrm{U11}}$	$\overline{U}_{\mathrm{U21}}$	$\widetilde{U}_{\mathrm{U21}}$	$\overline{W}_{\mathrm{W01}}$	$\widetilde{W}_{\mathrm{W01}}$	$\overline{W}_{\mathrm{W11}}$	$\widetilde{W}_{\mathrm{W11}}$	$\overline{W}_{\mathrm{W21}}$	$\widetilde{W}_{\mathrm{W21}}$
0	-2.3400	-15.624	0.0000	0.0000	0.0000	0.0000	0.0000	0.0000	-0.7101	-8.1690	0.0000	0.0000
0.1	-2.2048	-14.177	0.1836	0.6719	0.0245	0.0906	0.1000	0.4236	-0.6734	-5.0847	-0.1866	-8.6506
0.2	-1.8544	-10.564	0.3494	1.1870	0.0910	0.3128	0.2000	0.7799	-0.5713	-2.8733	-0.3218	-5.3541
0.3	-1.4006	-6.2335	0.4892	1.4695	0.1849	0.5626	0.3000	1.0240	-0.4235	-1.2175	-0.3788	-3.4354
0.4	-0.9343	-2.3003	0.6037	1.5173	0.2931	0.7473	0.4000	1.1324	-0.2475	0.0643	-0.3546	-1.8481
0.5	-0.5017	0.7217	0.6976	1.3521	0.4075	0.8034	0.5000	1.0874	-0.0555	0.9671	-0.2578	-0.4599
0.6	-0.1179	2.6973	0.7759	0.9833	0.5243	0.6796	0.6000	0.8616	0.1460	1.4556	-0.0992	0.6306
0.7	0.2176	3.6197	0.8426	0.3842	0.6420	0.3074	0.7000	0.3980	0.3533	1.4579	0.1118	1.2469
0.8	0.5107	3.4231	0.9008	-0.5523	0.7603	-0.4541	0.8000	-0.4372	0.5650	0.8057	0.3681	1.1011
0.9	0.7691	1.6718	0.9527	-2.2176	0.8794	-2.0397	0.9000	-2.0689	0.7805	-1.0180	0.6651	-0.4868
0.95	0.8876	-0.4488	0.9768	-3.8845	0.3995	-3.7320	0.9500	-3.7638	0.8897	-3.0349	0.8279	-2.5859
0.96	0.9105	-1.1232	0.9815	-4.4146	0.9515	-4.2763	0.9600	-4.3084	0.9117	-3.6713	0.8616	-3.2665
0.97	0.9332	-1.9731	0.9862	-5.1024	0.9636	-4.9833	0.9700	-5.0069	0.9337	-4.4846	0.8956	-4.1384
0.98	0.9557	-3.1297	0.9908	-6.0538	0.9757	-5.9597	0.9800	-5.9837	0.9558	-5.5848	0.9300	-5.3145
0.99	0.9780	-5.0017	0.9954	-7.7231	0.9878	-7.6640	0.9900	-7.6329	0.9779	-7.4419	0.9648	-7.2752
1.00	1.0000	$-1/\epsilon$	1.0000	$-1/\epsilon$	1.0000	$-1/\epsilon$	1.0000	$-1/\epsilon$	1.0000	$-1/\epsilon$	1.0000	$-1/\epsilon$

3.6 Example

The turbulent flow in a pipe downstream of a single bend has been measured in detail by Mickan (1999) using Laser-Doppler anemometry. Figure 3.2 shows the velocity distributions ten diameters downstream of the bend. The thirty first characteristic parameters have been determined from these experimental data and are listed in Table 3.3. The velocity distributions reconstructed from these characteristic parameters are compared with the experimental data in Fig. 3.2 (solid lines). The dotted lines result from assuming that the first six eigensolutions are sufficient to describe the flow field (minimal program, $k = 0, 1, 2$; $i = 1$; U- and W-type).

Table 3.3. The first 30 characteristic parameters $A_{\mathrm{U}ki}$, $B_{\mathrm{U}ki}$, $A_{\mathrm{W}ki}$, $B_{\mathrm{W}ki}$ ($k = 0, 1, 2$; $i = 1, 2, 3$) for the pipe flow in the cross section 10 diameters downstream of a bend according to measurements by Mickan (1999). $Re = 1.9 \times 10^5$ ($\epsilon = 0.0445$).

k	i	$A_{\mathrm{U}ki}$	$B_{\mathrm{U}ki}$	$A_{\mathrm{W}ki}$	$B_{\mathrm{W}ki}$
	1	0.0541	-	-	0.0020
$k = 0$	2	0.0170	-	-	0.0027
	3	-0.0113	-	-	-0.0024
	1	-0.0290	-0.0138	-0.0065	0.0131
$k = 1$	2	-0.0046	-0.0036	0.0004	0.0063
	3	0.0140	-0.0045	0.0080	-0.0078
	1	-0.0115	0.0060	0.0190	0.0086
$k = 2$	2	0.0012	0.0032	-0.0034	-0.0025
	3	0.0045	-0.0046	-0.0053	-0.0045

3.7 Experimental Determination of the Characteristic Parameters (Minimal Program)

In case that the first eigensolutions are sufficient to describe the flow, the characteristic parameters can be determined quite easily from experiments as Gersten et al. (2001) and Wildemann (2000) have shown, see also Chap. 14 of this book. The authors measured the circumferential distributions of the wall shear stress components and could determine the coefficients by using the formulas derived from (3.1) and (3.2):

$$\frac{\tau_{\mathrm{w}x}}{\tau_{\mathrm{w}\infty}} = 1 + 2 \sum_{k=0,\ldots} [A_{\mathrm{U}ki} \cos k\theta + B_{\mathrm{U}ki} \sin k\theta] \tag{3.28}$$

$$\frac{\tau_{\mathrm{w}\theta}}{\tau_{\mathrm{w}\infty}} = \sum_{k=0,\ldots} [A_{\mathrm{W}ki} \sin k\theta - B_{\mathrm{W}ki} \cos k\theta] \ . \tag{3.29}$$

Figure 3.3 shows a typical result of the measured wall shear stress distributions from Wildemann (2000). The coefficients of the Fourier series of these distributions are

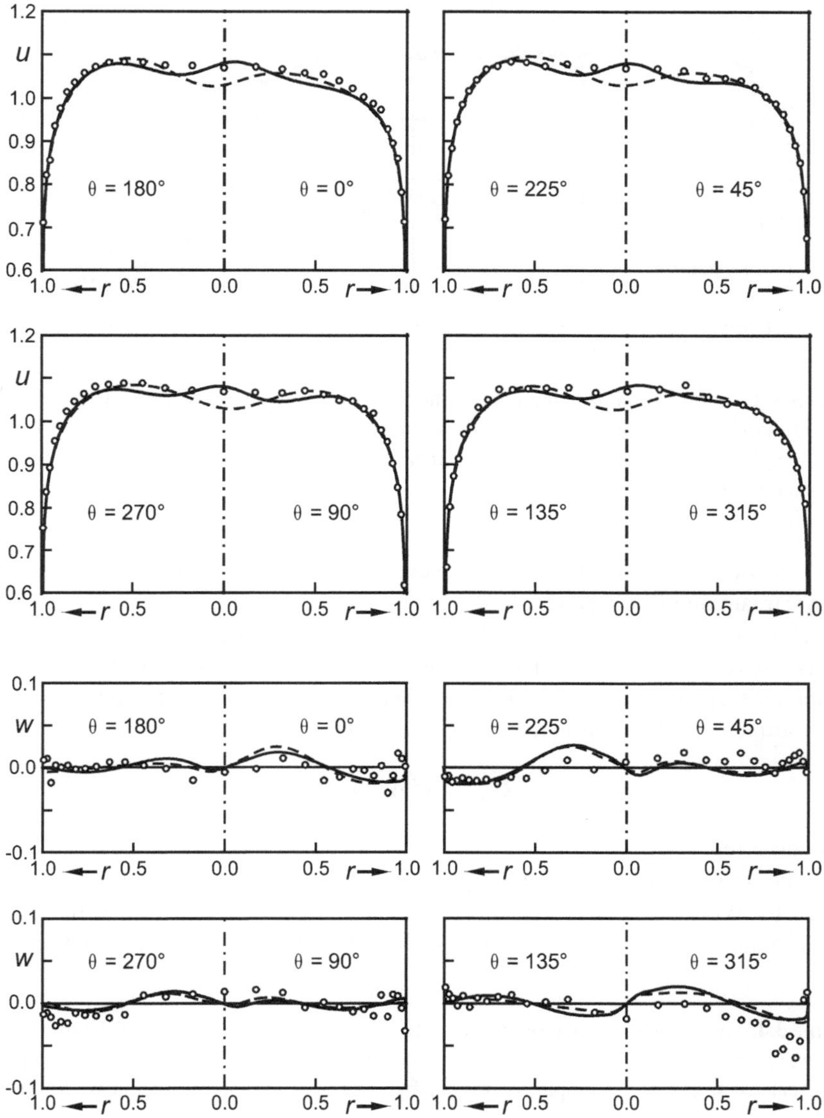

Fig. 3.2. Velocity distributions $u(r, \theta)$ and $w(r, \theta)$ 10 diameters downstream of a bend at $Re = 1.9 \times 10^5$. (○) Measurements after Mickan (1999); (——) Series expansions according to (3.1) and (3.2), $k = 0, 1, 2$; $i = 1, 2, 3$ for $x = 0$. The coefficients are listed in Table 3.3. (- - - -) Series expansions according to (3.1) and (3.2). Minimal program: $k = 0, 1, 2$; $i = 1$ for $x = 0$.

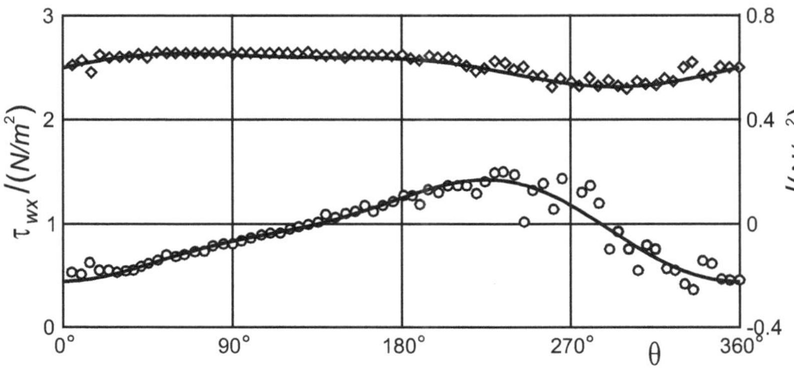

Fig. 3.3. Circumferential distributions of the wall shear stress components after Wildemann (2000). Position: 11 diameters downstream of a single bend. $Re = 2.2 \times 10^5$ ($\epsilon = 0.0439$). $\tau_{\mathrm{W}\infty} = 2.28 N/m^2$ (measured); (◇) $\tau_{\mathrm{W}x}$ measured; (○) $\tau_{\mathrm{W}\theta}$ measured; (—) Fourier series (terms up to $k = 2$).

the characteristic parameters. The measurements have been carried out 11 diameters downstream of a single bend.

Since the geometries and the Reynolds numbers of the experimental investigations are very similar, the characteristic parameters of Mickan (1999) and Wildemann (2000) are compared with each other in Table 3.4. Only the intensity of the three-dimensional flow structures characterized by

$$C_{\dots} = \sqrt{A_{\dots}^2 + B_{\dots}^2} \tag{3.30}$$

is considered.

The agreement is surprisingly good except for $C_{\mathrm{W}11}$. The value $C_{\mathrm{W}11}$ measured by Wildemann is probably the correct one because it agrees with other results found in the literature. Norman et al. (1989), for instance, measured the value

Table 3.4. The optimal characteristic parameters describing the intensities of the first six flow structures given in Table 3.1 for the turbulent pipe flow about 10 diameters downstream of bends at $Re \approx 2 \times 10^5$. Comparison of data by Mickan (1999) and Wildemann (2000).

	Mickan (1999) $Re = 1.9 \times 10^5$ $x/D = 10$	Wildemann (2000) $Re = 2.2 \times 10^5$ $x/D = 11$
$A_{\mathrm{U}01}$	0.054	0.053
$C_{\mathrm{U}11}$	0.032	0.032
$C_{\mathrm{U}21}$	0.013	0.012
$B_{\mathrm{W}01}$	0.002	0.013
$C_{\mathrm{W}11}$	0.015	0.070
$C_{\mathrm{W}21}$	0.021	0.021

$w_{\max}(= 0.12)$ at the axis 6.2 diameters downstream of a bend at $Re = 10^5$. This leads according to (3.23), (3.24) and (3.30) to

$$C_{W11} = \frac{|w_{\max}(0)|}{|W_{W11}|} = \frac{0.12}{1.1} = 0.11 \tag{3.31}$$

which is close to Wildemann's result.

References

Cebeci T, Chang KC (1978) A general method for calculating momentum and heat transfer in laminar and turbulent duct flows. Numerical Heat Transfer 1:39–68

Gersten K, Merzkirch W, Wildemann C (2001) Verfahren und Vorrichtung zur Korrektur fehlerhafter Messwerte von Durchflussmessgeräten infolge gestörter Zuströmung. Patent No. 197 24 116

Klein A (1981) REVIEW: Turbulent developing pipe flow. J. Fluids Eng. 103:243–249

Mickan B (1999): Systematische Analyse von Installationseffekten sowie der Effizienz von Strömungsgleichrichtern in der Großgasmengenmessung. Doctor thesis, University Essen

Morrison GL, DeOtte RE, Beam EI (1992) Installation effects upon orifice flow meters. Flow Meas. Instrum. 3:89–93

Morrison GL, Hauglie I, DeOtte RE (1995) Beta ratio, axisymmetric flow distortion and swirl effects upon orifice flow meters Flow Meas. Instrum. 6:207–216

Norman RS, Mattingly GE, McFaddin SE (1989) The decay of swirl in pipes. Proceedings International Gas Research Conference 1989:312–321

Steenbergen W, Voskamp J (1998) The rate of decay of swirl in turbulent pipe flow. Flow Meas. Instrum. 9:67–78

Sudo K, Sumida M, Hibara H (1998) Experimental investigation of turbulent flow in a circular-sectioned 90-degree bend. Exp. in Fluids 25:42–49

Wildemann C (2000) Ein System zur automatischen Korrektur der Messabweichung von Durchflussmessgeräten bei gestörter Anströmung. Doctor thesis, Universität Essen. See also: Fortschritt-Berichte VDI, Reihe 8, Nr. 868

Zagarola MV, Smits AJ (1998) Mean-flow scaling of turbulent pipe flow. J. Fluid Mech. 373:33–79

4

Measurement of Velocity and Turbulence Downstream of Flow Conditioners

Wei Xiong, Kathrin Kalkühler, Wolfgang Merzkirch

Institut für Strömungslehre, Universität Essen, 45117 Essen, Germany

Particle-image velocimetry and hot-wire anemometry are used for experimentally studying the flow downstream of three conditioners, a tube bundle and two perforated plates at Reynolds numbers of the order 10^5. The conditioners are exposed to the flow disturbed by two different installations: a 90° single bend and a $2 \times 90°$ out-of-plane double bend. Velocity profiles, turbulent fluctuations and Reynolds' stress are measured. The jets issuing from the holes and tubes of the conditioners are visualised in the near field, which extends to approximately four pipe diameters downstream of the conditioners. At that position the disturbance imposed on the flow by the conditioners disappears, while the decay of the disturbance caused by the installations takes place in the far field. The decay rate in the far field depends on the specific installation. While the velocity profiles match the profile for fully developed flow at a position of approximately 25 diameters downstream of the conditioners, the turbulent equilibrium state is not reached even at 50 diameters. The results also show that the perforated plates have a higher efficiency than the tube bundle in conditioning the disturbed flow.

4.1 Introduction

Flow conditioners serve for homogenising a velocity profile in pipe flow that is disturbed by an installation. A certain class of conditioners is intended to accelerate the formation of a fully developed turbulent velocity profile, i.e., a definite state of flow that must exist at the position of a flow meter whose calibration was done in fully developed flow. Such conditioners, a number of which are defined in national or international standards, e.g., DIN1952 (1982), allow to reduce the length of straight pipe required upstream of a flow meter. This obvious advantage must be paid for by an additional pressure decrease, expressed, e.g., by a pressure "loss" coefficient $\xi = (2\,\Delta p)/(\rho\, u_\mathrm{b}^2)$, with Δp being the pressure decrease caused by the conditioner, ρ the fluid density, and u_b the bulk velocity in the pipe (volume flow divided by the cross-sectional area). Many attempts have been made in optimising the design of a conditioner such that the desired reduction in straight pipe length is maximized and the inevitable pressure decrease Δp minimised. This led to a variety of different approaches in the design of flow conditioners that are currently in practical use.

Here we report on measurements of the distribution of flow velocity and turbulence downstream of three different conditioners: tube bundle (see (DIN1952, 1982)), and two perforated plates according to the design of Akashi et al. (1978) and Laws (1990). The applied experimental methods are hot-wire and particle-image velocimetry (PIV). The PIV measurements are an extension of experiments reported earlier by Schlüter and Merzkirch (1996) with considerably improved experimental capabilities: The tracer particles being much smaller than those used in the earlier experiments allow to resolve velocity fluctuations; time-averaged velocity profiles can be determined from a total of several hundred digitally recorded PIV images; PIV recordings can be taken in measurement planes normal to the pipe axis, as specified in detail in Xiong and Merzkirch (1999). The independent use of two different experimental methods and the comparison of the data acquired with them improves the confidence in the measurement results.

The aim of the measurements is to characterize the flow field downstream of the conditioners that are exposed to pipe flow disturbed by two specific installations: a 90° bend and a $2 \times 90°$ out-of-plane double bend. It is of interest to see how quickly the flow approaches the fully developed state, regarding time-averaged velocity profiles and turbulence characteristics. This equilibrium state is formed in the far field of the conditioners. The speed of its formation should depend on the rate at which the fluid jets issue from the tubes or openings mix in the near field of the conditioners. Therefore, the near field is also investigated and visualised with PIV.

The Reynolds number Re_{D} in the experiments is of the order of 100.000, based on the pipe diameter D. This magnitude of Re_{D} is of interest for many technical applications, and it permits to compare the results with other data reported in the literature. A great number of these results were obtained for the combination of a flow conditioner with a specific flow meter. The investigations reported by Morrison et al. (1997) are a typical example of respective research for this order of magnitude of the Reynolds number. Extensive descriptions of many details of the experimental programme reported in this chapter are given in the dissertations of Kalkühler (1998) and Xiong (2000).

4.2 Experiments

The test rig is a steel pipeline with $D = 100\,\mathrm{mm}$ inner diameter and hydraulically smooth internal surfaces. Several straight pipe sections can be assembled to provide a total length of straight pipeline of $120\,D$, if so desired. Air as the working fluid is sucked through the pipe by means of a ventilator positioned at the downstream end of the line. The highest achievable Reynolds number, approx. 300,000, is determined by the power of the ventilator. The installations are positioned $23\,D$ downstream of the intake nozzle, while the flow conditioners are at a distance of $2\,D$ downstream of the installations (Fig. 4.1). The distance of straight pipe between installation and ventilator can be up to $80\,D$. The test rig is equipped with a device for taking continuous reference measurements of the volumetric flow rate with an uncertainty of $\pm0.3\%$: further details of the test rig and the reference measurements can be found in Kalkühler (1998) and Xiong (2000).

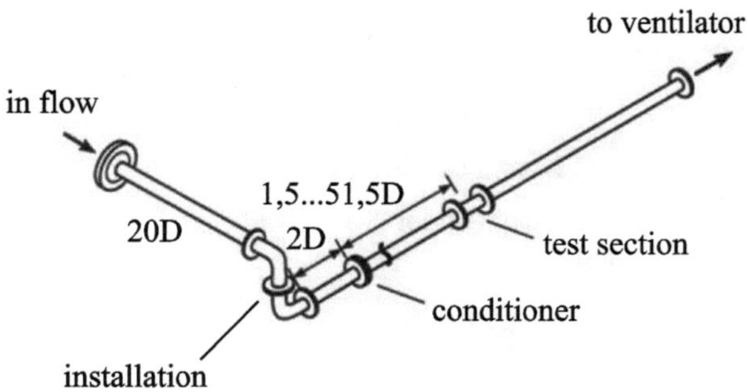

Fig. 4.1. Pipe line test rig.

Like in many comparable experimental investigations two installations are used in the experiments: A 90° bend with $D = 100\,\mathrm{mm}$ i.d. and bend radius $R_\mathrm{B} = 2\,D$, and a $2 \times 90°$ out-of-plane double bend, composed of two 90° single bends. The three flow conditioners are depicted in Fig. 4.2.

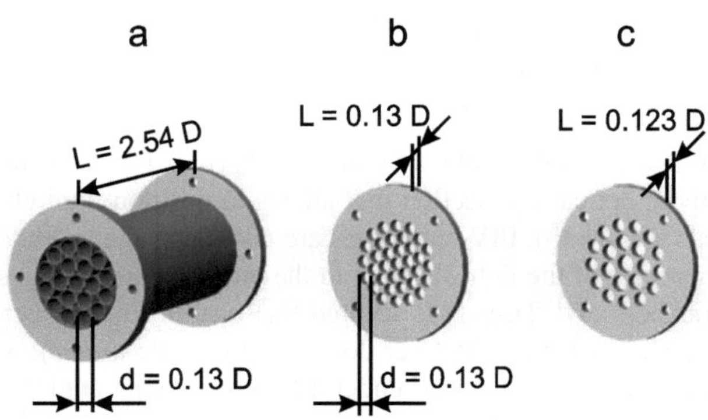

Fig. 4.2. The three conditioners investigated: **(a)** tube bundle , **(b)** perforated plate according to Akashi et al. (1978), **(c)** perforated plate according to Laws (1990).

An optical test section for taking PIV measurements can be installed at selected distances downstream of the conditioners. Two types of PIV test sections are used that allow different spatial orientation of the light sheet plane. For measuring the axial and radial components of the flow velocity in the pipe, the light sheet is vertical and includes the (horizontal) pipe axis. The motion of the tracer particles is observed and recorded by a CCD camera that is directed normal to the plane of the light sheet, i.e., in horizontal direction. Optical access for the observation is provided through the test section's transparent wall made of $0.3\,\mathrm{mm}$ thick acetate sheet so that optical distortions are minimised (Fig. 4.3a).

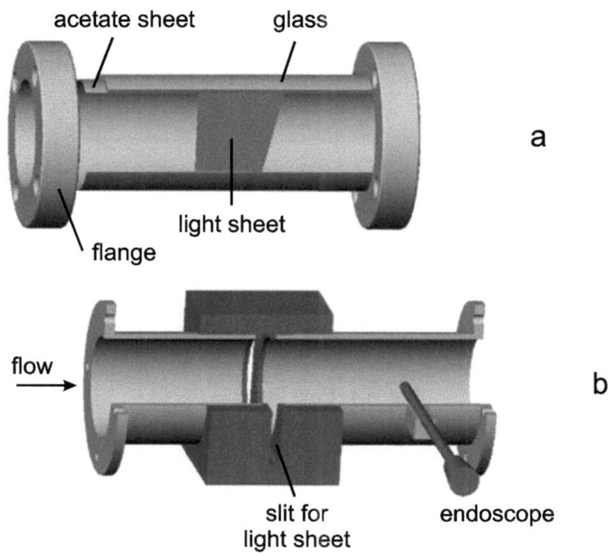

Fig. 4.3. Test sections used for the PIV experiments: **(a)** axial and radial velocity components are measured in the vertical light sheet that is observed through the acetate sheet forming part of the pipe wall; **(b)** radial and azimuthal velocity components are measured in the light sheet arranged normal to the pipe axis; observation is through the endoscope in upstream direction.

The radial and azimuthal velocity components (e.g., for determining swirl) are measured with an optical test section that allows to position the light sheet normal to the pipe axis (Fig. 4.3b). PIV recordings are taken through an endoscope located at $1.5\,D$ downstream of the light sheet, with the viewing direction upstream. With a light sheet thickness of $1.2\,\mathrm{mm}$ it is secured that a sufficient number of tracer particles is illuminated twice by the laser double pulse while moving within the light sheet. Since the particles are imaged under different viewing angles, an optical correction is necessary for determinig the true values of the radial and azimuthal velocity components (Xiong and Merzkirch, 1999).

Oil droplets as tracer particles are added to the air at the inlet nozzle of the test rig. Phase Doppler measurements show that these oil droplets have a size distribution with a maximum between 3 and $4\,\mu\mathrm{m}$ (Xiong, 2000). Illumination for the PIV measurements is provided by a double-pulsed Nd:YAG laser. The two successive particle-images are recorded with a CCD camera on separate frames, thus allowing an evaluation by means of the cross-correlation technique. The synchronisation of the laser's pulse repetition rate and the camera's frame rate results in a total of 6 PIV recordings that can be taken per second, practically at any desired length of measurement period. This allows to determine time-averaged velocity values from a total of several hundred individual and statistically independent PIV recordings.

A DANTEC constant-temperature hot-wire anemometer with a crossed-wire probe is used for velocity and turbulence measurements. Calibration and measurments are controlled by a PC and the software “Streamline”. In the calibration unit the probe is exposed to 13 different values of the flow angle between $-30°$ and $+30°$, and to 15 different values of the flow velocity ranging from $0.5\,\mathrm{m/s}$ to $55\,\mathrm{m/s}$. The results of the calibration are conserved in a look-up table which is also used, with respective interpolations, for the evaluation of the measurements.

The hot-wire measurements are performed using a special test section that can be inserted into the pipeline at selected positions along the pipe axis. The probe is traversed along radial paths separated by constant angular intervals, and measurements are taken at steps of $2\,\mathrm{mm}$ along each path and with the probe at three different orientations (0°, 45°, 90°) relative to the probe axis. This way, the three Carthesian velocity components can be determined and then converted into the axial, radial and azimuthal components. Two orientations of the probe (0°, 90°) are, in principle, sufficient for this purpose; the third orientation at 45° serves for increasing the measurement accuracy. The anemometer system and the applied software allow to determine mean (time-averaged) values of the velocity, RMS values of the turbulent fluctuations, and Reynolds stresses; further details are reported by Kalkühler (1998).

4.3 Results and Discussion

4.3.1 Nearfield Downstream of the Conditioners

The conditioner is a partial obstruction in the pipe cross section, therefore causing the mentioned pressure decrease, and the fluid is released to the downstream side through the openings or tubes from which the fluid issues in form of jets. Desirable is an intensive and quick mixing of these jets. We consider the near field to extend in downstream direction to an axial position where the jets have mixed completely and individual jets can no longer be discriminated. The measurement results presented in the following show that, for the range of Reynolds numbers investigated, this axial extension of the near field is approximately $4\,D$.

Profiles of the time-averaged axial velocity, measured with PIV in the near field of the three conditioners, are shown in Fig. 4.4. The vertical light sheet includes the pipe axis, and its orientation with respect to the exit plane of the conditioners is indicated on top of this figure. The Reynolds number is $Re_{\mathrm{D}} = 100{,}000$. The flow is disturbed, upstream of the conditioner, by the $2 \times 90°$ out-of-plane double bend. The velocity profiles exhibit two patterns characteristic for this configuration: the peaks indicating the position of the jets whose origin is obvious for the given orientation of the light sheet, and an asymmetry (skewness) caused by the double bend. For the purpose of comparison and reference, the diagrams also include the velocity profiles that would exist at this position in the case without conditioner, and the fully developed profile for this Reynolds number, calculated with the theory of

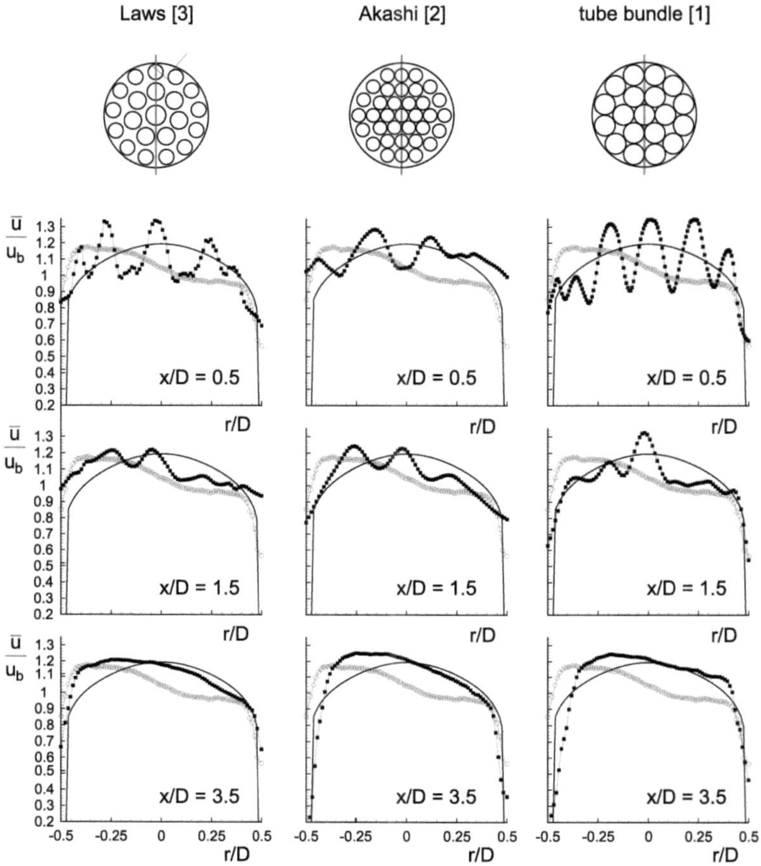

Fig. 4.4. Time-averaged profiles of the non-dimensional axial velocity component u in the near field of the three conditioners that are exposed to the flow disturbed by the $2 \times 90^{\circ}$ out-of-plane double bend; u_{b} = bulk velocity, r = radial coordinate, D = pipe diameter. Orientation of the light sheet with respect to the conditioner plane is indicated at the top. Fully developed profile (*solid line*) and the profile that would exist at the measurement position in the absence of a conditioner (*open circles*) are shown as reference. $Re_{\mathrm{D}} = 100{,}000$.

Gersten and Herwig (1992). The jets can be considered as a disturbance imposed to the flow by the conditioner, and it is visible that this disturbance decays rapidly, in contrast to the disturbance caused by the installation, the skewness, that persists clearly at the downstream end of the near field where the jets are no longer detectable. Thus, the self-generated disturbance of the conditioner disappears at the end of the near field, while the disturbance caused by the installation must decay in the far field, i.e., for $x/D > 4$. At the end of the near field there is no significant difference in the form of the velocity profiles, i.e., in the performance of the three conditioners investigated.

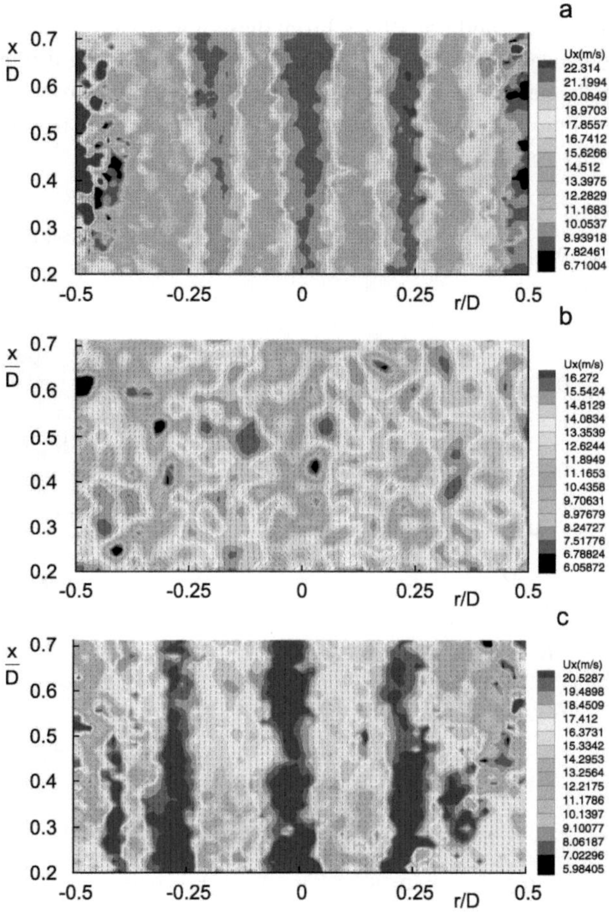

Fig. 4.5. Instantaneous distribution of the axial and radial velocity components in the near field of the three conditioners: **(a)** tube bundle, **(b)** perforated plate according to Akashi et al. (1978), **(c)** perforated plate according to Laws (1990). Orientation of the light sheet and installation upstream of the conditioner as in Fig. 4.4; $Re_{\mathrm{D}} = 100{,}000$. Velocities are depicted by arrows and velocity ranges by shades of gray.

Instantaneous velocity distributions measured immediately downstream of the conditioners, in the axial range from $0.2 \leq x/D \leq 0.7$, are presented in Fig. 4.5. The orientation of the PIV light sheet is indicated in Fig. 4.4; the left side of the velocity plots in Fig. 4.5 corresponds to the upper half of the light sheet plane as traced in Fig. 4.4. Form and position of the visible jets can be related easily to the positions of the openings of the conditioners from which these jets issue. Figure 4.5 provides an impression on the mixing of the jets in the regime covered by this PIV recording. This mixing is governed by interactions of the initially parallel jets. Until now, such interaction processes have been investigated only for two or three parallel jets (e.g., Nasr and Lai 1997). From these investigations it is known that the jet interaction causes an unsteadiness of the whole regime, and it must be expected that the jets in the near field of the conditioners tend to tumble around their axes. The applied PIV technique, however, does not provide the temporal resolution necessary for verifying this assumption.

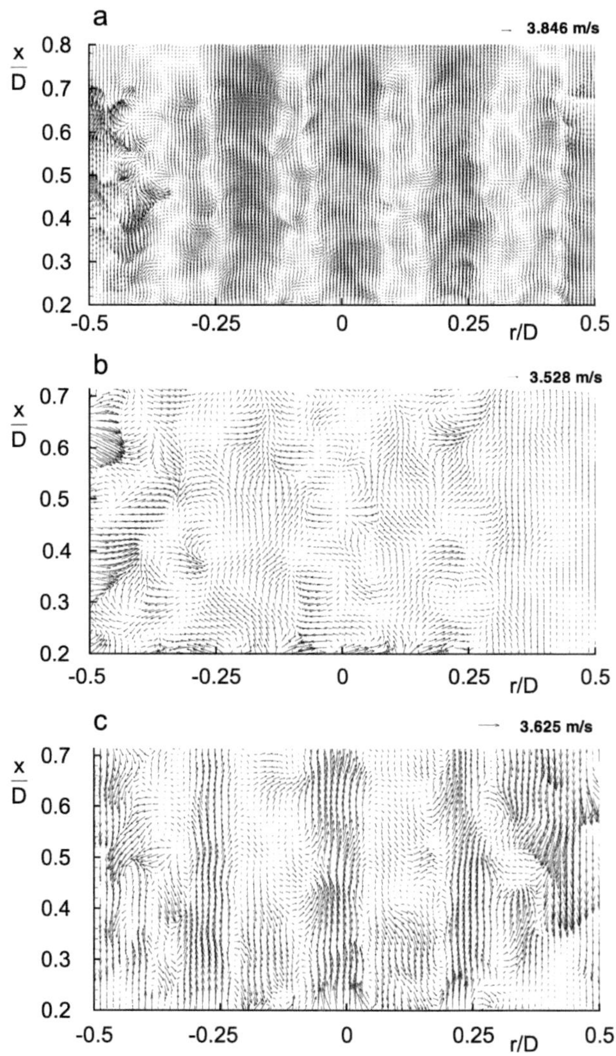

Fig. 4.6. Instantaneous pattern of the turbulence structure for the case shown in Fig. 4.5. A constant axial velocity of 15 m/s is subtracted from the distributions in Fig. 4.5.

Visualisation of the instantaneous distribution of the turbulent fluctuations in the measurement plane would require to first determine time-averaged values of the velocity, and then to subtract this distribution from an instantaneous velocity record, like the one in Fig. 4.5. Because of the mentioned unsteadiness of the whole flow, such time-averaging does not make sense, since both turbulence and unsteadiness would be averaged together. Instead, the constant value of the bulk velocity, $15\,\mathrm{m/s}$ (in axial direction), is subtracted from the velocity patterns of Fig. 4.5, in order to obtain a rough impression of the distribution of the turbulent fluctuations. The result is presented in Fig. 4.6 which gives an indication of the strong fluid motion normal to the jet axes. This motion which is a measure of the lateral mixing is veiled in the velocity plots of Fig. 4.5 by the dominating axial velocity component. A comparison of the rates of mixing caused by the three different conditioners appears not possible due to the different association of the light sheet plane to the distribution of openings in the exit plane of the conditioners.

4.3.2 Redevelopment of the Flow in the Far Field

The further decay of the flow disturbances in the far field of the conditioners is measured with hot-wire anemometry and PIV. Profiles of the time-averaged axial velocity component, measured in the vertical plane through the pipe axis (cf. Fig. 4.4), with the hot-wire anemometer are presented in Fig. 4.7 for the axial range $5.5 \leq x/D \leq 51.5$ downstream of the conditioners. As in the case of the previous PIV examples the installation upstream of the conditioners is the $2 \times 90°$ out-of-plane double bend. The profiles for $x/D = 5.5$ and 11.5 are measured with both PIV and hot-wire, and the patterns are practically identical. The skewness in the profiles, as it is visible at the end of the near field, disappears at approximately $x/D = 20$ downstream of the two perforated plates, while a residual skewness is still visible downstream of the tube bundle.

A comparison of the performance of the investigated conditioners, based on the measurement of the axial velocity component as results shown in Fig. 4.7, would require to compare the profiles for not only one orientation of the measurement plane. Four planes are used in the measurements, each including the pipe axis: $0°$(horizontal), $45°$, $90°$(vertical), $135°$. For the purpose of comparing the conditioners we determine mean values of the deviation (in percent) of the measured velocity from the value for fully developed flow along each measurement path (diameter according to the orientation of the measurement plane). The result applying to the $2 \times 90°$ out-of-plane double bend as the installation upstream of the conditioners is shown in Fig. 4.8. At $x/D = 11.5$ the relative deviation is approximately 3% for the two perforated plates (Laws and Akashi), while it is between 3% and 5%, depending on the measurement plane, for the tube bundle. The dependence of the deviation on the orientation of the measurement plane as it is obvious for the tube bundle again shows, that the skewness of the velocity profile has not yet disappeared. The values given here in $\%$ do not indicate the possible deviation of a flow meter's reading from the true value of the volumetric flow rate if the meter would be placed at that axial position: it is not known how a specific meter reacts on the disturbance of the velocity profile.

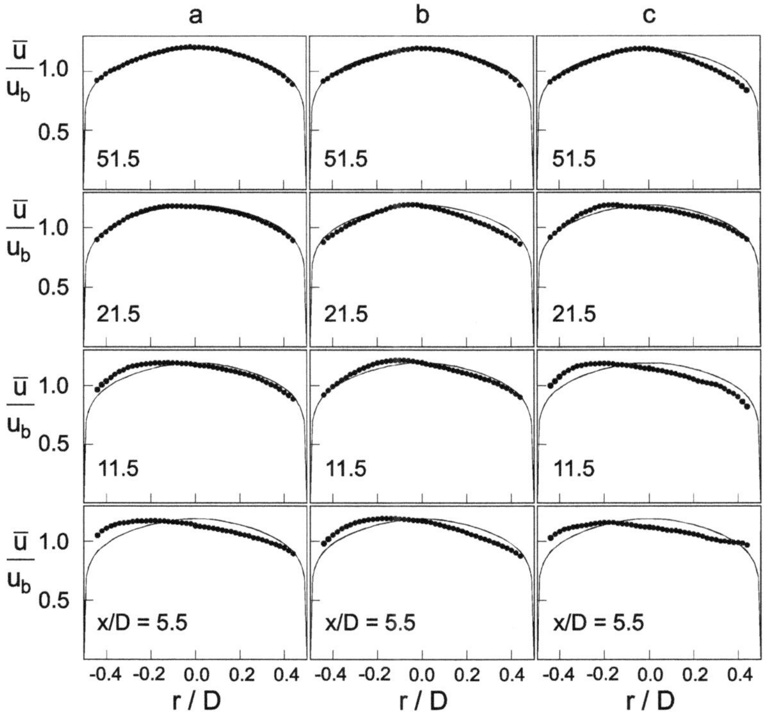

Fig. 4.7. Profiles of the time-averaged, non-dimensional axial velocity component u/u_b in the far field of the three conditioners that are exposed to the flow disturbed by the $2 \times 90°$ out-of-plane double bend (hot-wire data); u_b = bulk velocity in the pipe (volumetric flow rate/area of pipe cross section), r = radial coordinate, D = pipe diameter. Orientation of the light sheet with respect to the conditioner plane is as indicated in Fig. 4.4. Fully developed profile is shown for comparison (*solid line*); Re_D = 100,000. **(a)** perforated plate according to Laws (1990), **(b)** perforated plate according to Akashi et al. (1978), **(c)** tube bundle.

Figure 4.8 also indicates the relative deviation of the axial velocity profile from the fully developed state for the case that no conditioner is used. At $x/D = 11.5$ this value is approximately 6%. The hot-wire measurements show that, downstream of the three conditioners, the mean deviation is smaller than 2% for $x/D \geq 25$. It is believed that the accuracy in determining this value is 1% to 2%, so that the given value indicates practically the redevelopment of the undisturbed state, at least regarding the velocity profiles.

It is known that the $2 \times 90°$ out-of-plane double bend produces swirl. As stated above, the swirl components of the velocity field can be measured by means of the PIV setup with the light sheet perpendicular to the pipe axis (Xiong and Merzkirch, 1999), and also by hot-wire anemometry. As an example the instantaneous distribution of the radial and azimuthal velocity components at a position $x/D = 1.5$ downstream of the bend and for the Reynolds number Re_D = 100,000 is shown in Fig. 4.9. Besides the over-all swirl one recognises the two local vortices of the "sec-

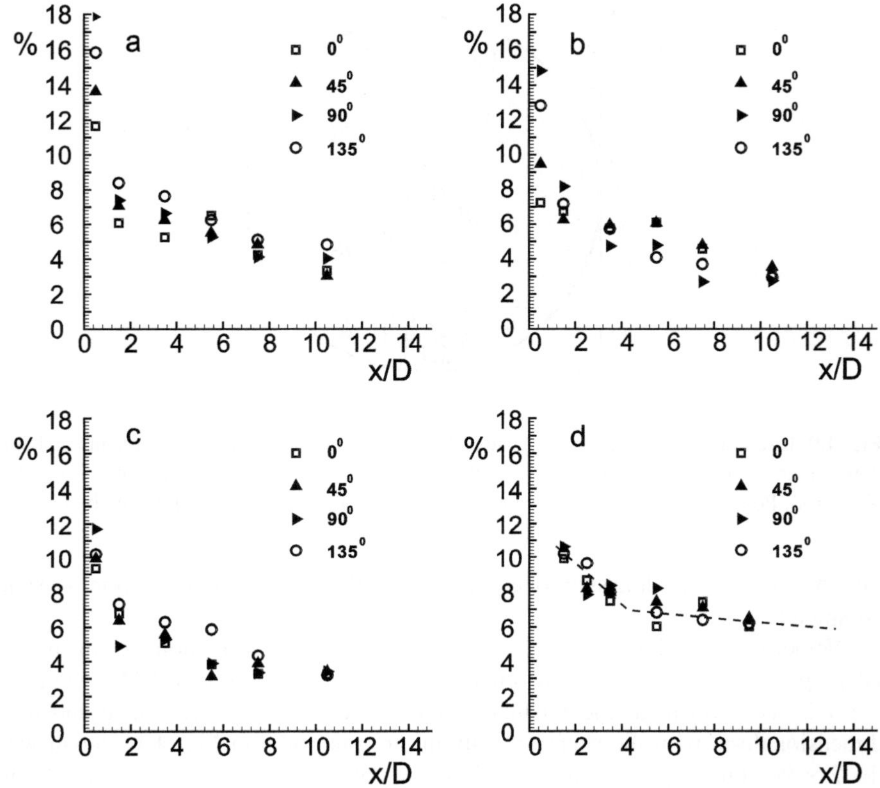

Fig. 4.8. Relative deviation in % (definition see text) from fully developed flow downstream of conditioners that are exposed to the flow disturbed by the $2 \times 90°$ out-of-plane double bend: **(a)** tube bundle, **(b)** perforated plate according to Akashi et al. (1978), **(c)** perforated plate according to Laws (1990); **(d)** values for the case without conditioner. Results for the measured axial velocity are shown for four orientations of the measurement plane, each including the pipe axis: 0°(horizontal), 45°, 90°(vertical), 135°. $Re_D = 100{,}000$.

ondary" flow generated in the bend. The positions of these secondary vortices are marked by circles in the figure. The degree at which a velocity distribution in the pipe is disturbed by swirl can be expressed by the "swirl angle", as defined by Mattingly and Yeh (1989). From both hot-wire and PIV measurements we determine the "mean" swirl angle, i.e., the swirl angle averaged over the respective cross section in the pipe, in the flow downstream of the three conditioners. As an example the result for the conditioner of Laws (1990) is shown in Fig. 4.10 together with the values of the mean swirl angle at the same axial positions as it is measured in the absence of any conditioner. The swirl is practically zero, even for $x/D = 1.5$, and this applies for all three conditioners investigated. In Fig. 4.10 the values of the swirl angle measured with conditioner are around 1°, and this is definitely below the value 2° that the norms, e.g., DIN1952 (1982), tolerate for measurements of the volumetric flow

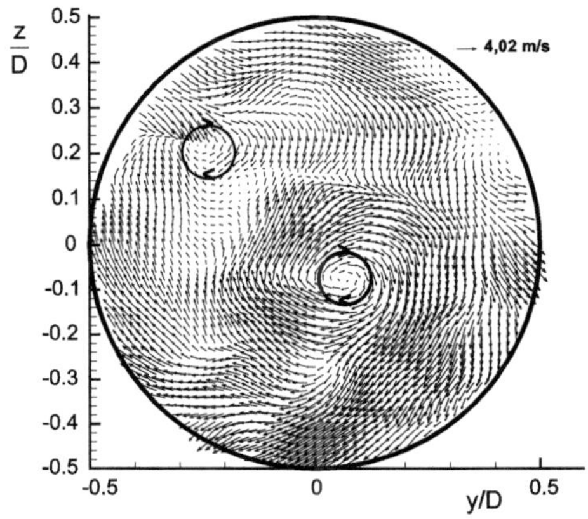

Fig. 4.9. Instantaneous distribution of the radial and azimuthal velocity components at $1.5\,D$ downstream of the $2 \times 90°$ out-of-plane double bend (PIV data in the y-z plane, normal to pipe axis). Positions of two vortices caused by the "secondary flow" are indicated by circles.

rate. As a result regarding swirl one can state that the three conditioners remove swirl completely.

Velocity fluctuations are determined with the hot-wire anemometer in the axial range $5.5 \leq x/D \leq 51.5$ downstream of the conditioners. RMS values of the axial component of the fluctuation, measured in the vertical plane and non-dimensionalised with the shear velocity ut, are presented in Fig. 4.11, again with the $2 \times 90°$ out-of-plane double bend as installation and for $Re_{\mathrm{D}} = 100{,}000$. The distribution for fully developed flow according to Lawn (1971) is given for comparison. Obviously the profiles for the tube bundle approach this equilibrium value more

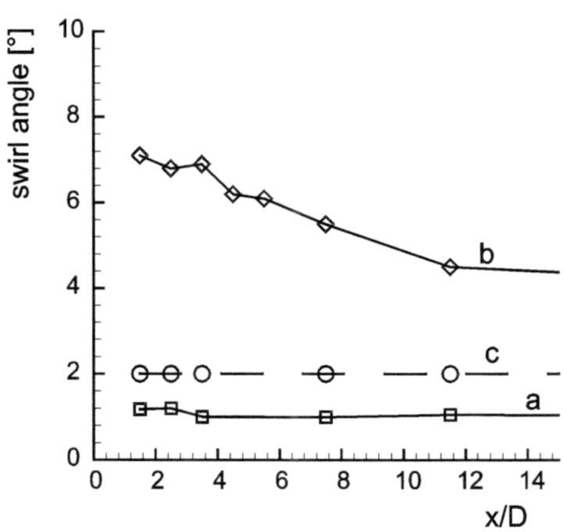

Fig. 4.10. Swirl angle as defined in Mattingly and Yeh (1989) downstream of position of conditioner; **(a)** with Laws (1990) conditioner, **(b)** without conditioner, **(c)** swirl angle 2° tolerated by norms. $Re_{\mathrm{D}} = 100{,}000$.

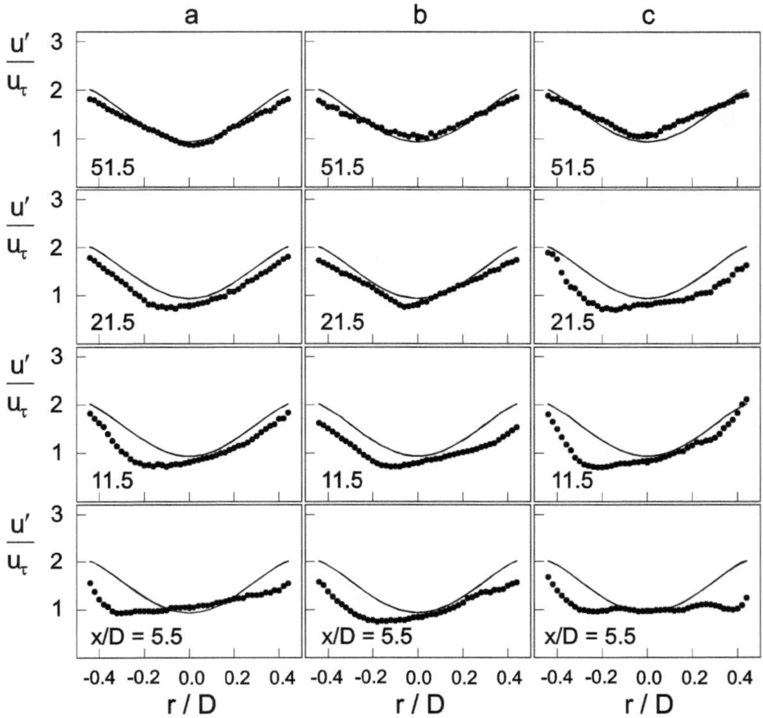

Fig. 4.11. RMS values of non-dimensional axial velocity fluctuations, u'/u_τ, downstream of the three conditioners exposed to the flow disturbed by the $2 \times 90°$ out-of-plane double bend (hot-wire data): **(a)** perforated plate according to Laws (1990), **(b)** perforated plate according to Akashi et al. (1978), **(c)** tube bundle. Vertical measurement plane, $Re_D = 100{,}000$. The profile for fully developed flow is shown as reference (*solid line*).

slowly than the profiles for the two perforated plates. Even at $x/D = 21.5$ there is a noticeable deviation from the equilibrium profile, and agreement with fully developed flow is only given, within the range of measurement accuracy, for $x/D = 51.5$. Similar results are found from the distribution of the Reynolds' stress formed with the axial and radial fluctuation (Fig. 4.12). Figures 4.11 and 4.12 indicate that the state of turbulence redevelops less rapidly than the velocity profiles, which can be explained with the additional production of turbulence by the conditioners. These results should be of interest for the modeling of the turbulence field in respective numerical computations, and also for the possible use of flow meters for which the generation of the signals depends on the turbulent state of the pipe flow, e.g., meters based on ultrasound.

Experiments are also performed with the 90° single bend as the installation upstream of the conditioners. Here, only two results typical for this series of measurements are presented. Further data can be found in Kalkühler (1998) and Xiong (2000). Mean values of the deviation of the time-averaged axial velocities from the

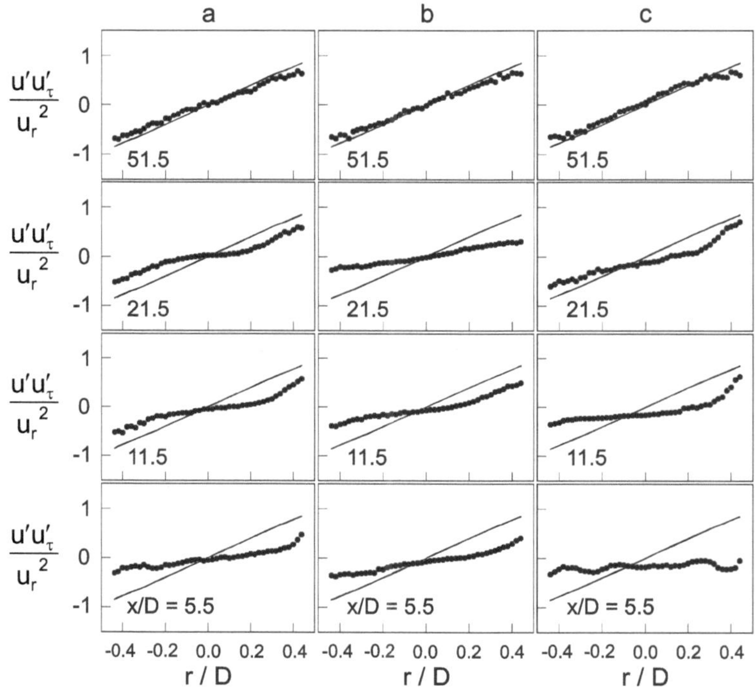

Fig. 4.12. Non-dimensional Reynolds' stress formed with the axial and radial fluctuation, u' and u'_r, downstream of the three conditioners. Flow, notation and reference as in Fig. 4.11.

fully developed profiles, defined above in connection with Fig. 4.8, are presented in Fig. 4.13. At $x/D = 11.5$ these deviations are somewhat higher than for the case with the $2 \times 90°$ out-of-plane double bend, between 3% and 4% for the two perforated plates, and around 6% for the tube bundle. The latter value is even not better than for the case without conditioner. In general, the disturbance in the velocity field caused by the single bend is damped to a lower degree by the conditioners than the disturbance produced by the double bend. Of course, this result applies here to the range of Reynolds numbers investigated.

As a second example for the single bend we show in Fig. 4.14 the profiles of the Reynolds' stress, equivalent to the case reported in Fig. 4.12. These data are determined with the hot-wire anemometer. They confirm the conclusion that the three conditioners are less efficient in removing the disturbance produced by the single bend in comparison to the case with the double bend. Even at $x/D = 51.5$, where the velocity profiles match well with the profile for fully developed flow (not shown here), there is a considerable difference to the Reynolds' stress profile for the equilibrium state.

Reynolds numbers higher than 300,000 cannot be produced due to the limitation in the power of the ventilator used in the experimentel setup. The experiments performed with $Re_{\mathrm{D}} = 300{,}000$ show no significant differences to the results already

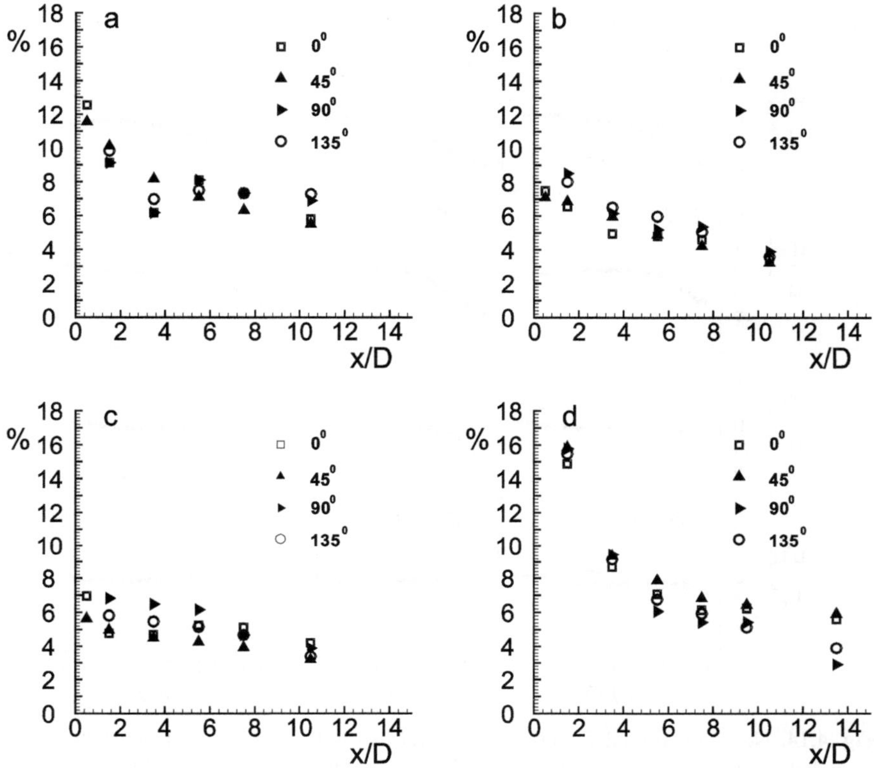

Fig. 4.13. Relative deviation (definition see text) from fully developed flow downstream of conditioners that are exposed to the flow disturbed by the 90° single bend. Notation and reference as in Fig. 4.8.

presented; for details the reader is again referred to Kalkühler (1998) and Xiong (2000).

4.4 Conclusion

The experimental results indicate, how rapid the flow that is disturbed by two specific installations redevelops under the influence of the three conditioners investigated, the perforated plates according to Akashi et al. (1978) and Laws (1990), respectively, and the tube bundle as defined in technical norms, e.g., DIN1952 (1982). The experiments are performed at Reynolds numbers of the order 10^5. They show that the flow downstream of the conditioners can be separated in two regimes: The disturbance imposed on the flow by the conditioner disappears at the end of the near field that extends to approximately four pipe diameters downstream of the conditioner. In this regard, the conditioners, particularly the perforated plates, can be compared with a turbulence grid. Such grids are often used for generating homogeneous and quasi-

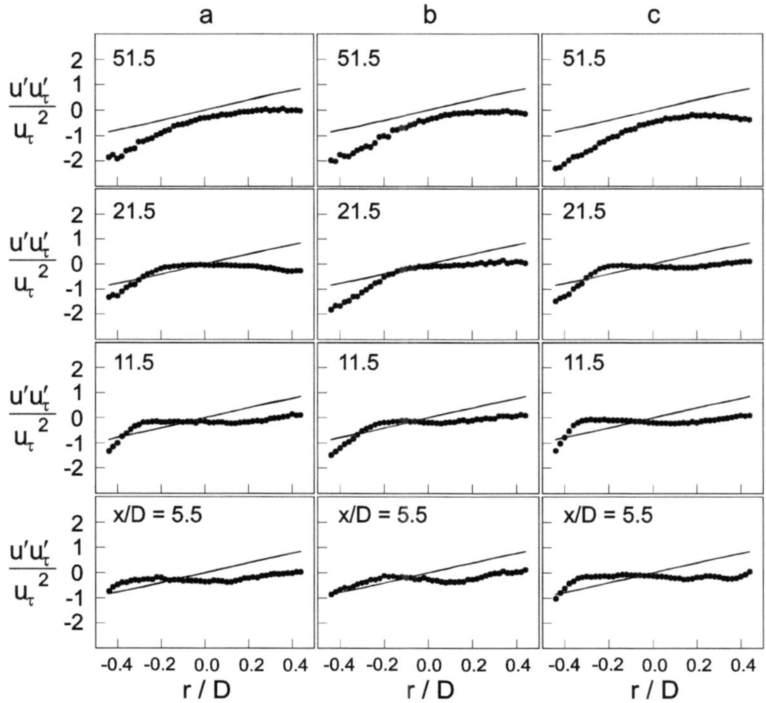

Fig. 4.14. Non-dimensional Reynolds' stress formed with the axial and radial fluctuation, u' and u'_r, downstream of the three conditioners with the 90° single bend as the upstream installation. Notation and reference as in Fig. 4.10.

isotropic turbulence which is assumed to establish about 30 mesh sizes downstream of the grid (see, e.g., Hinze (1975)). From the geometry of the perforated plates (Fig. 4.2) one can derive an equivalent mesh size of approximately 15 mm, and the end of the near field then coincides with the establishment of quasi-isotropic turbulence, i.e., a flow state in which any information on the characteristic length scales of the grid or plate has disappeared due to dissipation.

The decay of the disturbance caused by the installation takes place in the far field. The decay rate is different for the two installations: the decay is more rapid for the double bend. The velocity profiles match the profile for fully developed flow approximately at 25 diameters downstream of the conditioners, which is in agreement with results reported by Morrison et al. (1997). However, the turbulence field, as expressed e.g., by the Reynolds stress, does not reach the equilibrium state even at 50 diameters. This must be considered when describing this flow numerically. A comparison of the three conditioners shows that the efficiency of the two perforated plates is higher than that of the tube bundle.

The experiments do not yield direct information on how the reading of a flow meter is affected by the deviation of the flow from the fully developed state. If it is concluded that a meter must be positioned at 25 diameters or more downstream of the conditioner, an alternative to a conditioner for avoiding very long lengths of straight pipeline might be the correction of the meter's reading as proposed by Wildemann et al. (2002); see also Chap. 14 of this volume.

References

Akashi K, Watanabe H, Koga K (1978) Flow rate measurement in pipe line with many bends. Mitsubishi Heavy Ind 15:87–96

DIN 1952 (1982) Durchflussmessung mit Blenden, Düsen und Venturirohren in voll durchströmten Rohren mit Kreisquerschnitt. Beuth-Verlag, Berlin

Gersten K, Herwig H (1992) Strömungsmechanik - Grundlagen der Impuls-, Wärme- und Stoffübertragung aus asymptotischer Sicht. Vieweg-Verlag, Braunschweig

Hinze JO (1975) Turbulence, 2nd edition. McGraw-Hill, New York

Kalkühler K (1998) Experimente zur Entwicklung der Geschwindigkeitsprofile und Turbulenzgrößen hinter verschiedenen Gleichrichtern. Dissertation, Universität Essen; also published as VDI Fortschritt-Bericht, Reihe 7, Nr. 339, 1998, VDI-Verlag, Düsseldorf

Lawn C (1971) The determination of the rate of dissipation in turbulent pipe flow. J Fluid Mech 48:477–505

Laws EM (1990) Flow conditioning – a new development. Flow Meas Instrum 1:165–170

Mattingly G, Yeh T (1989) Flowmeter installation effects - single and double elbow installations. VDI Berichte 768:65–73

Morrison GL, Hall KR, Holste JC, Ihfe L, Gaharan C, DeOtte Jr RE (1997) Flow development downstream of a standard tube bundle and three different porous plate flow conditioners. Flow Meas Instrum 8:61–76

Nasr A, Lai J (1997) Two parallel plane jets: mean flow and effects of acoustic excitation. Exp. Fluids 22:251–260

Schlüter T, Merzkirch W (1996) PIV measurements of the time-averaged flow velocity downstream of flow conditioners in a pipeline. Flow Meas Instrum 7:173–179

Wildemann C, Merzkirch W, Gersten K (2002) A universal, nonintrusive method for correcting the reading of a flow meter in pipe flow disturbed by installation effects. J Fluids Eng 124:650–656

Xiong W, Merzkirch W (1999) PIV experiments using an endoscope for studying pipe flow. J Flow Visual & Image Processing 6:167–175

Xiong W (2000) PIV-Untersuchungen im Nahfeld von Strömungsgleichrichtern. Dissertation, Universität Essen; also published by Shaker-Verlag, Aachen (ISBN 3-8265-7575-X)

5

Signal Processing of Complex Modulated Ultrasonic Signals

Volker Hans

Institut für Mess- und Regelungstechnik, Universität Essen, 45117 Essen, Germany

Ultrasound is a natural choice for the measurement of turbulent flow. The ultrasonic signal is complex modulated by the structures in the fluid. The modulation frequency is low resulting in narrow sidebands to the high carrier frequency. The modulated signal can be separated by digital undersampling. By means of hardware- or software-applied Hilbert-transform the complex modulated signal can be determined for cross-correlation or vortex measurements. Phase angles can exceed the limited range of the arctan function caused by offset or strong vortices in the fluid. The density-distribution function of the phase angle admits the reconstruction of the phase angle. Extended Kalman filters have been proved to be suitable for phase demodulation. Also, analog signal processing will be explained for an electronic Hilbert-transform and amplitude demodulation in vortex measurement. Measurement results are represented for cross-correlation and vortex measurements at low and high flow velocities.

5.1 Introduction

On examining the propagation of small pressure fluctuations in a fluid a high similarity between flow turbulences and ultrasound can be determined (Lighthill, 1972; Poppen, 1997). This effect makes ultrasound a natural choice for the measurement and analysis of turbulent structures. An ultrasonic beam crossing a fluid perpendicularly will be complex modulated by the turbulent structures in the fluid. The modulation is connected with the size of structures and the wavelength of the ultrasound. Usually the carrier frequency of the ultrasonic wave is in the range of some hundred kHz whereas the bandwidth of the modulation frequency comes up to some kHz. This means that the sidebands of the modulated signal are very close to the carrier frequency. Therefore, signal processing is of great importance for ultrasound measurements.

Ultrasonic measurements in a gas flow aim at the determination of the mean flow velocity. Two principles are described in Chaps. 6 and 7 in detail. Vortex measurements are based on the well-known effect of the Kármán vortex street. The frequency of vortices in the wake of a bluff body is directly proportional to the mean flow velocity. For signal processing the modulation frequency is to be determined.

Correlation measurements as described in Chap. 7 require the determination of cross-correlation functions of two amplitude- or phase-modulated ultrasonic signals. In both cases the information needed is in the sidebands of the complex modulated signal. For signal processing the demodulation of the signal is necessary as may be shown by the example of cross-correlation measurements.

5.2 Cross-Correlation Measurements

The principle of cross-correlation flow measurement is based on two parallel ultrasonic barriers sending continuous beams radially through the pipe with a flowing fluid as shown in Fig. 5.1.

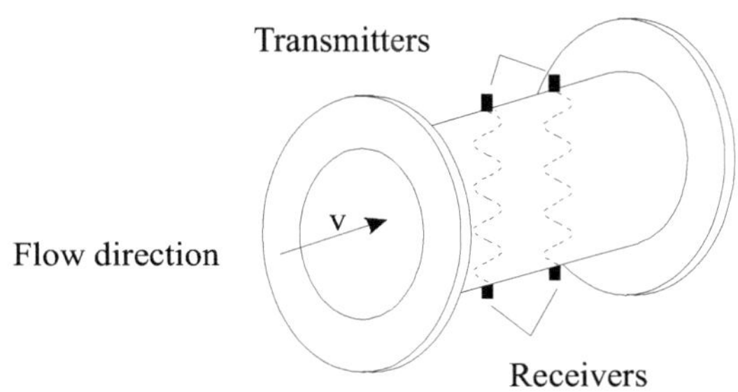

Fig. 5.1. Measurement setup.

The distance of the barriers is in the range of the pipe diameter.

Inhomogeneous structures in the fluid, such as turbulences, lead to a modulation pattern in the ultrasonic signal like a fingerprint. These patterns are similar in both signals but time shifted according to the traveling time of the modulating structures between the barriers. The similarity of the two patterns is compared by the cross-correlation function

$$\Phi(\tau) = \lim_{T \to \infty} \frac{1}{2T} \int_{-T}^{+T} s_1(t) s_2(t+\tau) \mathrm{d}t \; , \tag{5.1}$$

leading to a characteristic maximum of highest similarity at the traveling time. $s_1(t)$ and $s_2(t)$ are the signals of the two barriers, τ is a time shift. But as the modulation depth is relatively low the self-similarity of the sinusoidal carrier signal is predominating the cross-correlation function in a higher peak. There is a risk of the modulation signal being suppressed. Figure 5.2 shows the cross-correlation function of a modulated signal (left) with carrier signal and without carrier signal (right).

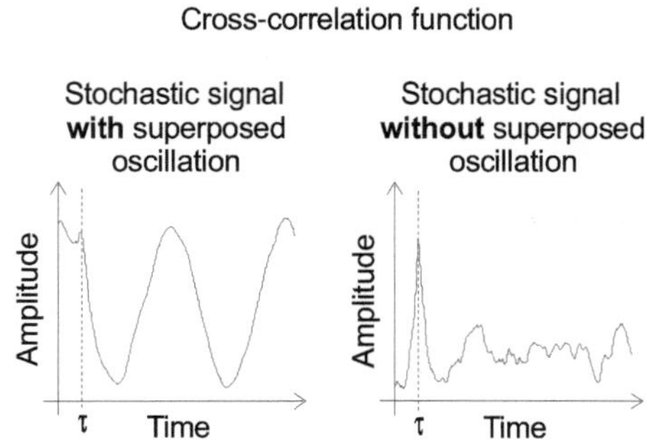

Fig. 5.2. Cross-correlation functions of stochastic signals with (*left*) and without (*right*) superposed oscillation such as a carrier or pulsation.

The figures demonstrate clearly the necessity of signal demodulation. In the following, various kinds of signal-processing methods are summarized that are described in detail in (Filips, 2003; Niemann, 2002; Poppen, 1997; Skwarek, 2000; Windorfer, 2001).

5.3 Demodulation by Digital Undersampling

The demodulation is based upon a synchronous sampling at zero-point crossings with integer submultiples of the carrier frequency taking advantage of the repetition of the spectrum of a sampled signal. This can be illustrated by the Fourier-transformation of a sampled signal from the time to the frequency domain. In the time domain sampling represents a multiplication with Dirac impulses in intervals T_s of the sampling frequency

$$s(t) = \sum_{-\infty}^{+\infty} s(nT_s)\delta(T + nT_s) \ . \tag{5.2}$$

In the frequency domain it can be described by a repeating spectrum

$$S(f) = \sum_{-\infty}^{+\infty} s(f - nf_s) \ , \tag{5.3}$$

with f_s the sampling frequency between two spectra. The gap between the repeating bands is equivalent to the sampling frequency. A subdivision of this frequency by an integer number shifts one of these bands to $f = 0$ Hz, Fig. 5.3. Then the redundant spectra can be suppressed by high-pass filtering and the signal is demodulated to its original band (Skwarek, 2000; Skwarek and Hans, 2000). Naturally, the Nyquist theorem must be taken into consideration with respect to the sidebands that are now shifted to a low-frequency range. From experience this range is less than 5 kHz. Suppose a 200-kHz ultrasound carrier frequency is undersampled by a factor 10, which means 20 kHz. This sampling frequency fullfills the Nyquist theorem up to

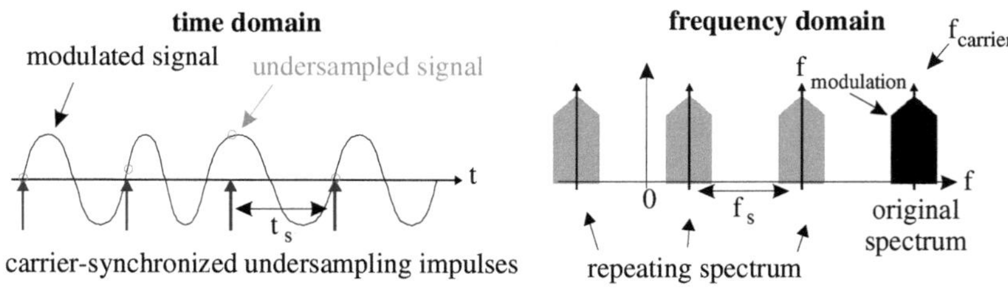

Fig. 5.3. Undersampling in time to frequency domains.

sideband frequencies of 10 kHz. This kind of digital processing is very time saving, memory saving and cost saving. It can be readily applied for the determination of correlation functions.

The demodulation of simultaneously amplitude- and phase-modulated signals is well-known from telecommunications engineering as quadrature-amplitude demodulation (QAD) or complex demodulation (Unbehauen, 1997). It can be demonstrated in the complex domain, Fig. 5.4. The modulated signal is represented by a

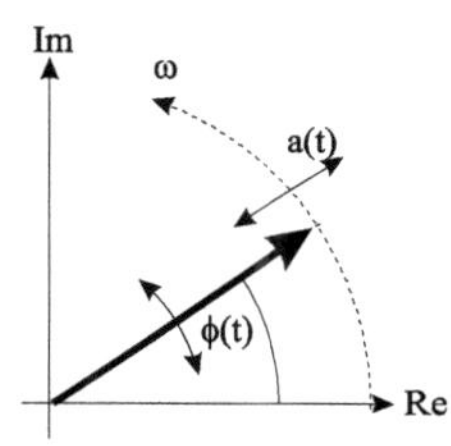

Fig. 5.4. Amplitude- and phase-modulated signal in complex domain.

phasor with changing amplitude and phase. Undersampling stops the rotation of the phasor such as is observed with a synchronized stroboscope and turns it into a standing pointer. Only local phase and amplitude vary according to modulation. Now, amplitude and phase can be obtained from the real and imaginary parts of the signal as

$$a(t) = \sqrt{\mathrm{Re}\{\underline{s}(t)\}^2 + \mathrm{Im}\{\underline{s}(t)\}^2}\,, \tag{5.4}$$

$$\varphi(t) = \arctan \frac{\mathrm{Im}\{\underline{s}(t)\}}{\mathrm{Re}\{\underline{s}(t)\}}\,. \tag{5.5}$$

But in practical applications the imaginary part cannot be measured directly. Therefore, the Hilbert-transform has to be used.

5.4 Digital Hilbert-Transform

5.4.1 Undersampled Hilbert-Transform

Each causal signal as the modulated carrier signal is convertible into an analytical signal (Bachmann, 1992; Otnes and Enochson, 1978). The Hilbert-transform defines the connection between the real and imaginary parts for analytical signals. Turning the phasor at t_1 by 90 degrees to t_2 the projection of the phasor on the imaginary axis at t_1 corresponds to the projection on the real axis at t_2, Fig. 5.5.

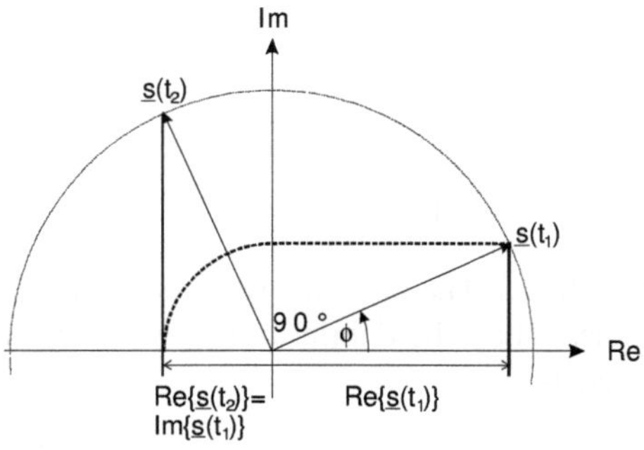

Fig. 5.5. Reconstruction of imaginary part by Hilbert-transform with 90° phase shifted samples.

For digital sampling (5.4) and (5.5) can be written as

$$a(t) = \sqrt{\mathrm{Re}\{s(t_1)\}^2 + \mathrm{Re}\{s(t_2)\}^2}\ , \tag{5.6}$$

$$\varphi(t) = \arctan \frac{\mathrm{Re}\{s(t_2)\}}{\mathrm{Re}\{s(t_1)\}}\ . \tag{5.7}$$

In principle, this theoretical procedure leads to deviations because the fluid moves, the structures are dissipating and the modulation changes in the time interval from t_1 to t_2. But a numerical example will show that this effect is negligible. Suppose a carrier frequency of $f_c = 220\,\mathrm{kHz}$ as applied in practical experiments and the time period is $4.545\,\mathrm{ms}$. A quarter period is $1.136\,\mathrm{ms}$. During this time the fluid moves only $5.7\,\mu\mathrm{m}$ at a mean flow velocity of $20\,\mathrm{m/s}$. It is not to be expected that this very small motion of the fluid influences the measured result.

Sampling the 220-kHz signal shifted by 90 degrees leads to a sampling frequency of 880 kHz resulting in expensive signal processing. It is obvious to apply the same undersampling procedure to the Hilbert-transform as to the carrier frequency. This undersampled Hilbert-transform must fulfill the Nyquist theorem with regard to the undersampled carrier frequency. In this time the flow will move about $31\,\mu\mathrm{m}$ for the example above.

5.4.2 Software-Based Hilbert-Transform

The Hilbert-transform of a real signal $s(t)$ is strictly defined as

$$\hat{s}(t) = H\{s(t)\} = \frac{1}{\pi} \int_{-\infty}^{+\infty} \frac{s(t)}{t-\tau} \mathrm{d}t \ , \tag{5.8}$$

$$\hat{s}(t) = s(t) * h(t) \ , \tag{5.9}$$

with

$$h(t) = \frac{1}{\pi t} \tag{5.10}$$

and its Fourier-transform

$$F\left\{\frac{1}{\pi t}\right\} = -\mathrm{j}\,\mathrm{sign}(\omega) \ , \tag{5.11}$$

with the sign function for ω. The pure imaginary spectrum is constant for negative and positive frequencies. Thus, the Fourier-transform of the Hilbert-transform of the signal can be described as

$$\hat{F}(\omega) = \hat{H}\left[s(t) * h(t)\right] == -\mathrm{j}\,\mathrm{sign}\left[\omega F(\omega)\right] \ , \tag{5.12}$$

which indicates that a Hilbert-transform can be determined when processing the signal shifted by 90 degrees.

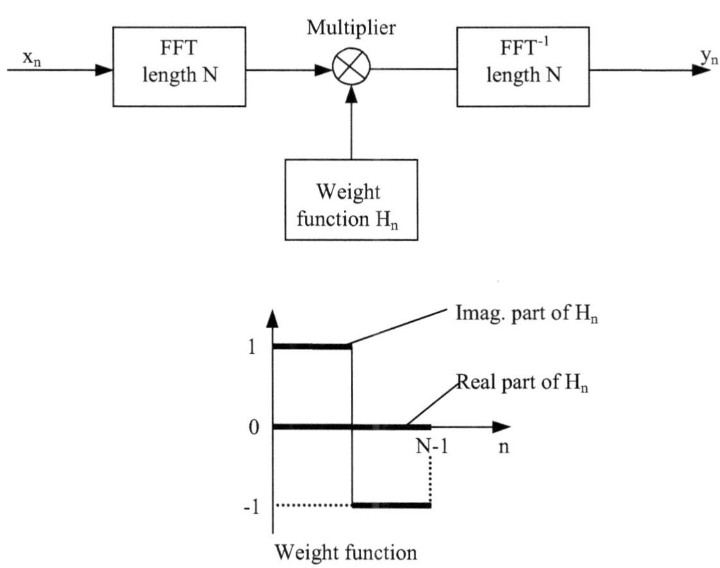

Fig. 5.6. Diagram of Hilbert-transform using FFT.

The Hilbert-transform can be calculated by the convolution of the discrete Fourier-transformed signal with the weight function H_{n}, which can be described as

$$H_{\mathrm{n}} = \begin{cases} 0 + \mathrm{j} * 0 & \text{for } \mathrm{n} = 0, N/2 \\ 0 + \mathrm{j} * 1.0 & \text{for } 1 \leq \mathrm{n} \leq N/2 - 1 \\ 0 - \mathrm{j} * 1.0 & \text{for } N/2 \leq \mathrm{n} \leq N - 1 \,, \end{cases}$$

as shown in Fig. 5.6 (Lin, 2003). The input x_{n} of the system is the measured real part, while the output y_{n} is the Hilbert-transform of x_{n}.

5.4.3 Measurement Results

Examples of measured and calculated imaginary parts of the demodulated signal are shown in Fig. 5.7 for flow velocities of 2 m/s (left) and 10 m/s (right).

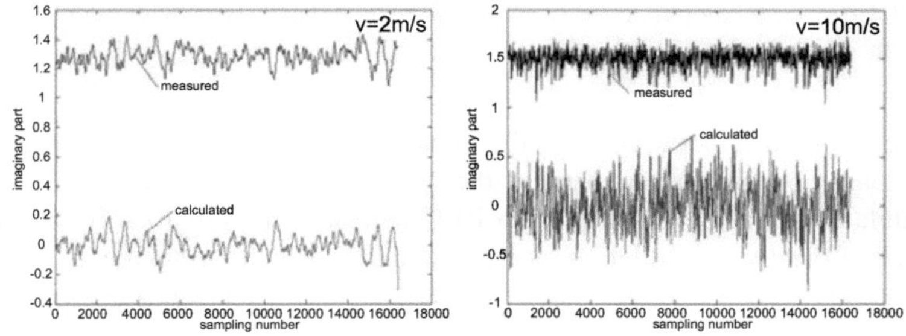

Fig. 5.7. Measured and calculated imaginary parts of the demodulated signal.

As explained in Chaps. 6 and 7 the effect of amplitude modulation is small in comparison to the phase modulation. Therefore, in further signal processing for determining the cross-correlation function usually the phase-demodulated signal is applied. Figure 5.8 shows a cross-correlation function of an amplitude-demodulated signal for two flow velocities.

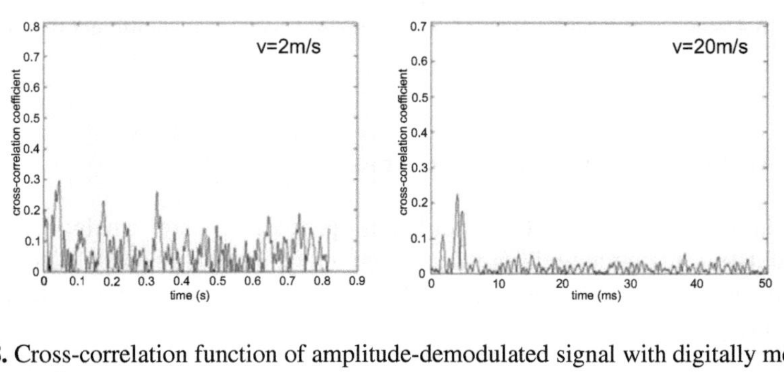

Fig. 5.8. Cross-correlation function of amplitude-demodulated signal with digitally measured Hilbert-transform.

The peaks indicating the traveling time of the modulating structures are low and sometimes not clear, involving the risk of misinterpretation.

The results using the calculated Hilbert-transform for the same signals are shown in Fig. 5.9.

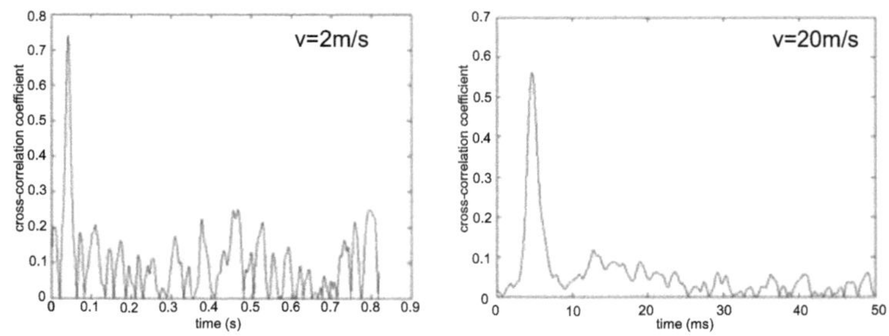

Fig. 5.9. Cross-correlation function of amplitude-demodulated signal with calculated Hilbert-transform.

The difference between digitally measured and calculated imaginary parts is obvious. For further comparison the cross-correlation function of the phase-demodulated signal is shown in Fig 5.10 . The results are remarkably similar. Only the peak of the cross-correlation function at higher velocity is smaller.

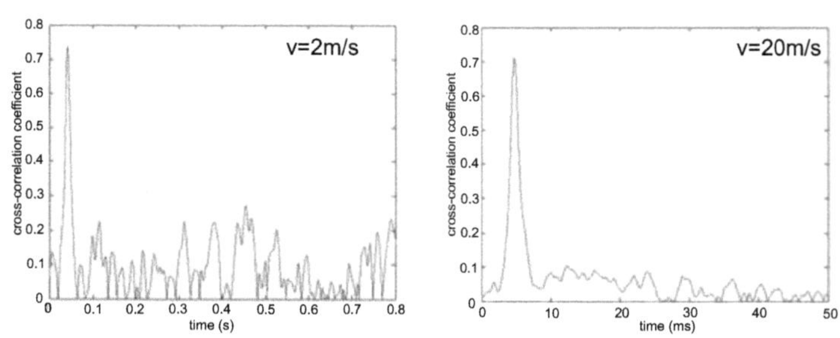

Fig. 5.10. Cross-correlation function of phase-demodulated signal at different flow velocities.

5.4.4 Analog Electronic Hilbert-Transform

An alternative to digital-demodulation procedures are analog electronic-demodulation methods widespread in communication techniques (Mäusle, 1988). Bachmann (1992) and Otnes and Enochson (1978) suggest a principle as shown in Fig. 5.11.

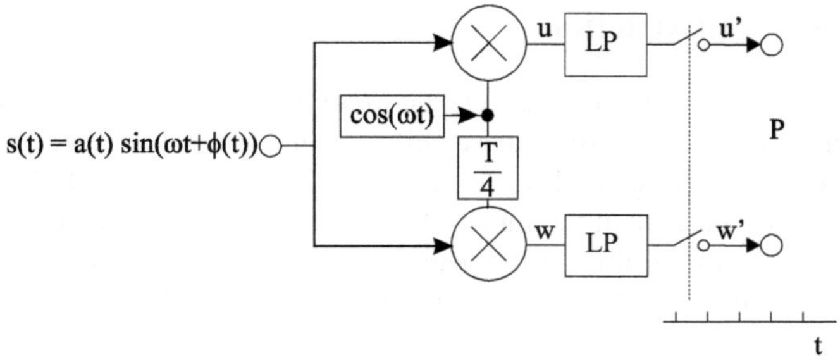

Fig. 5.11. Principle of an electronic Hilbert-transform.

Here, a modulated signal

$$s(t) = a(t)\ \sin(\omega t + \varphi)$$

is multiplied by the carrier and its 90°-shifted picture, respectively.

They form the signals

$$u(t) = a(t)\ \sin\left(\omega t + \varphi(t)\right)\ \cos(\omega t) \tag{5.13}$$

and

$$w(t) = a(t)\ \sin\left(\omega t + \varphi(t)\right)\ \sin(\omega t)\ , \tag{5.14}$$

which are transformed by trigonometric relations to

$$2u(t) = a(t)\left[\sin\left((\varphi(t)\right) + \sin\left(2\omega t + \varphi(t)\right)\right] \tag{5.15}$$

and

$$2w(t) = a(t)\left[\cos\left((\varphi(t)\right) - \cos\left(2\omega t + \varphi(t)\right)\right]\ . \tag{5.16}$$

The low-frequency portion is separated with a low-pass filter so that

$$u'(t) = \frac{a(t)}{2}\cos\left(\varphi(t)\right) \tag{5.17}$$

and

$$w'(t) = \frac{a(t)}{2}\sin\left(\varphi(t)\right) \tag{5.18}$$

remain without any influence of the carrier. Both parts, also called inphase and quadrature parts, respectively, represent a Hilbert-couple of a complex measure and can be further processed to (5.6) and (5.7). The already-mentioned problem becomes obvious again due to the calculation of tangent and squareroot functions with sufficient accuracy. At this point the signals are sampled and computers are needed for ongoing calculations.

An improved method for hardware-based demodulation is described in Skwarek and Hans (2001).

5.5 Phase Reconstruction

On installing the ultrasonic transducers a correct adjustment is necessary, considering the phase offset. Small zero-phase shifts are unavoidable. Especially in vortex measurements, high phase angles can occur and in connection with the offset they can have angles greater than the range of the definition of the arctan function from -90 to $+90$ degrees. In this case, special algorithms for the reconstruction must be applied (Bachmann, 1992; Otnes and Enochson, 1978; Skwarek, 2000; Tribolet, 1977).

A new kind of reconstruction has been developed by Niemann (2002) based on the analysis of the probability density distribution of the phase angle. The stochastic modulation of the ultrasonic beam by a streaming fluid is normally distributed. But assuming no adjustment, the density function is shifted. The position of the maximum of the distribution function represents the zero-adjustment point of the receiving sensor. Figure 5.12 shows a demodulated phase signal with zero-adjustment error (left) and the corresponding distribution function (right). The zero-phase angle is about 150 degrees.

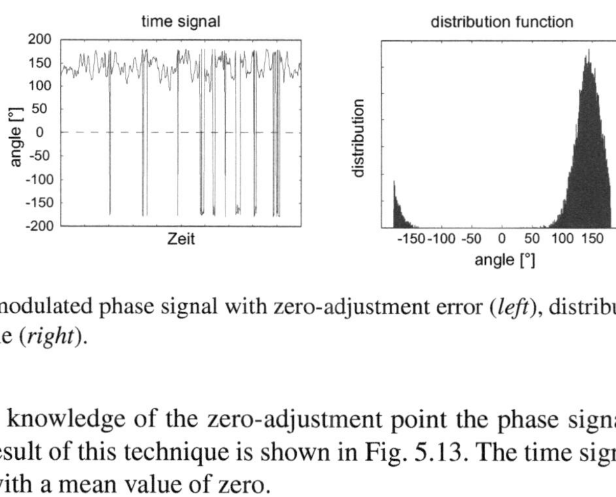

Fig. 5.12. Demodulated phase signal with zero-adjustment error (*left*), distribution function of the phase angle (*right*).

With the knowledge of the zero-adjustment point the phase signal can be relocated. The result of this technique is shown in Fig. 5.13. The time signal is normally distributed with a mean value of zero.

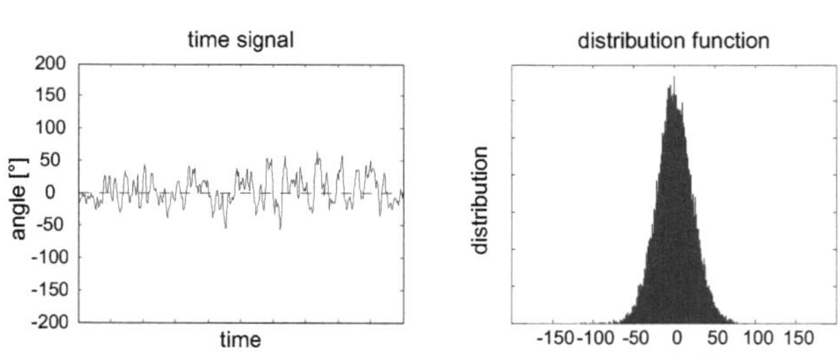

Fig. 5.13. Demodulated phase signal after reconstruction (*left*), distribution function of the phase angle (*right*).

The procedure results in the correct determination of the zero-adjustment point and the signal can be reconstructed correctly. The method is independent of the mechanical adjustment of the transducers and allows the measurement of flow velocity even with disturbed profiles behind single and double elbows (Niemann and Hans, 2002).

5.6 Phase Demodulation with Kalman Filter

The phase modulation of the ultrasonic wave by stochastically distributed structures in the streaming fluid is a random process especially in the case of cross-correlation measurement. In vortex measurements the modulation is caused primarily by well-defined vortices resulting in sinusoidal phase modulation superposed by random noise from other structures in the fluid. Therefore it is obvious to process the modulated signal with recursive filters considering process noise and measurement noise. Kalman filters are well-known for this task describing linear dynamic systems in terms of state-space concepts. Nonlinearities can be taken into account by extended Kalman filters (EKF) (Haykin, 1996; Loffeld, 1990). In addition, the EKF has the great advantage that it can dispense with the zero-point adjustment of the sensors. An example for low modulation degree is given in Fig 5.14 (Filips, 2003).

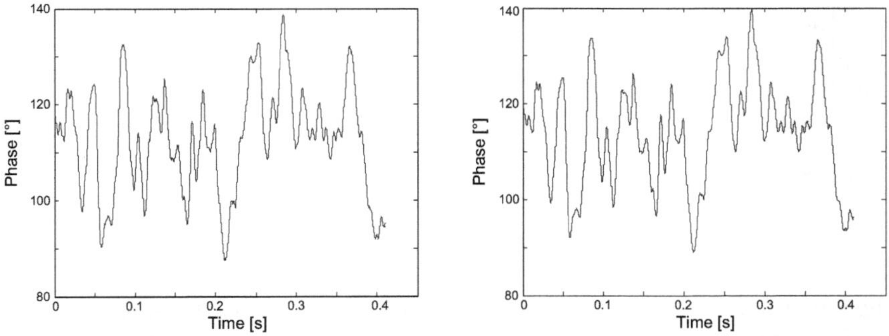

Fig. 5.14. Phase reconstruction by QAD (*left*) and prediction EKF (*right*) at low modulation.

The phase reconstruction by EKF is compared with the QAD algorithm. There is very good similarity, confirming the theory of EKF application. With increasing modulation the advantages of EKF become more obvious. Figure 5.15 shows an example.

The phase reconstruction by QAD is superposed by high noise and must be processed by digital filters, whereas the result of extended Kalman filter is free of noise and can be directly processed.

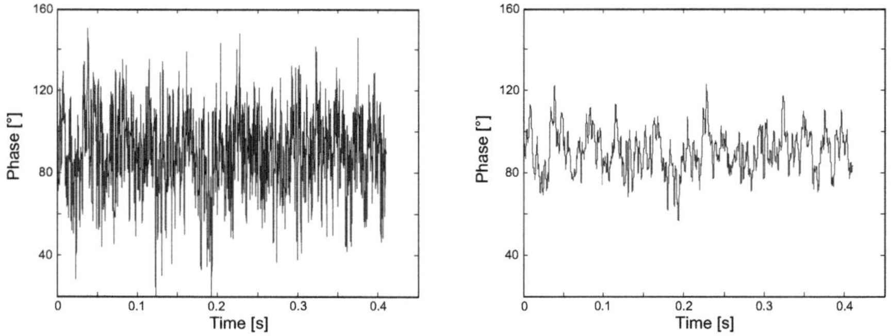

Fig. 5.15. Phase reconstruction by QAD (*left*) and prediction EKF (*right*) at high modulation.

To demonstrate the power of EKF, two complex modulated ultrasonic signals have been cross-correlated. The maximum of the cross-correlation function (CCF) represents the traveling time of structures between the two barriers. Figure 5.16 shows the cross-correlation function of a disturbed flow at high flow velocity of $20\,\mathrm{m/s}$ behind a single elbow.

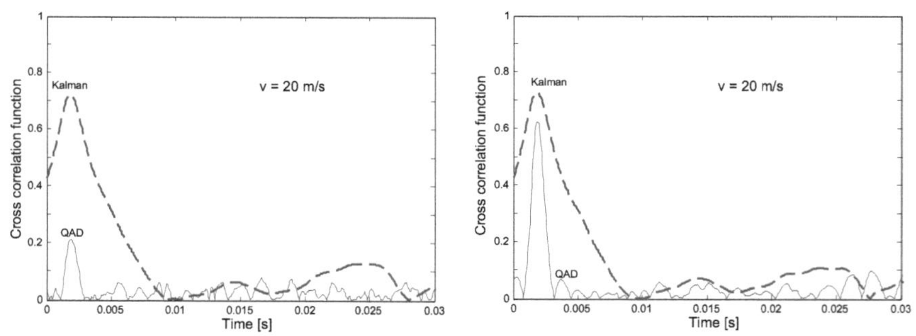

Fig. 5.16. Cross-correlation function of two demodulated ultrasonic phase signals by QAD and EKF without (*left*) and with (*right*) zero-point adjustment of the sensors.

In the left figure the QAD method shows a very low maximum of CCF in comparison to EKF. In this case, measurement was made without zero-point adjustment of the sensors. The right figure shows the same measurement with zero-point adjustment of the sensors. Comparing the two figures the advantage of the extended Kalman filter is evident.

Similar results for undisturbed flow as well as for disturbed flow after double elbows could be obtained (Filips, 2003).

Even vortex measurements could be performed successfully. The demodulated phase signal shows a periodical shape with noise in the time domain. But the phase angle often exceeds the range of the definition of the arctan function. Figure 5.17

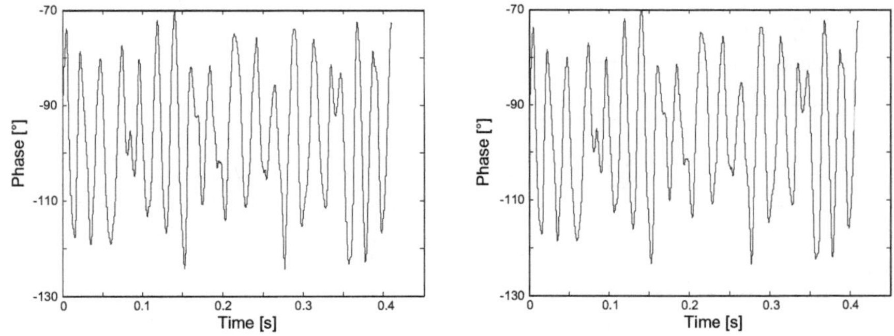

Fig. 5.17. Phase reconstruction by QAD (*left*) and prediction EKF (*right*) at low velocities.

shows the phase reconstruction for a M10 threaded control rod as a bluff body at $2\,\mathrm{m/s}$ flow velocity.

For low modulation grade the results of QAD and EKF algorithms are similar. Increasing flow velocity leads to high modulation grades caused by increasing the rate of vortices and superposed by increasing noise, Fig. 5.18.

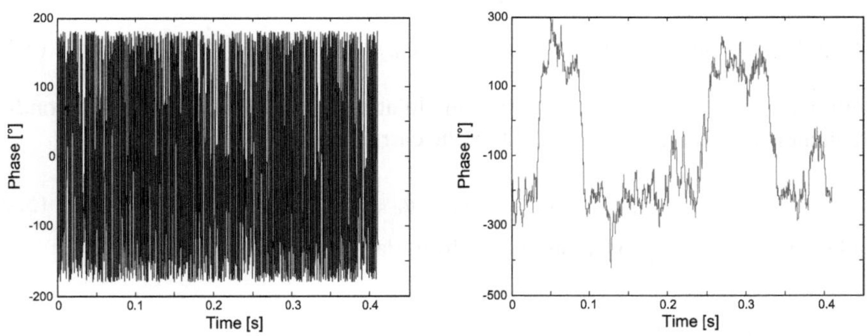

Fig. 5.18. Phase reconstruction by QAD (*left*) and prediction EKF (*right*) at high velocities.

The resolution using QAD gives no explanation. Phase jumpings characterize the illustration. The reconstruction using an extended Kalman filter demonstrates a noise-reduced phase shape and yields better signal processing for the determination of the vortex frequency.

5.7 Analog Signal Processing

Various analog demodulation techniques have been proved with more or less good results for disturbed or undisturbed flow in correlation as well as in vortex measurements (Filips and Hans, 2002; Filips, 2003). A simple but very powerful method

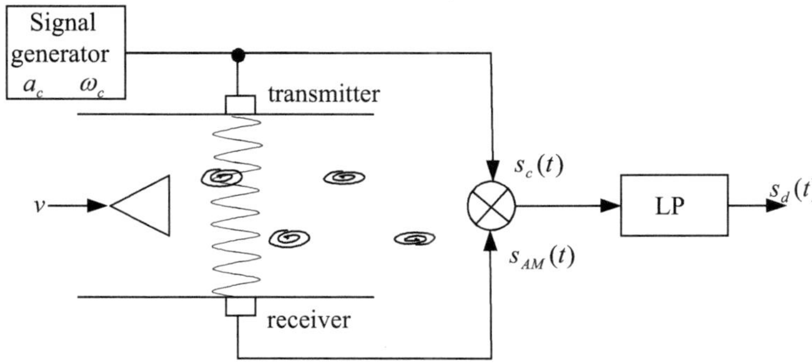

Fig. 5.19. Analog signal processing in vortex measurement.

of amplitude demodulation especially for vortex measurements is pointed out in the following.

The modulated receiver signal is multiplied by the feeding carrier signal, Fig. 5.19. For reasons of simplification only an amplitude-modulated signal may be considered, which is expressed by

$$s_{\mathrm{AM}}(t) = a_{\mathrm{c}} \sin(\omega_{\mathrm{c}} t) + \frac{a_{\mathrm{m}}}{2} \cos((\omega_{\mathrm{c}} - \omega_{\mathrm{m}})\, t) - \frac{a_{\mathrm{m}}}{2} \cos((\omega_{\mathrm{c}} + \omega_{\mathrm{m}})\, t) \ , \quad (5.19)$$

with a_{c}, a_{m} amplitudes of carrier and modulating signal and ω_{c}, ω_{m} corresponding frequencies. This signal, multiplied by the carrier signal, is

$$s(t) = s_{\mathrm{AM}}(t)\, a_{\mathrm{c}} \sin(\omega_{\mathrm{c}} t) \quad (5.20)$$

and results, according to trigonometric formulae, in

$$\begin{aligned} s(t) = & \frac{{a_{\mathrm{c}}}^2}{2} [1 - \cos(2\omega_{\mathrm{c}} t)] + \frac{a_{\mathrm{m}} a_{\mathrm{c}}}{2} \sin(\omega_{\mathrm{m}} t) \\ & + \frac{a_{\mathrm{m}} a_{\mathrm{c}}}{4} [\sin((2\omega_{\mathrm{c}} - \omega_{\mathrm{m}})\, t) - \sin((2\omega_{\mathrm{c}} + \omega_{\mathrm{m}})\, t)] \ . \end{aligned} \quad (5.21)$$

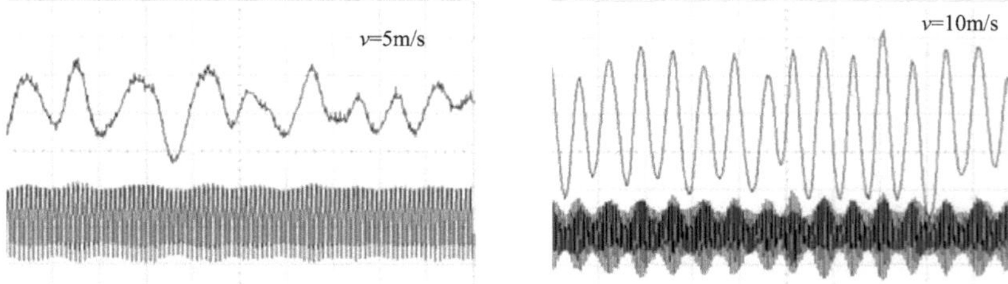

Fig. 5.20. Receiver signal (*lower signal*) and signal after multiplication and filtering (*upper signal*). Bluff-body width 4 mm, length 12 mm, tip to the inflow.

The signal contains the low-frequency modulating part with the modulating frequency ω_m and high-frequency parts with the double carrier frequency that can be separated by a low-pass filter (LP). Figure 5.20 shows oscillograms from the modulated receiver signal (lower signal) and the demodulated signal after multiplication and filtering that is caused directly by the vortices. The figure shows the signals for two velocities of $5\,\mathrm{m/s}$ and $10\,\mathrm{m/s}$ in the wake of a bluff body facing the tip to the inflow. The signals are nearly sinusoidal in shape. The vortex frequency as measured for the flow velocity can be determined directly. The same procedure can be applied in the cross-correlation measurement, see Chap. 7. An example of the amplitude-modulated signal of one channel is shown in Fig. 5.21 for two velocities. On account of the weak signal the cross-correlation function is not so effective as that with phase-demodulated signals.

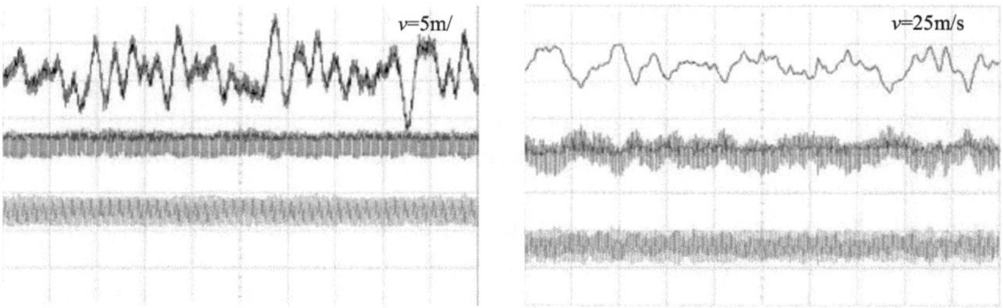

Fig. 5.21. Signals of one channel for cross-correlation measurements. *Lower signal*: receiver signal; *middle signal*: product; *upper signal*: analog output.

5.8 Conclusion

Processing of high-frequency complex modulated ultrasonic signals requires special procedures of demodulation. In the case of flow-velocity measurement the ultrasonic signal is modulated by the structures in the fluid. The bandwidth of modulation is up to $5\,\mathrm{kHz}$. Digital undersampling methods have been proved to be suitable to separate the modulated signal from its carrier. The amplitude and phase angle of the complex signal can be measured by normal Hilbert-transform and undersampled Hilbert-transform. Special attention must be directed to the offset and range of the phase angle. The density distribution of the phase angle is a helpful tool to compensate for offset. Extended Kalman filters are best suited for further signal processing, i.e., the development of cross-correlation functions. Even analog methods such as the electronic Hilbert-transform or the multiplication of the modulated signal with the carrier signal, lead to very good results for the determination of the flow velocity.

References

Bachmann W (1992) Signalanalyse. Vieweg Verlag, Braunschweig

Filips Ch, Hans V (2002) Comparison of analog and digital demodulation methods of modulated ultrasonic signals in vortex flow metering. IEEE Conf. IMTC 2002, Proceedings:1657–1660

Filips Ch (2003) Ultraschallsignalverarbeitung bei Korrelations- und Vortexverfahren zur Durchflußmessung. Dissertation, Universität Essen, also published by Cuvillier Verlag Göttingen (ISBN 3-89873-711-X)

Haykin S (1996) Adaptive Filter Theory. Prentice Hall Information and System Sciences Series

Lighthill J (1972) The propagation of sound through moving fluids. Journal of Sound and Vibration, 24(4):471–492

Lin Y (2003) Analog signal processing in flow measurement. Techn. Report, Universität Essen, Mess- und Regelungstechnik

Loffeld O (1990) Estimationstheorie II, Anwendungen Kalmann-Filter. Oldenbourg Verlag, München

Mäusle R (1988) Analoge Modulationsverfahren. Hüthig Verlag, Heidelberg

Niemann M, Hans V (2002) New signal processing methods of ultrasound phase recontruction for measurement of disturbed flow. Systemics, Cybernetics and Informatics Conference Proceedings, Vol. III:345–349

Niemann M (2002) Signalverarbeitung in der Ultraschall-Durchflussmessung. Dissertation, Universität Essen, also published by Shaker Verlag Aachen (ISBN 3-8322-0340-0)

Otnes RK, Enochson L (1978) Applied Time Series Analysis. Wiley, New York/USA

Poppen G (1997) Durchflußmessung auf der Basis kreuzkorrelierter Ultraschallsignale. Dissertation, Universität Essen, also published by Shaker Verlag Aachen (ISBN 3-8265-3108-6)

Skwarek V (2000) Verarbeitung modulierter Ultraschallsignale in Ein- und Mehrpfadanordnungen bei der korrelativen Durchflußmessung. Dissertation, Universität Essen; also published by Shaker Verlag Aachen (ISBN 3-8265-7589-X)

Skwarek H, Hans V (2000) The ultrasonic cross-correlation flow meters - new insights about the physical background. FLOMEKO 2000, Proceedings

Skwarek H, Hans V (2001) An improved method for hardware-based complex demodulation. Measurement 28/2:87–93

Tribolet JM (1977) A new phase unwrapping algorithm. IEEE Trans. on Acoustics, Speech and Signal Processing 25 (2):170–177

Unbehauen R (1997) Systemtheorie. Oldenbourg Verlag, München

Windorfer H (2001) Optimierung von Wirbelfrequenzmeßgeräten mit demodulierten Ultraschallsignalen. Dissertation, Universität Essen; also published by Shaker Verlag Aachen (ISBN 3-8265-9261-1)

6

Vortex-Shedding Flow Metering Using Ultrasound

Volker Hans[1], Ernst von Lavante[2]

[1]Institut für Mess- und Regelungstechnik, Universität Essen, 45117 Essen, Germany
[2]Institut für Strömungsmaschinen, Universität Essen, 45117 Essen, Germany

The coincidence of vortices generated by a bluff body in a gaseous flow, which is well-known as the Kármán vortex street, with an ultrasonic beam crossing these vortices raises many questions concerning physics and signal processing. Usually the vortices are detected by pressure sensors in the pipe wall or inside the bluff body. Using ultrasound for vortex detection is a powerful advantageous alternative combination. The ultrasonic beam is modulated by the velocity components of the vortices and varying fluid density. The kind of modulation depends on the structure of the vortices that are influenced by the bluff-body geometry and size. Measurements and simulations show that conventional triangular bluff bodies with the large size used for pressure measurements are unsuitable for ultrasonic measurements because of the lack of knowledge about secondary vortices. They can be avoided by turning the bluff body so as to face the tip to the inflow. As ultrasound is very sensitive to the smallest influences, the bluff-body size can be reduced considerably. Even the shape can be adapted. The best results are obtained with a 3-mm threaded rod as the bluff body in a pipe of 100 mm diameter. Pressure losses behind such small bluff bodies can be neglected. Independent of the bluff-bodies shapes and sizes vortex meters are sensitive to disturbed profiles and pulsation.

6.1 Introduction

Commercial vortex flow meters are based on the well-known relationship between the vortex frequency in the wake of a bluff body and the mean flow velocity. Usually, piezoelectric or piezoresistive pressure sensors, strain gauges or thermal sensors are applied for detecting the vortex frequency. Pressure sensors are installed in the wall of the pipe or inside the bluff body for the detection of separating vortices. Various bluff-body shapes have been designed creating regular and well-defined vortex structure and pressure signals at the sensor. Large bluff-body dimensions are required for strong pressure fluctuations. A width of 24 to 28% of the pipe diameter is recommended in the literature (Baker, 2000; Bentley, 1983; Breier and Gatzmanga, 1998) covering an area of more than 30% of the pipe profile and causing immense pressure losses.

Signal processing presupposes well-defined vortices at only one dominant frequency. Often the triangular bluff bodies used do not fulfil this condition. They generate secondary vortices that are not so well-known, complicating the signal processing. Recent research has studied vortex flow meters that detect the vortex frequency as a modulation of ultrasonic signals (Filips, 2003; Perpeet, 2000; Windorfer, 2001). Special analog and digital signal procedures have been developed for the demodulation of the ultrasonic signal, as shown in Chap. 5. Ultrasonic signals show a higher sensitivity for the detection of vortices. Therefore much smaller bluff-body sizes can be realised with negligible pressure losses. Various bluff-body geometries and kinds of installation have been investigated that were assisted by numerical simulations. This chapter is a summary of research work described in detail in (Filips, 2003; Perpeet, 2000; Poppen, 1997; Windorfer, 2001).

6.2 Physical Background

The interaction of streaming fluid and its vortices with an ultrasonic wave results in a complex modulation of amplitude and phase of the ultrasound signal (Poppen, 1997). Different physical effects in the fluid lead to different kinds of modulation (Windorfer, 2001).

The ultrasonic wave is damped by acoustic absorption as a consequence of the conversion of mechanical into thermal energy. The intensity of sound varies on changing temperature, density and pressure. As a vortex structure passing an ultrasonic beam represents an area of lower pressure and density than the surrounding fluid, the ultrasonic wave is modulated in amplitude. Each vortex causes one minimum in the time signal of the transmitted amplitude. Experiments and simulations have shown that the modulation degree is about 0.4%.

The vortex structures also represent areas of different refraction indices. They are influenced by the same physical parameters as temperature, density and pressure. Refraction changes the direction of ultrasound propagation and results in interference of the sound. The amplitude is modulated and attenuated. Experimental investigations and measurements have shown a refraction angle of about $\pm 3^\circ$, resulting in an amplitude modulation of about $\pm$ 10%.

Finally, the ultrasonic beam is deflected and elongated by high axial velocity components of the fluid from its perpendicular propagation direction. Because of the cone characteristics, sound waves with lower amplitude are transmitted to the receiver. The drift angle is dependent on the flow velocity and could be found in measurements of about 5% at a flow velocity of $30\,\mathrm{m/s}$. The resulting modulation degree of the amplitude was about 2%.

The change of the parameters mentioned above further varies the local sound speed, resulting in phase shift. It could be found that each vortex generates one complete oscillation phase angle with an amplitude of 20%. The phase modulation additionally is caused by the superposition of flow-velocity vectors and sound-speed vectors. The phase modulation can reach values of some hundreds of degrees depending on the flow velocity and the size of vortex structures, which in turn depend on the geometry of the bluff body.

6.3 Measurement Principle and Test Arrangement

Vortex flow meters use the well-known effect of a Kármán vortex street. Vortices separate periodically from both sides on the back of a bluff body that is flowed around. Special shaping of the bluff body has an influence on the linearity of the characteristic of vortex frequency f versus mean flow velocity u_{m} expressed by the dimensionless Strouhal number

$$S_r = \frac{fd}{u_{\mathrm{m}}}, \tag{6.1}$$

where d is the height of the bluff body. If the Strouhal number can be kept constant over the interest velocity range of the mean flow velocity can be determined directly by measurement of the vortex frequency.

All measurements were performed in a test arrangement with a pipe diameter of $100\,\mathrm{mm}$. The length of the pipe from the inflow to the test chamber with bluff bodies was $5\,\mathrm{m}$ to ensure a fully developed flow profile. The flow velocity was controlled in a range 2–30 $\mathrm{m/s}$ corresponding to Reynolds numbers 13000–195000 with a deviation of max. $2\,\mathrm{cm/s}$. A turbine gasmeter was used as reference with a deviation within 1% of the mean flow. The test fluid for all experiments was $1\,\mathrm{bar}$ static pressure. The ultrasound transducers were usually operated with a carrier frequency of $220\,\mathrm{kHz}$. The location of the ultrasonic beam was selected by the demand for the most sinusoidal time signal of the demodulated amplitude in a range of 10–80 mm behind the bluff body. Various bluff-body shapes were tested, such as triangular, rectangular, T-shape, special developed shapes, circular and threaded control rods.

Only the most interesting sizes and shapes will be presented in the following. Signal processing of the complex modulated ultrasonic signals has been performed as described in Chap. 5.

6.4 Simulations

The numerical algorithm employed uses the three-dimensional, time-dependent full Navier–Stokes equations describing the conservation of mass, momentum and energy of the flow (Perpeet, 2000). The program is based on the finite-volume formulation, using a cell-centered organisation of the control volumes. The spatial discretisation is carried out with the help of Roe's flux difference scheme, a Godunov-type method providing an approximate solution of the Riemann problem on the cell interfaces. The method has been demonstrated to be very accurate and effective in the simulation of low Mach number viscous flows. Upwind-biased differences are used for the convective terms, central differences for the viscous fluxes.

Starting with a constant initialisation of the scalar variables and body-fitted velocity components, the integration in time is carried out by a modified explicit Runge–Kutta time stepping as well as, optionally, an implicit approximate-factorisation method (AF) or symmetric Gauss–Seidel (SGS) scheme.

The simulations were carried out on structured grids. The mesh points were arranged according to an algebraic distribution and clustered at the solid walls to ensure enough gridpoints in the boundary layers. The domain was divided into several blocks to make the formulation of the boundary conditions and handling a complex geometry easier. Furthermore, the multiblock structure was necessary to compute the flow on parallel computers. For the inlet and outlet planes, the subsonic one-dimensional nonreflecting boundary conditions were implemented. They were based on the Riemann invariants normal to the pipe cross section. These conditions made the pressure waves and other disturbances run out of the domain without reflection. The interzonal boundary conditions provided the data exchange among the blocks; at the walls, the no-slip solid viscous wall boundary conditions were used.

6.5 Comparisons of Pressure and Ultrasonic Signals

For basic investigations a conventional triangular bluff body of 24 mm height and 48 mm length was used. The pressure signal was measured with a miniaturised probe of only 1.6 mm diameter 30 mm downstream behind the tip of the bluff body and 20 mm above the centerline of the bluff body. The ultrasonic barrier was also 30 mm behind the bluff body. Both signals were measured at a flow velocity of 15 m/s. Figure 6.1 shows in the upper picture the averaged signal of the demodulated amplitude of the ultrasonic signal in the time domain. In the lower picture the pressure signal is presented. Two points are remarkable. The ultrasonic signal shows in a

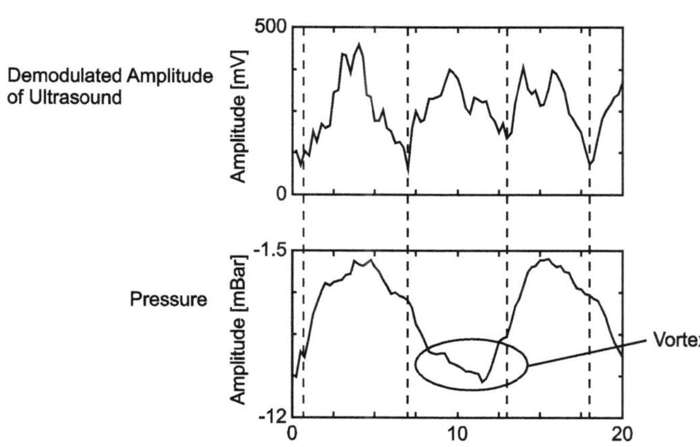

Fig. 6.1. *Top*: Time signal of the demodulated amplitude of the ultrasonic signal. *Bottom*: Time signal of pressure.

higher resolution the influence of secondary and stochastic effects. Secondly, the frequency of the ultrasonic signal is twice as high as the measured pressure signal. This is effected by the two different measurement principles.

The ultrasonic signal detects along the beam, through the whole pipe diameter, all structures in the fluid, especially vortices both in the upper and lower half of the pipe, whereas the pressure sensor measures only the pressure fluctuations passing the measurement location. The same effect is measured with a pressure sensor in the pipe wall. The signal is generated only by vortices in the upper half of the pipe wall. Due to the low sensitivity of pressure sensors in comparison to ultrasound, secondary and stochastic effects are not detected. This means that vortex measurements with ultrasound from the beginning have double sensitivity in comparison to arrangements with a pressure sensor (Hans et al., 1997).

6.6 Experiments

6.6.1 Large Triangular Bluff Body

In Fig. 6.2 the simulated pressure plot of a vortex street behind a triangular bluff body is shown (Hans et al., 1998, 2000). The pressure plot was selected for the presentation of the simulation results because of the best visualisation. The velocity field circulates clockwise and anticlockwise around the pressure structures. The vortex at the lower side has fully separated but a secondary, smaller vortex develops on the same side. This secondary vortex leads to a secondary maximum in the simulated time signal (Fig. 6.3). The same secondary effect can be observed at the measured phase shift of the ultrasonic signal. It can be readily detected in the frequency-domain measurements. The pressure signal measured in the tube wall is not affected by the secondary vortex (Fig. 6.4).

Vortex detection by a pressure sensor shows a very good sinusoidal signal in the time domain. It is superposed by noise. The determination of the signal frequency is very easy in the frequency domain. The vortex frequency is significant and overtops the noise with a high signal-to-noise ratio.

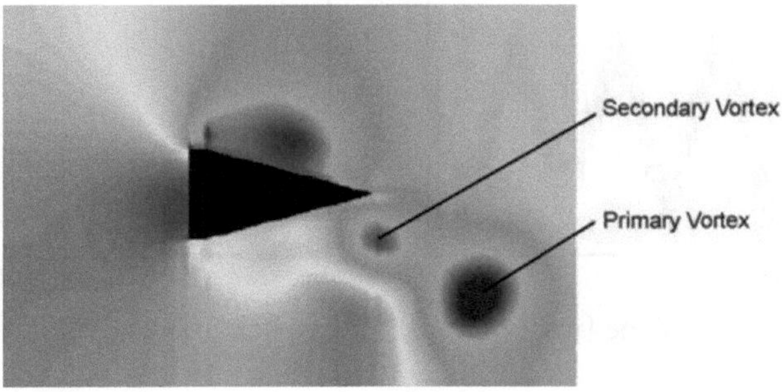

Fig. 6.2. Pressure field of a simulated vortex street of a triangular bluff body with a width of 24 mm used in the usual way.

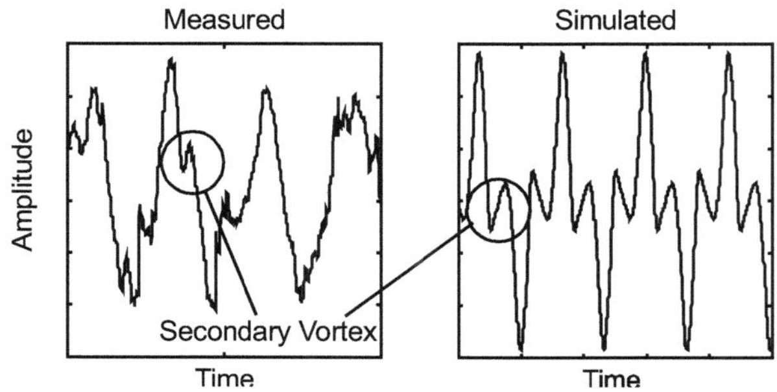

Fig. 6.3. Measured ultrasonic phase shift and simulated time signal of a triangular bluff body used conventionally.

The shape of the triangular bluff body facing the flat side to the inflow is optimised for many commercial vortex-shedding flow meters using pressure sensors for the detection of the vortex frequency. The same shape used for the measurement system combined with ultrasound leads to worse signals. Secondary effects on the ultrasonic signal prevent simple digital signal processing.

Using the same shape rotated for vortex generation, measurement and simulation lead, as expected, to very different results with an advantage for the ultrasound measurement method.

Figure 6.5 shows a pressure plot of the triangular bluff body facing the edge to the inflow (Hans et al., 1998). The vortices separate at the backside without the development of secondary vortices. The vortex street is fully developed in the middle of the pipe. The pressure fluctuations at the wall are very small, so that a detection of the vortex frequency by pressure sensors becomes impossible. In the simulated time signal there is no secondary effect visible. The measured time signal of the demodulated ultrasound signal also shows very sinusoidal behavior without the influence of

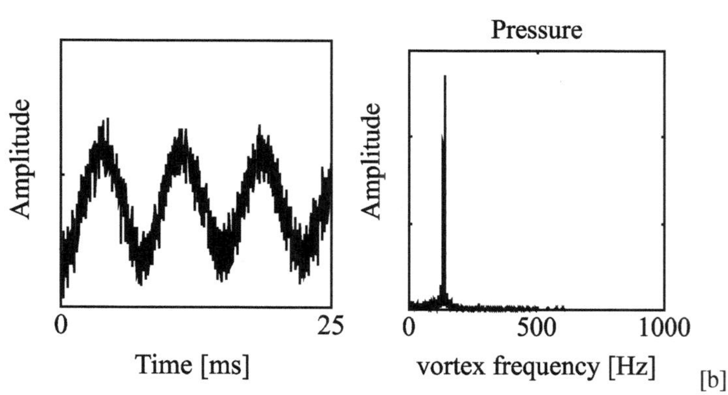

Fig. 6.4. Pressure signal in time (*left*) and frequency domain (*right*) of a 24 mm triangular bluff body.

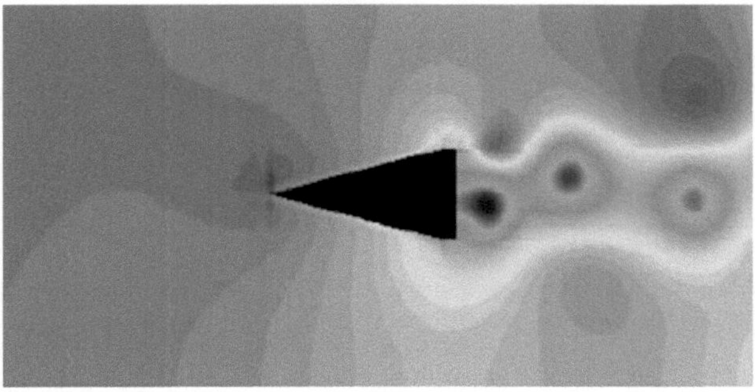

Fig. 6.5. Pressure field of a simulated vortex street of a triangular bluff body with a width of 24 mm facing the edge to the inflow.

secondary effects, but the amplitude varies strongly (Fig. 6.6). The spectrum shows only the smallest secondary effects. The use of the triangular bluff body in the two different directions shows the influence of ultrasound and pressure sensors on both detection methods. The pressure sensor requires a strong pressure signal, and secondary effects have no influence on the signal. An ultrasonic barrier is much more sensitive to secondary vortices, it requires well-defined structures without any secondary effects.

Not only the time signals and spectra of the two different arrangements of bluff bodies are different, but also the characteristics, as shown in Fig. 6.7.

The sensitivity in the conventional arrangement with the flat side to the inflow is 6.7 Hz per m/s, corresponding to 6.7 primary vortices per meter. Each primary vortex occupies a distance of 15 cm. The Strouhal number is $S_r = 0.16$.

On turning the tip of the bluff body to the inflow, the sensitivity more than doubles to 14 Hz per m/s. The distance between successive vortices is 7.1 cm and the Strouhal number is 0.34. It is evident that secondary vortices lead to decreasing sensitivity. For ultrasonic measurements the conventional arrangement of triangular bluff bodies should be avoided.

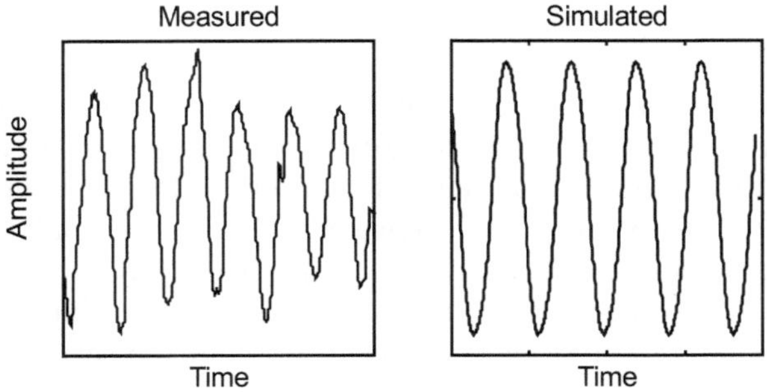

Fig. 6.6. Measured and simulated time signal and frequency spectrum of the ultrasonic phase shift of a triangular bluff body facing the edge to the inflow.

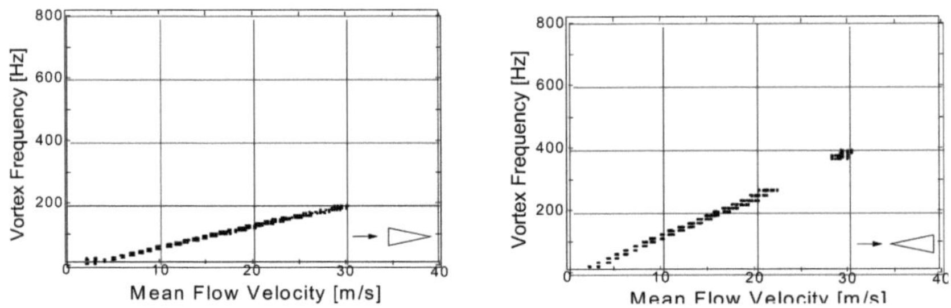

Fig. 6.7. Vortex frequency versus mean flow velocity of conventional arrangement (*left*) and turned around bluff body (*right*) of 24 mm height with ultrasonic measurement.

6.6.2 Small Triangular Bluff Body

Large bluff bodies may be good for vortex detection with pressure sensors, but because of the much higher sensitivity of ultrasonic waves for disturbances it is self-evident to investigate smaller bluff bodies. In the first instance, the triangular shape will be maintained. Experiments have shown that the smallest dimension with best results is a bluff body of 4 mm height and 8 mm length. This is a drastic reduction to 4% of the pipe diameter.

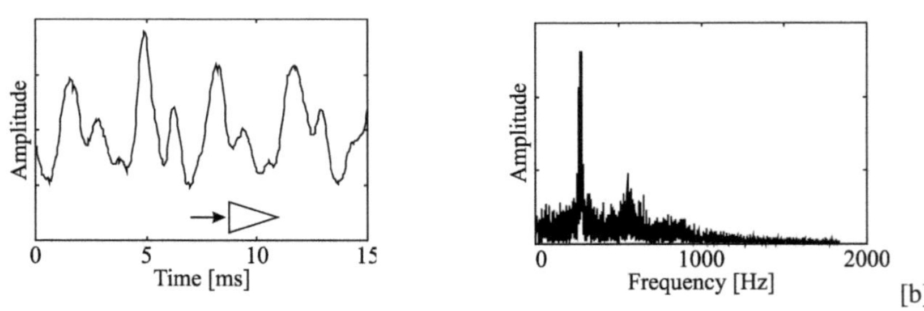

Fig. 6.8. Signal in time and frequency domains for a 4-mm bluff body in conventional arrangement.

Simulations and measurements have shown for a conventional arrangement that secondary vortices are generated at the tip of the bluff body following the main vortex and resulting in a second, small amplitude of double frequency in the frequency domain (Fig. 6.8). Physically, it is the same effect as at a large bluff body. The main frequency can be detected definitely. The 4-mm bluff body shows a much higher sensitivity of 25 Hz per m/s with a corresponding Strouhal number of $S_r = 0.1$. It is most remarkable that pressure losses behind the bluff body decrease substantially by a factor of about 5.

On rotating the small triangular bluff body by 180° so that the gas flows against the tip of the triangle it can be observed that secondary vortices dissolve. Even here

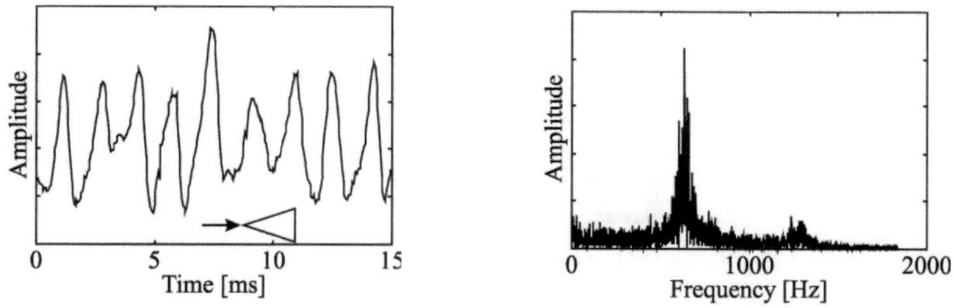

Fig. 6.9. Ultrasound signal of "turned-around" 4-mm triangular bluff body in time and frequency domain.

the same effect occurs as at the large bluff body. In the time domain they are not recognizable, with only a small effect observed in the frequency domain (Fig. 6.9).

This kind of arrangement has an additional advantage. The sensitivity increases to 60 Hz per m/s and is more than twice as high as in the conventional arrangement with the same very small pressure losses. The Strouhal number is 0.24. Figure 6.10 shows a comparison of the characteristics of conventional and "turned-around" arrangements.

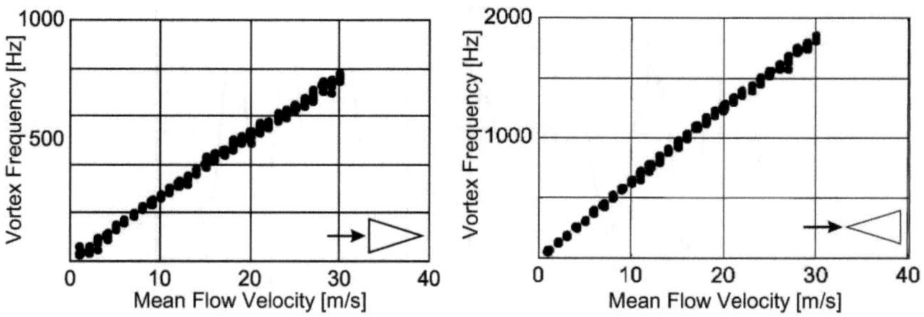

Fig. 6.10. Vortex frequency versus mean flow velocity of conventional arrangement (*left*) and a turned-around bluff body (*right*) of 4 mm height with ultrasonic measurement.

6.6.3 T-Shaped Bluff Body

Another type of bluff body used in commercial vortex-shedding flow meters combined with pressure sensors is the T-shape form. For ultrasonic measurements the width of the body can be reduced to only 10 mm, while maintaining signal quality.

Figure 6.11 shows the density plot of a T-shaped bluff body used in the conventional way facing the flat side to the inflow. The picture is zoomed into the central part of the pipe for better visualisation. The simulation was performed for the whole diameter, but the flow near the wall is hardly affected by the vortices. Also, parallel to the primary vortex a secondary vortex of very low amplitude is separating. The

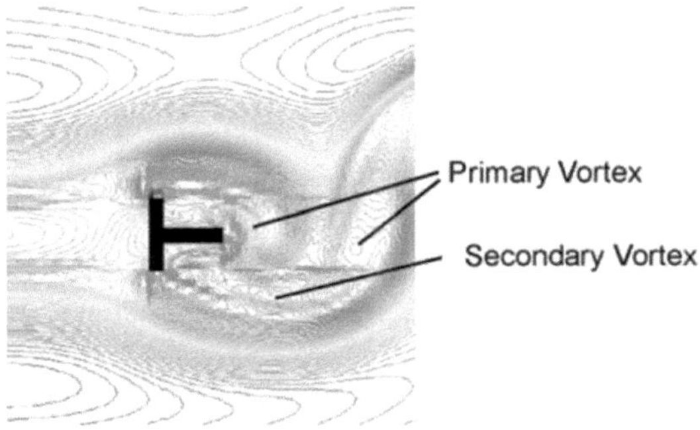

Fig. 6.11. Density plot of a simulated vortex street of a T-shaped bluff body with a width of 10 mm used in the conventional way. Only the central part of the pipe is displayed.

measured time signal of the phase shift of the ultrasonic signal shows that the influence of the secondary vortex is similar to the signal of the triangular bluff body (Fig. 6.12). The simulated time signal is not affected by the secondary vortex. The small influence can be explained by the simplification of the ultrasonic transmission.

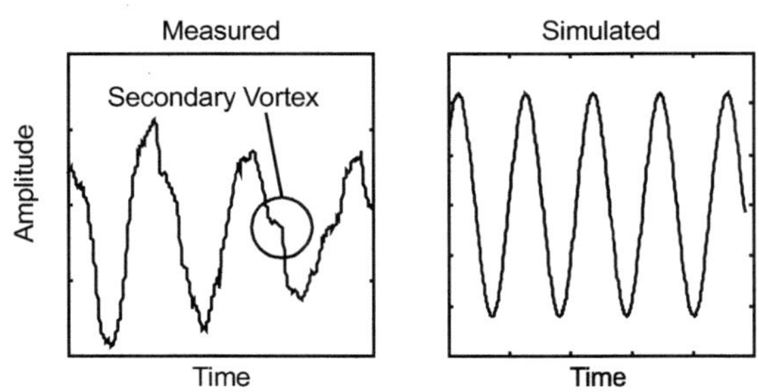

Fig. 6.12. Measured and simulated time signal of the ultrasonic phase shift of a T-shaped bluff body used in the conventional way.

As for the triangular bluff body, the T-shaped form generates well-defined signals for the ultrasonic detection method if it is used the other way round. The density plot (Fig. 6.13) shows only primary vortices with strong restriction. The simulated and the measured time signals (Fig. 6.14) show a very well-defined sinusoidal behavior that is not disturbed by any secondary vortex. The amplitude modulation of the phase shift is much smaller than for the triangular bluff body. A pressure signal at the pipe wall could not be measured because of the very small dimensions of the body.

On comparing the sensitivity of the two arrangements, 12 Hz per m/s could be detected for the conventional arrangement with the flat side to the inflow, while the turned-around arrangement resulted in 20 Hz per m/s for the 10-mm bluff body. The corresponding Strouhal numbers are $S_r = 0.12$ and $S_r = 0.2$, respectively.

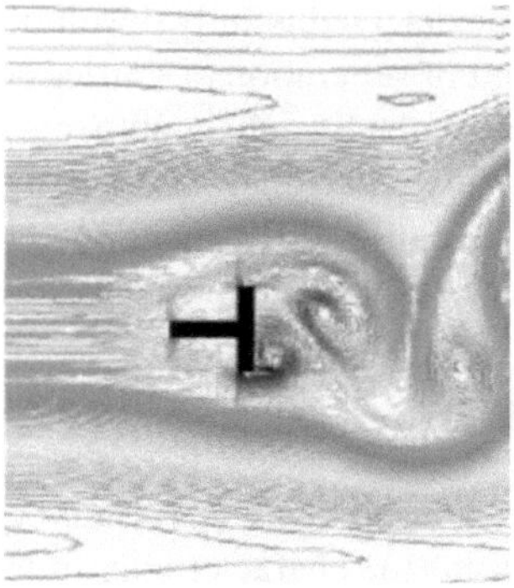

Fig. 6.13. Density plot of a simulated vortex street of a T-shaped bluff body with a width of 10 mm facing the bar to the inflow. The central part of the pipe only is displayed.

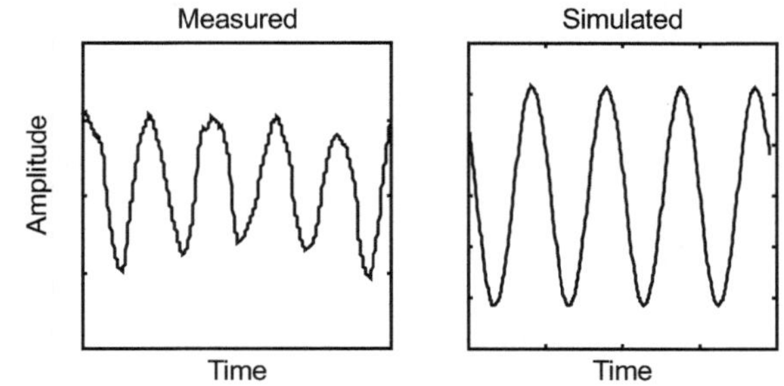

Fig. 6.14. Measured and simulated time signal of the ultrasonic phase shift of a T-shaped bluff body facing the bar to the inflow.

6.6.4 Circular Form

In order to decrease pressure losses, a circular form of only 3 mm diameter has been tested. Figure 6.15 shows the result in the time and frequency domains.

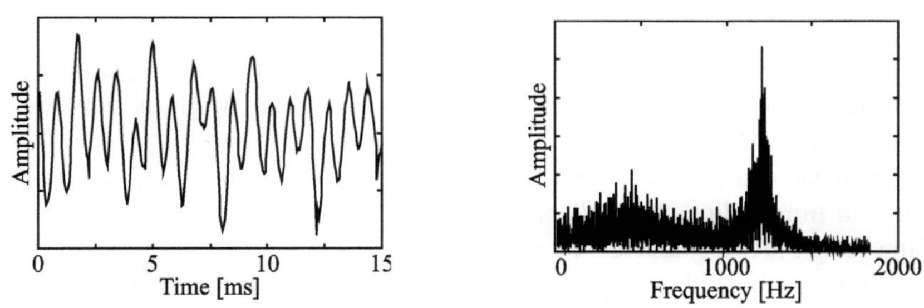

Fig. 6.15. Demodulated ultrasound signal in the time and frequency domains of a 3-mm circular bluff body.

The main frequency to be detected is fully developed, but amplitudes vary over a wide range because the point of separation of vortices on the round and smooth surface is not stable. Nevertheless, the Strouhal number is $S_r = 0.22$ and the sensitivity reaches 73 Hz per $\mathrm{m/s}$. This is more than 10 times that of a triangular form in the conventional arrangement. Because of the mentioned instabilities of the separation points of vortices, rough surfaces were supposed to lead to better stability.

6.6.5 Threaded Control Rod

Best results with very good sinusodial signals could be obtained with threaded control rods. Figure 6.16 represents the characteristics of threaded control rods of different sizes.

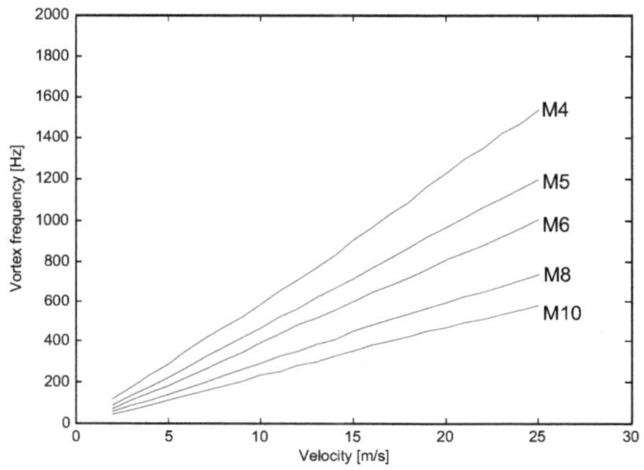

Fig. 6.16. Characteristics of different bluff-body sizes of threaded control rods.

It is evident that the sensitivity of the vortex meter depends on the bluff-body size. The sensitivity is given by

$$E = \frac{\mathrm{d}f}{\mathrm{d}v} \text{ in Hz per m/s} . \tag{6.2}$$

This refers to the number of vortices per meter and represents the periodic length of vortices. The sensitivity as a function of the bluff-body height is shown in Fig. 6.17 (Filips, 2003; Hans, 2003).

The mean characteristic can be determined by curve fitting as $E = 208d^{-0.93}$, with d in mm (Filips, 2003). This very simple relation permits the determination of the vortex meter characteristic as a function of the dimension of the bluff body. Measurements have shown that similar relations are applicable to other bluff-body sizes, too.

The best results of various bluff-body shapes were obtained using a M3 threaded control rod. The generated vortices are very stable and no secondary effects could be observed. The frequency is distinct with a good signal-to-noise ratio and lower

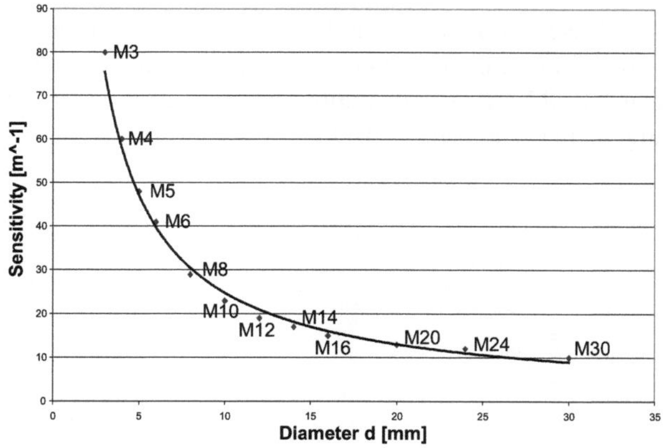

Fig. 6.17. Sensitivity versus size of threaded control rod for undisturbed flow.

frequencies can not be detected (Fig. 6.18). The rough surface of the bluff body increases the friction on the bluff-body surface and clearly defines the separating point of vortices.

The sensitivity increases to 80 Hz per m/s with a Strouhal number of $S_r = 0.24$. Pressure losses decrease to a negligible minimum. The uniform vortex street leads to reliable and sinusoidal demodulated ultrasonic signals that can readily be handled in signal processing.

The reduction of the bluff body from 24 mm to 3 mm leads to an enlargement of the cross-sectional area in the pipes resulting in drastically decreasing pressure losses.

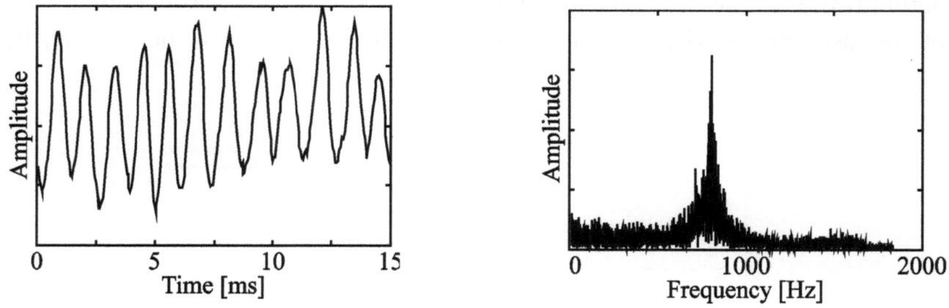

Fig. 6.18. Demodulated ultrasound signal in time and frequency domains of a 3-mm threaded control rod.

6.6.6 Pressure Losses

Figure 6.19 shows the pressure losses for different bluff-body sizes. The classical triangular size of 24% bluff-body height related to the pipe diameter causes rapidly increasing pressure losses with increasing flow velocity. By minimization of the bluff-body size pressure losses can be reduced to more than one tenth. Pressure losses of threaded control rod M3 are only negligibly higher than those in a free pipe in the same test arrangement (Windorfer, 2001).

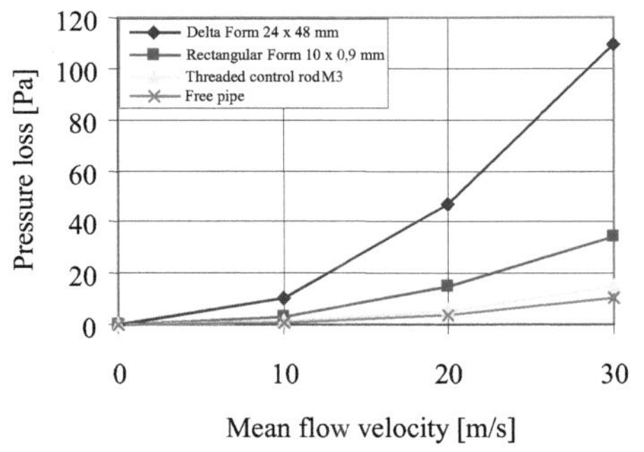

Fig. 6.19. Pressure losses of different sizes and arrangements of bluff bodies.

6.7 Influence of Disturbances

6.7.1 Single and Double Elbows

In industrial applications the necessary requirements of installation for a fully developed profile can not be met in any case. Single and double elbows influence the velocity profiles of the flow. The unsymmetrical velocity distribution behind a single elbow is based on centrifugal forces displacing the maximum of the velocity in the

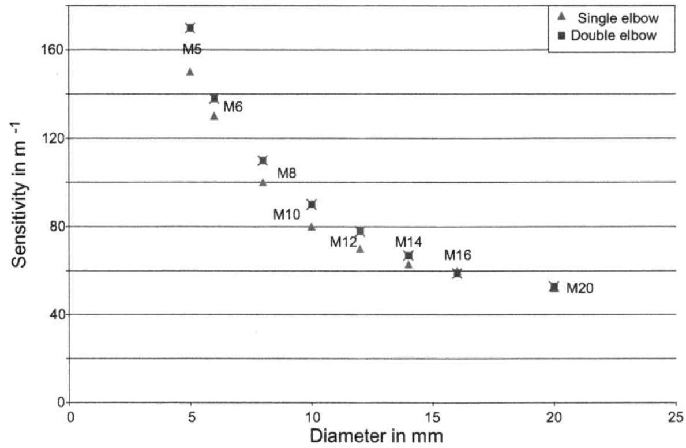

Fig. 6.20. Sensivity versus size of threaded control rod for disturbed flow.

outer range of the pipe. Additionally the main flow is superposed by a cross-flow resulting in a double vortex. Double elbows, however, result in a flow profile with spin.

Investigations on the influence of unsymmetrical and spin-affected flow have been made with threaded control rods M3 to M30 in a distance of the double pipe diameter $2D$. Sizes smaller, M5, and larger, M20, are too sensitive on disturbed profiles, especially behind a single elbow (Filips, 2003).

It is remarkable that the vortex frequency is nearly twice as high as for the undisturbed flow. Figure 6.20 shows the sensitivity as a characteristic of the threaded-rod size for disturbed flow.

Measurements behind a single elbow show high deviations independent of the kind of signal processing. Obviously they must be explained with the physics of vortex generation. Measurements behind a double elbow, however, lead to more reliable results. The vortex street is more stable. Nevertheless, the deviations are higher than for the undisturbed profile.

6.7.2 Pulsation

Two types of pulsation generators have been applied to the vortex meters. The first was an industrial lobed-impeller flow meter (lif) generating a discontinuous flow of a defined volume by the rotating piston. The second was a rotating plate inside the pipe with a plate diameter of $0.7\,D$ (D = pipe diameter).

Measurements have been made with triangular bluff bodies and threaded control rods in a distance of $2\,D$, $20\,D$ and $40\,D$ to the pulsation generator. For the lobed-impeller flow meter, in general, it can be stated that the energy of the pulsating flow is higher than the energy of vortices generated by bluff bodies. Therefore, mostly the pulsating frequency was determined with high deviations independent of the bluff-body shapes and sizes even in a distance of $40\,D$.

The pulsation frequency of a rotating plate inside the pipe could be adjusted independent of the flow velocity. A lock-in effect could be detected only if the natural vortex frequency was very close to the adjusted pulsating frequency. The turned-around triangular bluff body with the tip to the inflow was less sensitive to pulsation effects than the conventional bluff-body arrangement with the flat side to the inflow.

All measurements have shown large deviations.

It can be stated, in general, that vortex flow meters are very sensitive to any kinds of pulsation.

6.8 Conclusion

The application of ultrasound in vortex-shedding flow meters is a powerful combination. The ultrasonic signal is modulated in amplitude and in phase by the vortices. The demodulated amplitude signal has been evaluated. Compared with conventionally used pressure sensors, the ultrasonic method shows higher sensitivity. This method requires a new design for measuring systems. Simple signal processing presupposes sinusoidal signals that can be generated by triangular bluff bodies rotated 180° compared with conventional arrangements. The size can be drastically reduced to 4% of the pipe diameter, leading to negligible pressure losses behind the bluff body. The best results were obtained with an M3 threaded control rod with a sensitivity of more than ten times that of conventional, pressure-based measurements. Independent of the bluff-body shapes and sizes ultrasonic vortex meters are sensitive to disturbed flow profiles and pulsations.

References

Baker RC (2000) Flow Measurement Handbook. Cambridge University Press

Bentley JP (1983) Principles of Measurement Systems. Longman, London and New York

Breier A, Gatzmanga H (1998) Parameterabhängigkeit der Durchfluss-Frequenz-Kennlinie von Vortex Zählern im Bereich kleiner Reynolds Zahlen. Technisches Messen 62:22–26

Filips Ch (2003) Ultraschallsignalverarbeitung bei Korrelations- und Vortexverfahren zur Durchflußmessung. Dissertation, Universität Essen, also published by Cuvillier Verlag Göttingen (ISBN 3-89873-711-X)

Hans V, Poppen G, von Lavante E, Perpeet S (1997) Interaction between vortices and ultrasonic waves in vortex shedding flowmeters. FLUCOME 97, Proceedings Vol. 1:43–46

Hans V, Windorfer H, von Lavante E, Perpeet S (1998) Experimental and numerical optimization of acoustic signals associated with ultrasound measurement of vortex frequencies. FLOMEKO 98, Proceedings:363–367

Hans V, Windorfer H, Perpeet S (2000) Influence of vortex structures on pressure and ultrasound in vortex flowmeters. IMEKO 2000, TC9, Proceedings

Hans V (2003) Detection of vortex frequency in gas flow using ultrasound. Int. Symp. on measurement technology and intelligent instrumentation, Hongkong, Proceedings

Perpeet S (2000) Numerische Simulation von Strömungsfeldern um Durchfluß-Meßanordnungen. Dissertation, Universität Essen; also published by Shaker Verlag Aachen (ISBN 3-8265-7219-X)

Poppen G (1997) Durchflußmessung auf der Basis kreuzkorrelierter Ultraschallsignale. Dissertation, Universität Essen; also published by Shaker Verlag Aachen (ISBN 3-8265-3108-6)

Windorfer H (2001) Optimierung von Wirbelfrequenzmeßgeräten mit demodulierten Ultraschallsignalen. Dissertation, Universität Essen; also published by Shaker Verlag Aachen (ISBN 3-8265-9261-1)

7

Ultrasonic Gas-Flow Measurement Using Correlation Methods

Volker Hans

Institut für Mess- und Regelungstechnik, Universität Essen, 45117 Essen, Germany

Cross-correlation functions are in widespread use especially for the measurement of wave-propagation time and all kinds of velocity. The coincidence of turbulent structures in a gaseous flow with ultrasonic waves raises many questions concerning physics and signal processing, in particular on account of dissipating structures between the two ultrasound barriers. It appeared that the flow velocity measured by the cross-correlation function of the complex modulated signals is determined by the most frequent velocity components in the fluid. Besides, the ultrasonic beam measures along a line, whereas the average flow velocity is defined by an area integral. This velocity-dependent nonlinear effect necessitates a calibration of the measurement device. Uncertainties better than one percent are obtainable. Even disturbed flow by single and double elbows or pulsation can be measured in a qualified sense concerning uncertainty and scattering. Multipath arrangements can improve scattering. Furthermore, using special algorithms the profile of disturbed or undisturbed flow can be represented by tomography.

7.1 Introduction

In recent years an increasing market of flow-measuring instruments using ultrasound can be observed. Contactless measurements, no mechanical parts in the flow, no pressure drops are evident advantages of this technique.

Usually, the propagation time of ultrasound based on the Doppler effect is used to determine the flow velocity (Schrüfer, 1995a). Cross-correlation methods are well known but not so widespread on account of difficult signal analysis. The signals to be correlated are modulated only slightly by the flow and the sidebands are very close to the carrier frequency. In this case, signal processing with correlation functions is very difficult. Especially if the carrier frequency is very high, as for ultrasound in the range of some hundred kHz, oversampling techniques with some megahertz lead to problems in realtime signal processing on account of the limited computer storage, see Chap. 5.

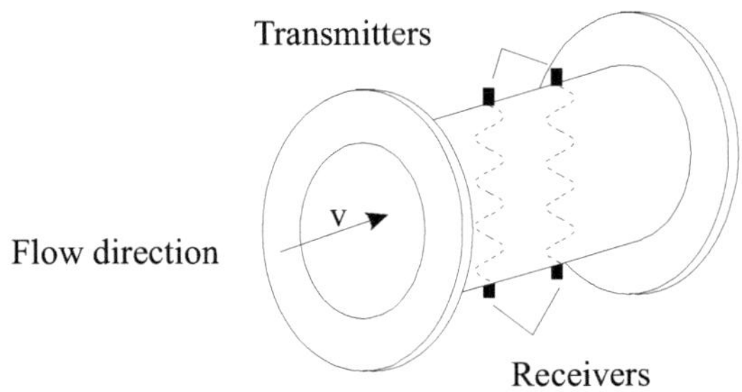

Fig. 7.1. Cross-correlation measurement device.

Modern principles of signal processing improve the possibilities of research in countless scientific areas. In this chapter, a practical application of such methods is presented for the explanation and the improvement of an ultrasound cross-correlation flow meter. It is used for the determination of the volume flow, which is proportional to the average axial flow speed. Although the basic theory of this measurement is simple and easy to understand, much effort was necessary to improve the instrument range from multiphase and particle flow to gaseous fluids. The main reason for these difficulties is based on the unsatisfactory knowledge of the properties of turbulent fluids and the interaction with ultrasound, see Chap. 6. Therefore, there are still unexplainable systematic deviations and unexpected behavior in the measured data. By performing a reverse processing of that data more information about the fluid can be gained and a further understanding and improvement of this device is possible. This chapter is a summary of research work described in detail in Filips (2003); Niemann (2002); Poppen (1997); Rettich (1999); Skwarek (2000).

The idea for this device was developed between 1935 and 1955 by several scientists, but no exact date and origin can be determined. A first patent was registered in 1967 with a corresponding PhD thesis by Beck in 1969 with a huge number of publications following by other scientists. For further references see Beck (1987). As a main intention the feedback on the fluid should be reduced by using a continuous ultrasonic wave transmitted between transmitter–receiver pairs in the radial direction perpendicular to the main flow. For multiphase and particle flows, the amplitude of the ultrasonic beam s_1 is modulated stochastically with the density and distribution of the particles crossing the beam. Each combination leaves a certain pattern like a fingerprint in the received signal. Assuming that the fluid and the particle distribution do not change significantly within a short distance, a second downstream transmitter–receiver pair s_2 strongly receives the same signal – only with a time delay according to the traveling time of the particles between the ultrasonic barriers (Fig. 7.1).

The point of maximum similarity between s_1 and s_2 is computed by using the mathematical cross-correlation function (CCF) Φ_{12}

$$\Phi_{12}(\tau) = \lim_{T\to\infty} \frac{1}{2T} \int_{-T}^{T} s_1(t) s_2(t-\tau) \mathrm{d}t \tag{7.1}$$

and the result can be directly read from the abscissa of the extremum of the CCF (Fig. 7.2).

With a known distance d between the barriers the average axial speed can be calculated by

$$\bar{v} = \frac{d}{t} \, . \tag{7.2}$$

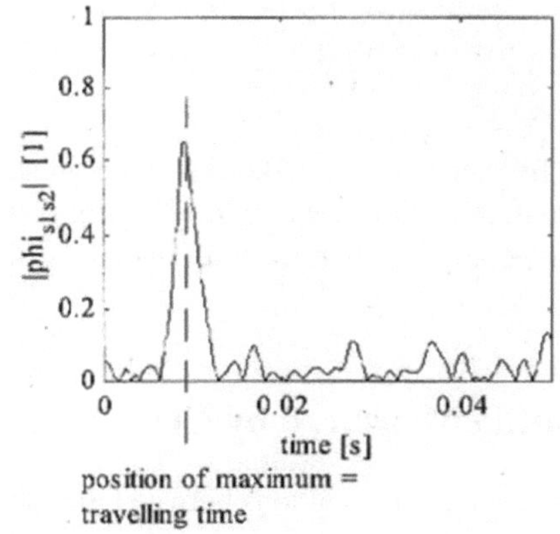

Fig. 7.2. Normalized result of the CCF between s_1 and s_2. Unique maximum at $t = 0.01\,\mathrm{s}$ represents traveling speed of structures $\bar{v} = 10\,\mathrm{m/s}$.

7.2 Determination of Traveling Time by Cross-Correlation Functions

For digital signal processing (7.1) the cross-correlation function can be written as

$$\begin{aligned}
\Phi_{12}(kT_\mathrm{a}) &= \frac{1}{T}\sum_{n=1}^{N} s_1(nT_\mathrm{a})s_2((n-k)T_\mathrm{a}) \, , \\
\Phi_{12}(kT_\mathrm{a}) &= \frac{1}{T}\sum_{n=1}^{N} s_1(nT_\mathrm{a})s_2((n-i-k)T_\mathrm{a}) \, , \\
\Phi_{12}(kT_\mathrm{a}) &= \Phi_{11}((k+i)T_\mathrm{a}) \, ,
\end{aligned} \tag{7.3}$$

with the sampling interval T_a, N number of samples and n sampling index. Further details of signal processing are described in Chap. 5. The cross-correlation function equals the autocorrelation function Φ_{11} shifted by the traveling time iT_a (Schrüfer, 1995b). As the magnitude of CCF depends on the root mean square of single-signal components usually the absolute value of the normalized CCF is applied

$$|\Phi_{12\text{norm}}(kT_\text{a})| = \frac{\Phi_{12}(kT_\text{a})}{\sqrt{\Phi_{11}(0)\Phi_{22}(0)}} . \tag{7.4}$$

The magnitude ranges between 0 and 1 as a measure of the similarity of both signals. Only signals with the same frequency can be correlated. On account of dissipation the similarity of the two signals decreases. The degree of similarity depends on the distance between the two ultrasound barriers. The exact determination of the maximum of CCF and the traveling time depends on the relative deviation given by the sampling frequency f_a and distance (Niemann, 2002),

$$\frac{\Delta v}{v} = \frac{v}{df_\text{a} - v} . \tag{7.5}$$

For a given sampling frequency small distances cause greater deviations especially for high velocities. Small deviations require long barrier distances leading to higher dissipation. Best results could be obtained with distances of 60 to 80% of the pipe diameter D.

7.3 Physical Quantity Measured by Cross-Correlation Functions

Many measurements have shown deviations between the measured flow velocity by CCF and the real mean flow velocity. As explained in Chap. 6 various physical effects regarding the interaction between the fluid and the ultrasonic wave cause amplitude and phase modulation of the ultrasound signal. The question arises if both kinds of modulation result in the same measured time, whereas the causal connections are different. The next question is which physical quantity of turbulent structures is measured in the correlation measurement. Many authors gave various answers (Braun, 1984; Niemann, 2002; Poppen, 1997; Schneider, 2001; Shu, 1987; Skwarek, 2000; Worch, 1998) developing interesting models. Interesting considerations confirming the systems theory model in Chap. 8 have been presented by Niemann (2002).

Both ultrasonic barriers register many turbulent structures with different velocities in each case. The amplitude- and phase-demodulated signals therefore are composed by single-signal components with different traveling times. The question is which of these components determines the maximum of the correlation function. In a first step it may be supposed that both signals s_1 and s_2 are composed by two components

$$\begin{aligned} s_1 &= s_{\text{x}1}(nT_\text{a}) + s_{\text{y}1}(nT_\text{a}) , \\ s_2 &= s_{\text{x}2}(nT_\text{a}) + s_{\text{y}2}(nT_\text{a}) , \end{aligned} \tag{7.6}$$

with different traveling times $t_1 = i_1 T_\text{a}$ and $t_2 = i_2 T_\text{a}$.

The CCF yields

$$\begin{aligned} \Phi_{12}(kT_\text{a}) &= \frac{1}{N}\sum_{n=1}^{N} [s_{\text{x}1}(nT_\text{a}) + s_{\text{y}1}(nT_\text{a})]\,[s_{\text{x}2}((n-k)T_\text{a}) + s_{\text{y}2}((n-k)T_\text{a})] , \\ \Phi_{12}(kT_\text{a}) &= \Phi_{\text{x1x2}}(kT_\text{a}) + \Phi_{\text{x1y2}}(kT_\text{a}) + \Phi_{\text{y1x2}}(kT_\text{a}) + \Phi_{\text{y1y2}}(kT_\text{a}) . \end{aligned} \tag{7.7}$$

The components $s_{x1}(nT_a)$ and $s_{y1}(nT_a)$ are supposed to be stochastically independent of each other resulting in $\Phi_{x1y2}(kT_a)$ and $\Phi_{y1x2}(kT_a)$ equal zero:

$$\Phi_{12}(kT_a) = \Phi_{x1x2}(kT_a) + \Phi_{y1y2}(kT_a) \; . \tag{7.8}$$

The CCF of both signals corresponds to the sum of both CCF of the single-signal components. Both components generate their own maximum superposing each other. The traveling time is to be determined by the highest maximum. The traveling time of the other component does not influence the result. If both maxima are close to each other both correlation functions are superposed to one characteristic. The position of this new maximum is not given by the arithmetic mean value of the position of the single maxima. From two components the general case with m signal components can be expressed as

$$\Phi_{12}(kT_a) = \sum_{l=1}^{m} \Phi_{xlyl}(kT_a) \; . \tag{7.9}$$

The CCF of the sum of signals results in the sum of single CCF of all signal components. This means that without dissipation the most frequent velocity component will be detected. The bell-shaped characteristics differ clearly in magnitude as well as in their width. The magnitude is determined by similarity and dissipation of components, while the width represents the bandwidth of the signals. Narrowband signal components generate wide shapes of CCF, wideband components lead to narrow shapes of CCF.

Accordingly, four different stochastically independent influences determine the position of the maximum of CCF:

- The number of recognized signal components
- The magnitude of amplitudes of signal components
- The width of summarized characteristics and thus the bandwidth of signal components
- The similarity of components in both barriers

Summing up, it may be said that by evaluating the CCF the most frequent velocity component will be measured generating a high degree of modulation of the carrier signal with the maximum rate of change and high amplitudes with low dissipation. The measured velocity by cross-correlation corresponds to the modal value of the probability density function (PDF) of velocity components. It does not represent any mean value of all detected velocity components. As pointed out in Chap. 8 measurements of fluid components by means of particle-image velocimetry have shown that the PDF of velocity components is skewed. The modal value of velocity components is given by the maximum of the skewed PDF (Braun, 1984; Schneider, 2001; Skwarek, 2000).

Correlation functions presuppose ergodic processes. A stochastic process is called ergodic if its ensemble averages equal appropriate time averages (Papoulis, 1984). As the modal value is a mean value within the meaning of statistics the theorem of ergodicity is fullfilled. These reflections are in agreement with the analysis by systems theory in Chap. 8.

7.4 Measurements

Experiments have been made in a test rig with a pipe diameter of 100 mm. The stationary turbulent gas flow was controlled by a calibrated turbinemeter for average flow velocities up to 30 m/s. The distance between the ultrasonic barriers was 100 mm, and the carrier frequency was 220 kHz. Signal processing was executed by digital undersampling and Hilbert-transform as described in Chap. 5. As the effect of phase modulation was more distinct than that of amplitude modulation the determination of CCF was related to the phase-demodulated signal.

Measurement results are represented as normalized characteristics according to (7.4). The result is calculated in units of T_a. However, with the time antiproportional to the velocity, the graphical results are not directly interpretable due to its nonlinear behavior. Increasing speed shifts the results not to infinity but to zero, and the width of the maximum peak becomes smaller instead of broader. This is a reason for rescaling abscissa. By applying (7.2) on each value of the abscissa the function $\Phi_{12}(kT_a)$ is converted into $\Phi_{12}(v)$ with equally spaced velocity units, Fig. 7.3 (Skwarek and Hans, 1999).

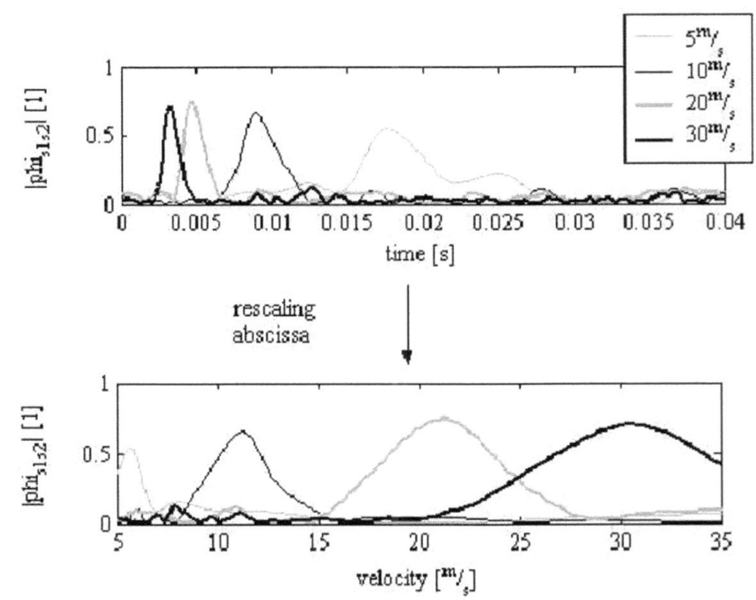

Fig. 7.3. Rescaling abscissa from time to velocity.

According to physical assumptions, amplitude modulation is caused by diffraction, reflecion and damping due to density changes as described in Chap. 6. Phase modulation, on the other hand, is caused by velocity components in the direction of the ultrasound de- and accelerating signal. Both effects are linked. Turbulent structures are a combination of local pressure, density and velocity distribution and move with the average velocity of the fluid. Furthermore, they dissipate due to the turbulent viscosity surrounding the structures and become larger with a decreasing energy density. Although the structures become larger the average velocity distribution stays the same. Nevertheless, the evaluation of amplitude modulation is not suitable for calculating the average flow velocity because the measurement does not refer to the same relative position in a turbulence (Skwarek and Hans, 1999).

7.4.1 Measurement Uncertainty

As mentioned above, the modal value of the flow-velocity components is measured instead of the average value. This modal value is dependent on the flow profile and thus on flow velocity. As the relationship between modal value and average value changes, a calibration of cross-correlation flow meter is necessary in any case.

The measurement principle itself causes further systematic deviations between the real mean flow velocity and the measured flow velocity by CCF. Unlike the definition of the average flow velocity by an area integral

$$\bar{v}_{\text{def}} = \frac{1}{A} \int_A v(A) \mathrm{d}A \,, \tag{7.10}$$

an ultrasonic beam represents a measurement along a line

$$\bar{v}_{\text{meas}} = \frac{1}{L} \int_L v(L) \mathrm{d}L \,, \tag{7.11}$$

with L and A sizes of the line and the area, respectively. The radial velocity distribution for a fully developed, undisturbed flow shows a speed-dependent exponential-like curvature due to the friction at the wall. This curvature causes a relative deviation of the measurements

$$r = \frac{\bar{v}_{\text{meas}}}{\bar{v}_{\text{def}}} = 1 + \frac{1}{2n} \tag{7.12}$$

with $n \in \{6, 7, 8...\}$ for increasing v and $n = f(v)$, see Chap. 8. Usually the theoretical values of r are in the range of 1.085 to 1.05 (Nikuradse, 1932). However, practical tests show further deviations between expected and real results necessitating a calibration of the flow meter. The probable reason for this deviation is the difference between the theoretical assumption of the ultrasonic beam as a line. In reality it covers a non-negligible volume of the test rig and measures not only the centerline of velocities but even slower velocities at the left and right of the centerline. Consequently, the real measured velocity is lower than expected.

7.4.2 Measurement of Disturbed Flow Profiles

Measurements Behind a Single Elbow

In industrial applications disturbed profiles in consequence of installations such as single or double elbows are often typical and unavoidable. Measurements have been

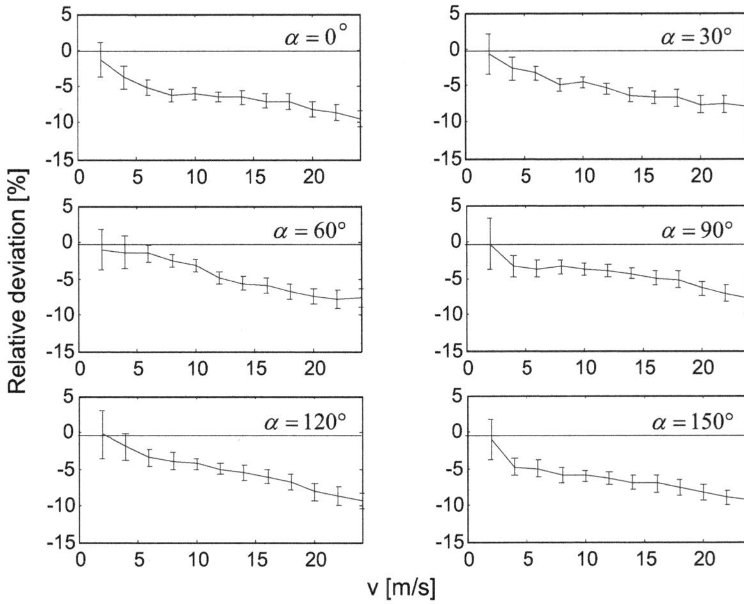

Fig. 7.4. Measurement results in ultrasound paths at different angles behind a single elbow.

made in a distance of $2D = 200\,\mathrm{mm}$ behind a 90-degree single elbow. In this case a secondary flow is superposed to the main flow leading to a transverse motion in the form of a spiral double eddy. Examining the influence of the asymmetric profile the ultrasonic path has been rotated by six angles. The results of relative deviations of the average velocity and the standard deviation of measurement are shown in Fig. 7.4 (Niemann, 2002). It is remarkable that all measured velocities are lower than the real velocity independent of the angle. On account of the unknown asymmetric profile no deviation factor r could be determined. To compensate the deviations the measurement device must be calibrated.

The results of the calibrated device are shown in Fig. 7.5. The relative deviations are smaller than one percent with standard deviations of about one percent. The angle of the ultrasound path has practically no effect on the uncertainty.

Measurement Behind a Double Elbow

Two single elbows are distorted to each other by 90 degrees in space. By this means a spin-affected flow is generated. The relative deviations of the uncalibrated device show characteristics from positive to negative values (Niemann, 2002). Best results could be obtained at an angle of 60 degrees. As for the single elbow the results can be improved by calibration. The average values of deviations are in the range of one percent. The standard deviation increases up to two percent with increasing velocity.

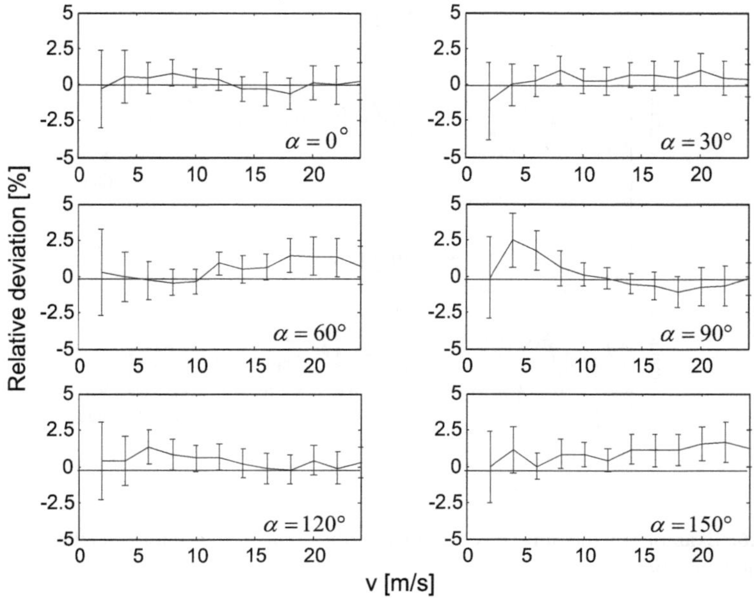

Fig. 7.5. Measurement results with calibrated measuring device behind a single elbow.

Pulsation Effects

Possible influences of pulsation on a flow meter are manifold so that they will not be completely described in this chapter. Alternatively, the main focus will be put on problems occurring in combination with an ultrasonic measurement device (Skwarek et al., 2001). Therefore, effects are generally divided into two subgroups:

- Direct effects on transmitter or receiver
- Modulation-like effects on ultrasonic wave

Direct effects are caused by shock-waves due to pulsation: Piezoelectric receivers are sensitive to every kind of pressure independent of their origin. Every pressure wave from pulsation is superposed onto the modulated ultrasonic signal at the receiver. The following estimation gives an impression about the magnitudes of the pulsation amplitude in comparison with the ultrasonic wave:

The transmitters are continuously driven with a sinusoidal supply voltage $u_{\max} = 10\,\mathrm{V}$. According to the datasheets of the sensor (Massa, 1999) a transmitting sensitivity of 20 dB versus 1 μbar per Volt has to be expected, which is equivalent to a maximum pressure of 10 Pa. Assuming a wave propagation without any damping losses these 10 Pa are superposed by pulsation. With the fans producing a pressure drop of about 6000 Pa in the pipe, theoretically a pressure pulse of the same magnitude appears in a pipe opened and closed for a short time. Consequently, ultrasonic signals are hardly detectable within the pulsation for disadvantageous boundary conditions concerning pulse frequency and speed of flow. An example of such poor

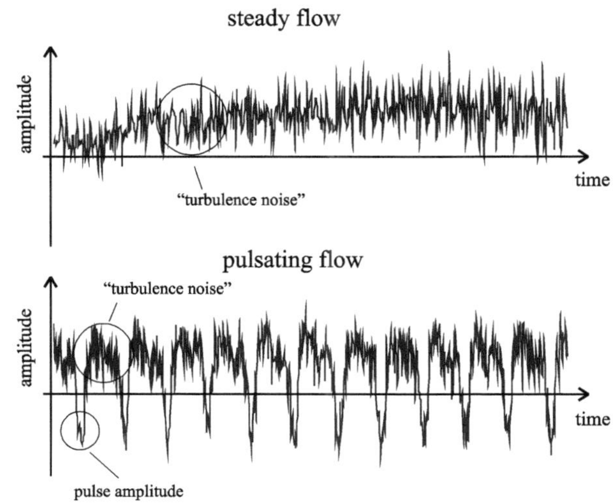

Fig. 7.6. Example of the superposition of the ultrasonic signal with modulated turbulence noise and pulsation.

conditions is given in Fig. 7.6. It is easy to imagine that cross-correlation as well as vortex flow meters using nonselective algorithms react sensitively to this pulsation.

For pulse amplitudes similar to or less than the modulation amplitude the second case mentioned above has to be taken into consideration. Not only do these impacts directly influence the sensors, they also influence the behavior of the fluid. As already explained, modulation properties depend on the boundary conditions given by the environment such as temperature, pressure, density and velocity fields. According to the magnitude of the pulse one or several quantities change significantly. Consequently these changing fields modulate the sinusoidal carrier as well. Their influence can possibly be higher than the turbulence-caused modulation itself. For more considerations about the interaction between ultrasound and fluid see Kolpatzik (1999); Poppen (1997); Rettich (1999); Skwarek (2000).

Beyond the general effects of pulsation on ultrasonic measurements further influences concerning the physical principle of a device can be thought of. Coherent structures at the correlation method are unstable and not very powerful. They could either be changed or even destroyed by the pressure, so they cannot be redetected at the second barrier or the pressure moving with sound speed produces such a high amplitude at the receiver that ultrasound cannot be received at all.

In industrial applications pulsations are usually generated by pumps, valves or measurement devices such as oval gear meters. For modeling all these disturbances a simple pulse generator was designed that imitated a wide variety of pulses. It consists of an actively driven rotating flap closing and opening the cross section of the pipe sinusoidally with a frequency between 5 and 40 Hz. By using flaps of different diameters and shapes the intensity and type of pulsation can be changed. The complete test rig consists of a straight pipe with a diameter of $D = 0.1\,\mathrm{m}$ and an overall length of $90.5\,D = 9.05\,\mathrm{m}$. After an inlet of $20D$ the measuring device is installed with the pulsation generator $10D$ further downstream. A turbine flow meter after $15D$ is used for reference although this reference gives only an estimation because the flow type exceeds the prescribed setup conditions. An expected, systematic deviation is not corrected yet. Finally, the test rig ends $32D$ after the reference with the underpressure fan. For air as fluid the mean velocity can be varied between $Re = 20000$–$200,000$ and stabilized on about 99–101% of the desired set point.

For the evaluation of the influence of technical devices and the comparability to the rotating-paddle measurements have been performed with an oval gear meter. In this device the rotating frequency of the oval gear depends linearly on the mean flow velocity so the pulsating frequency is proportional to the mean velocity. The amplitude of the pulsation is conjunct with the volume flow, too. Independent manipulation of the parameters is impossible.

For the evaluation of the results the lack of an exact reference has to be taken into consideration. As already mentioned, the turbine flow meter is only suitable for estimating the average flow speed. Therefore, the measurements cannot be evaluated absolutely but only checked for their plausibility.

The measurements were performed at velocities $\bar{v} = 5$, 10, 15 and $20\,\mathrm{m/s}$ with flap diameters varying between $60\,\mathrm{mm}$ and $100\,\mathrm{mm}$ and a rotating frequency between 0 Hz (open flap) and 40 Hz.

Obviously two states of measurements have to be considered: Either all or no measurements are correct for a given boundary condition. Flap diameters up to $80\,\mathrm{mm}$ and rotating frequencies up to 20 Hz have no influence on correct measuring results.

The cross-correlation method seems to be very sensitive to pulsation, independent of the average flow velocity. The only parameters defining the behavior of the device are the pulsation frequency and the flap diameter. For further interpretation the flap diameter is set equivalent to the amplitude of the pulse wave because the lower the diameter the higher the amount of bypassing fluid that is not influenced by pulsation.

For understanding the influence of pulsation amplitudes in detail formula (7.1) for the cross-correlation function of $s_1(t)$ and $s_2(t)$ has to be used. According to the Wiener–Chintschin theorem (Schrüfer, 1995b) this is equivalent to the cross-power spectrum in the frequency domain

$$\mathrm{CPS}(\omega) = \frac{1}{2T} S_1(\omega) S_2^*(\omega) \,. \tag{7.13}$$

The spectrum of a turbulent fluid contains frequencies from 0 to 3000 Hz with higher amplitudes at lower frequencies (Fig. 7.7). Generally, the power is distributed stochastically over the whole bandwidth. If the fluid starts pulsating with a high amplitude at both sensors for signals s_1 and s_2 pressure waves move with the speed of sound and appear nearly simultaneously at both ultrasonic barriers. This causes a very powerful periodic component that is squared in the CPS and superposes the maximum caused by the traveling time in the CCF (Fig. 7.8). Although the oscillation of the CCF in the right diagram looks very harmonic even a single bandstop filter

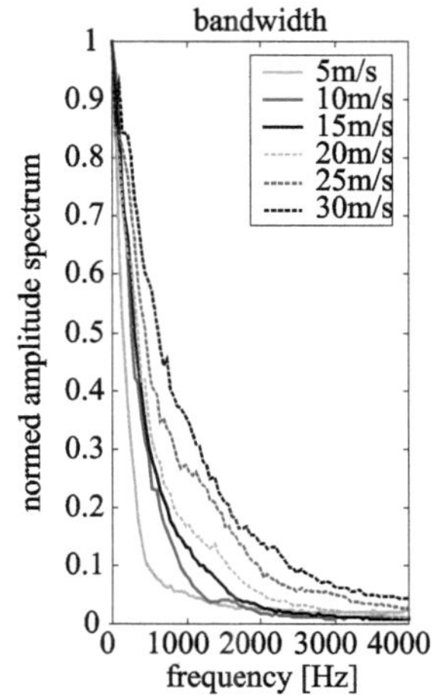

Fig. 7.7. Bandwidth of a demodulated ultrasonic signal of a steadily flowing fluid.

does not help to detect the maximum of the CCF. The vibrating fluid makes several coupled systems in the test room resonate with spectral lines that can also be found in Fig. 7.8. Only after filtering the frequency of several maxima and multiples of it up to 200 Hz with bandstop filters can a sufficient result be obtained. The difference between the filtered and the unfiltered signal is shown in Fig. 7.9. Here a unique maximum in the CCF can be evaluated. Although filtering seems to be a solution of the problem, two further difficulties arise:

- The main information of the turbulent fluid is found at frequencies below 200 Hz (Skwarek, 2000; Skwarek and Hans, 1999). Although the remaining, high-frequency part can still be evaluated, the "filtered maximum" differs from the "unfiltered maximum" by several sampling points. This is equivalent to some percent of the calculated speed.
- Furthermore the bandwidth and starting frequency and the number of bandpass filters is unknown and depends on the velocity of the fluid and pulsation parameters.

Lobed-Impeller Gasmeter

For the verification of the measurements on pulsating flow generated by an artificial pulsation generator a lobed-impeller gasmeter was also used. The comparison of the two different measurement setups with rotating paddle and lobed-impeller gasmeter shows a similar influence on the measurement system and the disturbing effects in the frequency domain. The limited independent variation of the pulsation parameters of the second setup shows only a small section of the variety of pulsation parameters. Quantification of the pulsation in flow parameters as time-dependent pressure and velocity components should be carried out for a better numerical comparison of different pulsation generators and the unsteady flow.

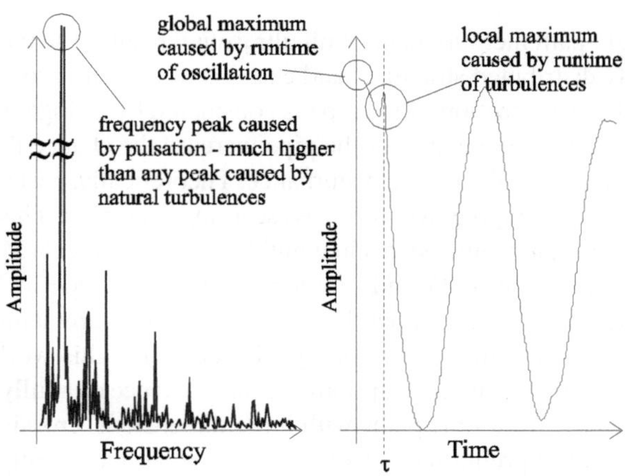

Fig. 7.8. Faulty calculated cross-correlation function caused by pulsating flow.

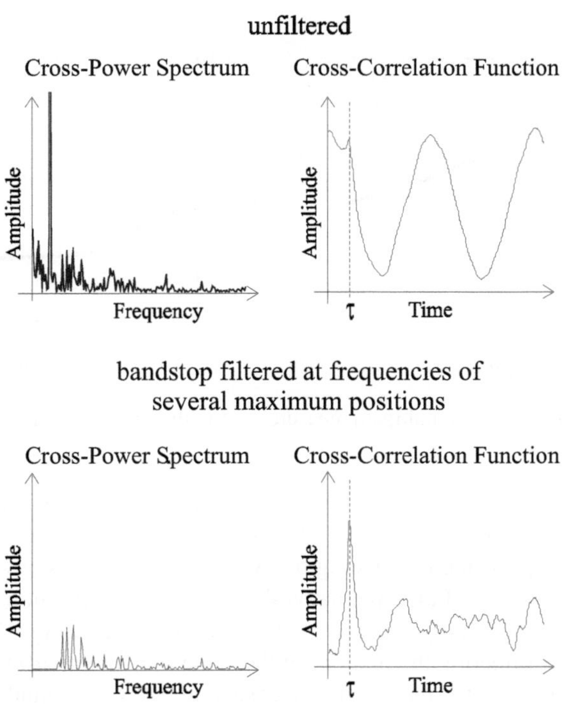

Fig. 7.9. Filtered and unfiltered signals.

7.5 Multipath Arrangement

As shown, single-path measurements with ultrasound result in scattering on account of stochastically distributed structures and changes of velocity profile. Furthermore, only a part of the cross section of the pipe is considered by a single ultrasonic beam. Measurements at another position of the pipe increase the chance that further structures are met, equalizing the profile disturbance. Theoretically, best results should be expected by spatial averaging over the cross section of the pipe. This can be approximated with several ultrasonic beams in a multipath arrangement (MPA).

First examinations of MPA and problems were introduced in Papoulis (1984) with some basic consideration about the feasibility of measurements and the design of the pipe. With better algorithms for signal processing, this work is particularly being continued. Simplifying assumptions are the existence of a fully developed flow profile and diametric ultrasonic paths with finite dimensions, crossing exactly in the center of the pipe. For preliminary examinations, a three-path configuration is taken into consideration.

Measurements using the three single paths and their common average are shown in Fig. 7.10 (Skwarek and Hans, 2000). The range of the set speed had to be limited for the introductory measurements to 20 m/s. As already mentioned, it is still not

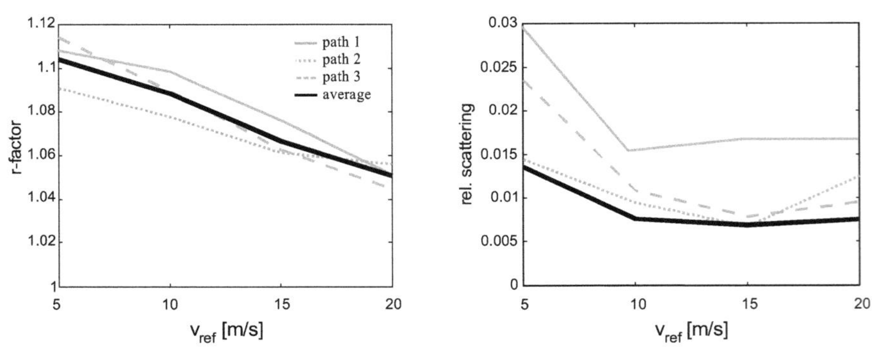

Fig. 7.10. Results of a multipath measurement with sensors in star arrangement.

possible to estimate the quality of the r-factor. All three paths as well as their average run in similar intervals, which are at least plausible compared with the expected value. The real improvement of averaging over multiple paths is obvious in the right diagram: The scattering of all three paths shows a different and nonmonotonous behavior, which is significantly reduced by their common averaging. This is an indication of oppositely effecting fluctuations at the same time but at different positions in the pipe. Consequently, the mean value over several paths will equalize and suppress these effects.

7.6 Tomographic Reconstruction of Flow Profile

Multipath measurements can be extended into tomographic representations of turbulent flow profiles. Only a short overview about ultrasound tomography with cross-correlation functions and the resultant problems will be given. More details are described in Filips (2001).

In the case of disturbed profiles, however, simple averaging is not sufficient due to missing rules about weighting the results over the single paths. Therefore, a tomographic representation of the profile could be helpful to determine the total average velocity of the fluid by an integration over this profile. The results of CCF as measures of the average velocities over the ultrasonic paths can be used for a reconstruction of the axial velocity distribution on the fluid.

Main problems arise with the definition of boundary conditions for tomographies. One parallel projection is ideally represented by the entity of an infinite number of parallel beams. This means an infinite number of projections.

Parallel beams can be realized in the test rig by a grid arrangement. With respect to crosstalk between the sensors and the macroscopic size of the beams only two perpendicular projections with three beams each are performed. Furthermore, the parallel expansion of the ultrasonic beam as a cone is not ensured. These six paths represent the only information of the flow. They are used for the Radom transforma-

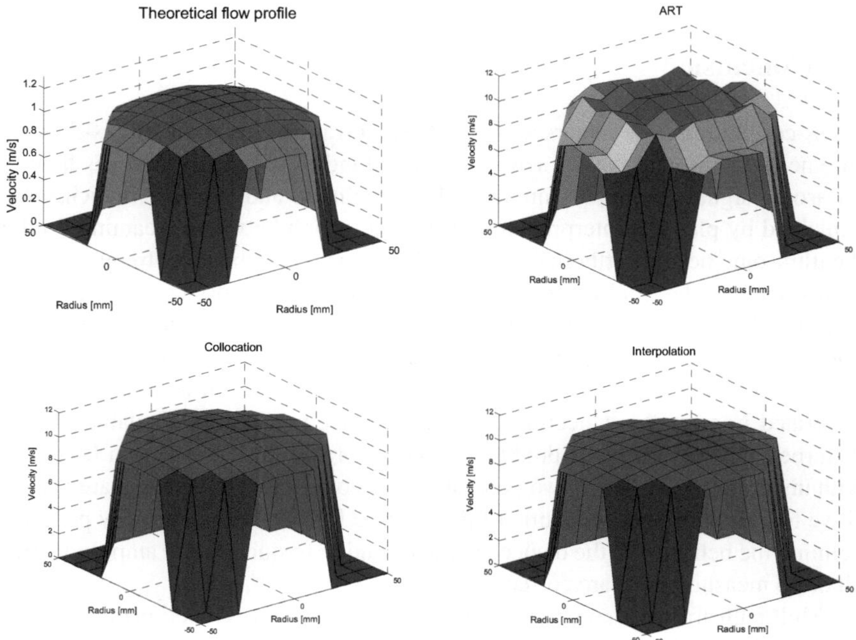

Fig. 7.11. Fully developed flow profile by different reconstruction algorithms. *Upper row left*: theoretical profile (Gersten and Herwig, 1992); *right*: reconstruction by ART. *Lower row left*: collocation method; *right*: interpolation method.

tion. Special procedures for minimal sets of data must be applied, which are known as the algebraic reconstruction technique (ART).

Using the same distribution of pixels as in ART a linear combination of several constant approach functions can be applied, which is known as the collocation method. The tomographic representation can be improved by a simple interpolation method with polynomial functions.

For the undisturbed profile a function is taken as a basis that can be changed by matched polynomial functions. The theoretical description of the profile is applied following Gersten and Herwig (1992) as reference for those reference polynomial functions that are to be reached in the reconstruction. The different kinds of tomographic reconstruction are shown in Fig. 7.11 (Filips, 2003, 2001).

For the tomographic representation only a limited number of measuring paths with small content of information is available. The collocation method results in a reconstruction with the smallest deviations for the undisturbed profile. This is caused by the rough partition of the cross-sectional area. The irregular reconstruction by ART was obtained by a simplified algorithm. The reconstruction by interpolation using polynomial functions shows remarkable improvements.

The reconstruction methods described have also been applied to disturbed flow after single and double elbows (Filips, 2003, 2001). For the disturbed profile no satisfactory results could be obtained with the collocation method. Best results could be realized with ART showing clearly local maxima and minima.

7.7 Conclusion

Cross-correlation flow meters are useful and advantageous tools in gas-flow measurement. Ultrasound is very sensitive to any kind of turbulence in a gas flow. The ultrasonic signal is complex modulated in amplitude and in phase, which may be explained by physical interpretation of the interaction between streaming fluid and the ultrasonic beam. Furthermore, it could be shown that the most frequent velocity components in the fluid determine the magnitude of the cross-correlation function and thus the measured flow velocity. Additionally, the average flow velocity is defined by an area integral over the cross section, whereas an ultrasonic beam represents a line integral. Therefore ultrasonic cross-correlation flow meters have to be calibrated in any case. Uncertainties less than one percent are available. Even for disturbed flow by single or double elbows calibrated flow meters can be installed with uncertainties of one to two percent. Cross-correlation flow meters are very sensitive to pulsation, independently of the average flow velocity. The only parameters defining the behavior of the device are the pulsation frequency and amplitude. Either all or no measurements are correct.

Multipath arrangements consider a larger area of the cross section of the pipe and therefore lead to lower scattering. But they also have to be calibrated. They can be used for tomographic representations of the turbulent flow profile. Special procedures for minimal sets of data can be applied to the view of disturbed or undisturbed flow profiles.

References

Beck MS (1969) Powder and fluid flow measurement using correlation techniques. PhD thesis, University of Bradford

Beck MS (1987) Cross correlation flowmeters: their design and application. IOP Publishers, Bristol

Braun H (1984) Statistik der Signale berührungsloser Strömungssensoren. Dissertation, Universität Karlsruhe

Filips Ch (2003) Ultraschallsignalverarbeitung bei Korrelations- und Vortexverfahren zur Durchflußmessung. Dissertation, Universität Essen, also published by Cuvillier Verlag Göttingen (ISBN 3-89873-711-X)

Filips Ch (2001) Tomographische Darstellung des Strömungsprofils mit wenigen Daten. Techn. Messen 68:226–233

Gersten K, Herwig H (1992) Strömungsmechanik - Grundlagen der Impuls-, Wärme- und Stoffübertragung aus asymptotischer Sicht. Vieweg-Verlag, Braunschweig, Wiesbaden

Kolpatzik SJ (1999) Numerische Simulation der Ausbreitung von Ultraschallwegen in Gasströmungen. Dissertation, Universität Essen, also Fortschritt-Berichte VDI, Reihe 7, 373

Massa (1999) Datasheet E-188. Hingham, MA, USA

Niemann M (2002) Signalverarbeitung in der Ultraschall-Durchflussmessung. Dissertation, Universität Essen

Nikuradse J (1932) Gesetzmäßigkeiten der turbulenten Strömung in glatten Rohren. VDI-Forschungsheft 356

Papoulis A (1984) Probability, Random Variables, and Stochastic Processes. Mc Graw Hill

Poppen G (1997) Durchflußmessung auf der Basis kreuzkorrelierter Ultraschallsignale. Dissertation, Universität Essen

Rettich T (1999) Korrelative Ultraschall-Durchflussmessung auf Basis turbulenter Strukturen. Dissertation, Universität Essen also VDI-Fortschr.-Ber., Reihe 7, Nr. 359, VDI-Verlag Düsseldorf

Schneider F (2001) Eine Analyse der Entstehung der Messsignale bei der korrelativen Ultraschall-Durchflussmessung in turbulenter Strömung. Dissertation, Universität Essen

Schrüfer E (1995a) Elektrische Messtechnik. Hanser-Verlag, München

Schrüfer E (1995b) Signalverarbeitung. Hanser-Verlag, München

Shu W (1987) Durchflussmessung in Rohren mit Hilfe von künstlichen und natürlichen Markierungen. Dissertation, Universität Karlsruhe

Skwarek V (2000) Verarbeitung modulierter Ultraschallsignale in Ein- und Mehrpfadanordnungen bei der korrelativen Durchflußmessung. Dissertation, Universität Essen

Skwarek H, Hans V (1999) Modern principles of signal processing for analysing properties of turbulent fluids. WISP 99, Proceedings:5–10

Skwarek H, Windorfer H, Hans V (2001) Measuring pulsating flow with ultrasound. Measurement 29/3:225–236

Skwarek H, Hans V (2000) Multipath cross-correlation flow meters. IMEKO XVI, Proceedings, TC 9

Worch A (1998) A clamp-on ultrasonic cross-correlation flow meter for one-phase flow. Meas. Sci Techn. 9:622–630

8

Ultrasound Cross-Correlation Flow Meter: Analysis by System Theory and Influence of Turbulence

Franka Schneider, Franz Peters, Wolfgang Merzkirch

Institut für Strömungslehre, Universität Essen, 45117 Essen, Germany

The physical principles of the ultrasound cross-correlation flow meter are described quantitatively by means of an approach based on system analysis. We extend an earlier theoretical model of Shu (1987) by accounting for the characteristics of the turbulent pipe flow and making use of characteristic turbulence parameters. The velocity obtained by cross-correlation of the phase-modulated ultrasound signals can be predicted with the theoretical model as a function of the Reynolds number. These theoretical predictions agree favorably with experimental results. It is shown that the velocity determined by cross-correlation is nearly the "modal value" of the turbulent velocity profile and, thereby, higher than the bulk velocity in the pipe. The investigations include studies of the influence of a number of parameters that can be varied in a practical setup. It is shown that a calibration of the system, as is done in practice, is only valid for a specific set of values of these parameters and restricted to fully developed flow in the pipe.

8.1 Introduction

The principles of flow metering by means of ultrasonic cross-correlation were communicated, e.g., by Mesch (1982) and Beck (1987); see also Chap. 7 of this book. This method of measuring the volumetric flow rate in a pipe is an interesting approach because the method is nonintrusive, does not need information on the speed of sound of the fluid, and it can be applied to single- and multiphase flow. In such a flow meter two ultrasonic waves (or rays), separated by an axial distance L, are transmitted continuously across the pipe, from wall to wall and normal to the pipe axis (Fig. 8.1). The two signals received continuously in time are cross-correlated. The position of the maximum of the cross-correlation function is interpreted as being the time needed by markers in the flow for being convected from the upstream to the downstream ultrasonic "ray". The distance L divided by this convection time gives a convection velocity that is calibrated in terms of the bulk velocity (volume flow rate divided by the pipe cross-sectional area), $\bar{u}_{\mathrm{b}}$.

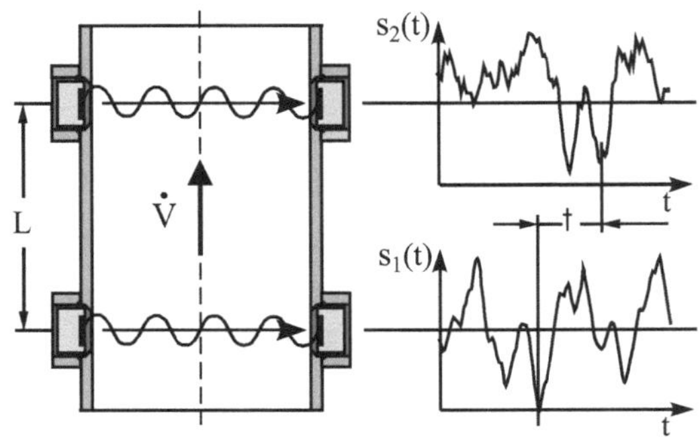

Fig. 8.1. Schematic view of a cross-correlation ultrasonic flow meter applied to pipe flow with the volumetric flow rate $\dot{V}$. The ultrasound signals $s_1(t)$, $s_2(t)$, received continuously with time t at the transducers separated by the axial distance L, are cross-correlated. The time $t = \tau$ determining the maximum of the correlation function is the delay for a particular pattern of the signal to appear at the downstream position (2), at a distance L from (1).

In multiphase flow the markers can be particles dispersed in the continuous phase. We restrict the investigations to single phase flow for which it is believed that the markers are vortical or coherent structures convected by the turbulent flow of the fluid. The existence of such vortical structures was verified by experiments using particle-image velocimetry (PIV) (Westerweel et al., 1996; Schneider and Merzkirch, 2001). When a vortex moves through an ultrasound "ray", the radial velocity component of the vortex is superimposed to the propagation velocity of the sound waves, thus resulting in a phase modulation of the ultrasound wave. Another possible mechanism of the sound wave's modulation is the pressure field associated to the vortical structure. This would result in an amplitude modulation of the wave. The vortical structures are convected in the flow with a velocity that depends on their radial position in the pipe, i.e., the (local) convection velocity is a function of the velocity profile of the turbulent pipe flow. The velocity determined by cross-correlation is associated with the "modal value" (abscissa of the maximum of the PDF) of the wall-to-wall probability density function of the local convection velocities. The final value of the volumetric flow rate found by calibration depends on a number of parameters, the most important being the state of flow, e.g., fully developed flow. When the meter is in practical use, these parameter values must coincide with those present at calibration.

First descriptions of the ultrasonic cross-correlation flow metering method were presented two decades ago, e.g., Mesch (1982), Beck (1987). Explanations of the measured results were given, e.g., by Manook (1981), Poppen (1997), Worch (1998), Rettich (1999); all these explanations are based on the idea, that the velocity determined by the cross-correlation is the result of an integration of the convection velocities of the vortical structures along the "ray" path. The discrepancy found between the bulk velocity and the so determined (integrated convection) velocity was explained with certain properties of the vortical structures. Because of the lack of statistical data on the vortical structures, e.g., the distribution of the vortical structures in the turbulent pipe flow, these explanations do not include a quantitative analysis of the measuring principle.

The principle of flow metering by ultrasonic cross-correlation can be considered as a technical system with the following main components:

- the pipe flow whose spatially and temporally averaged velocity is to be measured,
- the turbulent vortical structures serving as markers,
- the ultrasound waves interacting with the markers in the flow,
- the cross-correlation process used for determining the convection velocity of the markers.

Several theoretical approaches have been reported for describing the principle of this meter by means of system analysis (Kipphan and Mesch, 1978; Kronmüller, 1980; Braun, 1984; Shu, 1987; Shu and Tebrake, 1989). These investigations refer to the general principle of cross-correlation flow metering, independent of the choice of the markers and sensors. These theoretical models were not used for deriving quantitative results that would simulate measurements with the ultrasonic cross-correlation system based on natural markers.

In this chapter we describe the development of a model for the flow metering method by means of system analysis. Thereby we follow the strategy used by Shu (1987), we extend this model, and we make use of input data derived from experiments on turbulent pipe flow in order to arrive at quantitative results. In order to compare the analytically derived predictions with measured data, we also perform experiments with the ultrasonic cross-correlation method in pipe flow at Reynolds numbers in the order of 10^5. Measurements with particle-image velocimetry (PIV) performed in the same pipe flow serve for providing data on the vortical structures in the flow; the results of these measurements (Schneider and Merzkirch, 2001) are used as input data for the approach by means of system analysis.

In the following we first describe the ultrasonic flow metering system. Thereafter, the model based on system analysis is described, and this model is then evaluated quantitatively and compared with the flow metering results.

8.2 Experiments

Experiments with the ultrasound cross-correlation setup illustrated in Fig. 8.1 are performed with air flow in the pipeline (inner diameter $D = 100\,\mathrm{mm}$) that is mentioned in Chap. 4. The measurement position is at 90 diameters downstream of the inlet section, thus securing fully developed flow. The pipe Reynolds number $Re_\mathrm{D} = D \times \bar{u}_\mathrm{b}/\nu$ (ν = kinematic viscosity of air, $\bar{u}_\mathrm{b}$ = bulk velocity) is varied between 0.2×10^5 and 2.4×10^5. Reference measurements of the volumetric flow rate can be recorded continuously in time with an uncertainty of $\pm 0.3\%$ (see also Chap. 4).

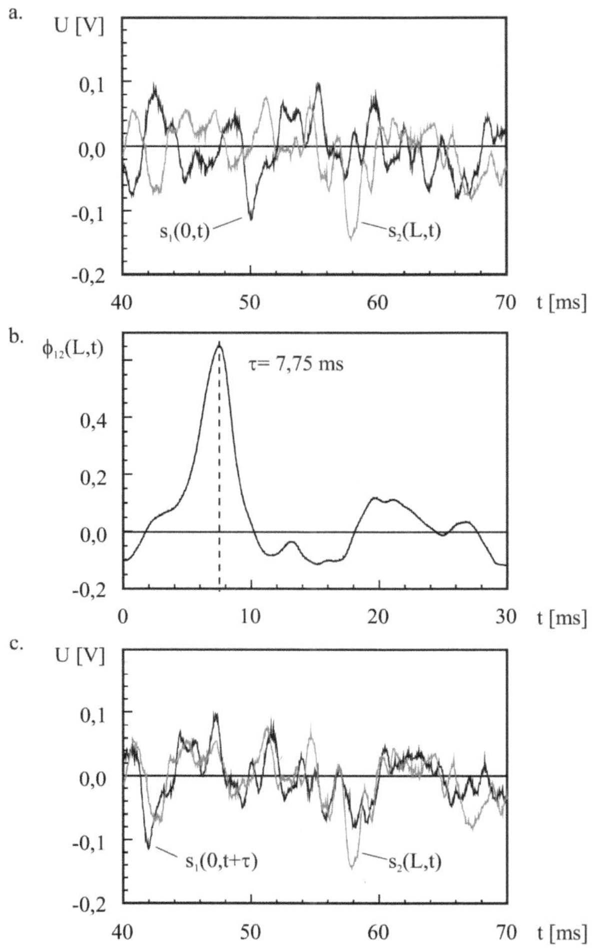

Fig. 8.2. **(a)** Example of ultrasound signals s_1, s_2 received by sensors (1) and (2), respectively, during a period of time t of 30 ms; dimension of signal amplitude is electric tension U in Volt. **(b)** cross-correlation function ϕ_{12} formed with the two signals s_1, s_2; **(c)** superposition of the two signals s_1, s_2 after shifting s_1 by $\tau = 7.75\,\text{ms}$; data from Rettich (1999).

For generating and receiving the ultrasonic waves we use piezo transducers with a resonance frequency of 220 kHz. The signals are processed by undersampling at a frequency of 20 kHz. The axial distance between the two ultrasonic "rays" (see Fig. 8.1) is chosen in the majority of the measurements as $L = 100\,\text{mm}(= 1\,D)$ according to the optimisation studies of Rettich (1999). As an example we show in Fig. 8.2a the two signals s_1 and s_2 received at the two sensors as function of time t for a period of time of 30 ms. The cross-correlation function ϕ_{12} with its maximum at $\tau = 7.75\,\text{ms}$ is seen in Fig. 8.2b, while the two signals, superimposed after a shift of s_1 by 7.75 ms are shown in Fig. 8.2c. For performing the cross-correlation we use the phase-modulated signals. Details of the signal processing are described in Chap. 5 of this book; see also Skwarek (2000). For the performance as well as results of the cross-correlation process we refer to Schneider (2001).

8.3 System-Theoretical Model

The measurements reported by Poppen (1997) and Rettich (1999) had shown that the velocity $\bar{u}_{\mathrm{corr}}$ determined from the maximum of the correlation function ϕ_{12} deviates systematically from the bulk velocity. For giving an explanation and an estimate of this deviation, a model is needed, that can describe the functional relationship between the correlation velocity $\bar{u}_{\mathrm{corr}}$ and the bulk velocity $\bar{u}_{\mathrm{b}}$. For this purpose, we interpret the measurement system as a linear time-invariant system. The model described here is based on earlier work of Shu (1987) and Shu and Tebrake (1989). We modify the model with regard to the assumed velocity profile, the diffusion term, and the influence of the natural flow markers.

8.3.1 Interpretation of the Measurement System as a Linear Time-Invariant System

A linear, time-invariant system is defined by time-invariant and linear transmission characteristics. It is postulated here that the ultrasonic cross-correlation flow meter has such properties. Assumption of a true linear, time-invariant system requires the existence of stationary flow between the two sensors, which is given for fully-developed, time-averaged turbulent pipe flow. However, in turbulent flow the velocities are time-dependent and stochastic. By forming the cross-correlation function ϕ_{12} of the two signals, s_1 and s_2,

$$\phi_{12}(\tau) = \frac{1}{T}\int_0^T s_1(t)s_2(t-\tau)\mathrm{d}t \tag{8.1}$$

(with T being the length of the measurement period) it is evident that a temporal average is provided, from which it follows that the system can be considered as being time-invariant in the average (Shu, 1987). The PIV measurements (Schneider, 2001; Schneider and Merzkirch, 2001) showed that the vortical flow structures, i.e., the markers, are relatively small (e.g., in comparison to the pipe diameter). Hence, they allow the application of the superposition criterion of a linear system.

For a linear, time-invariant system the signal $s_2(t)$ is linked to the signal $s_1(t)$ by

$$s_2(t) = s_1(t) * h(t) \tag{8.2}$$

with $h(t)$ being the impulse response of the system. $h(t)$ expresses the action of the flow on the markers, that are released to the flowing fluid at a given instant of time t_1 and at the defined position (1). If the markers are released along the line of a pipe diameter, they will be convected from their initial positions with different velocities, due to the given velocity profile and diffusion processes occuring in turbulent pipe flow. As a consequence, the impulse response changes its shape along the distance L separating the two measurement positions (1) and (2), which means that $h(t)$ depends also, to some extent, on the axial position x (see Fig. 8.3). With regard to the measurement position (2) at $x = L$, the impulse response defines the instants of time for which the markers, convected along different radial coordinates $r = const.$, arrive at position (2), provided they were released simultaneously to the flow at position (1).

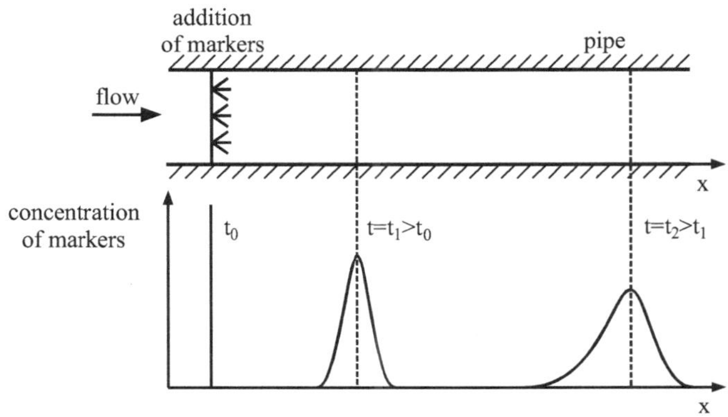

Fig. 8.3. Schematic representation of the changing shape of the impulse function along the axial coordinate in the pipe flow.

Before we investigate how $h(t)$ can be determined (Sect. 8.3.2), a few general properties of the impulse response will be discussed, mainly with concern to the use of natural markers (turbulent structures), as we consider them here. The following relationships apply to the linear time-invariant system:

$$\phi_{12} = \phi_{11} * h(t) \tag{8.3}$$

$$\phi_{22} = \phi_{11} * h(t) * h(-t) \tag{8.4}$$

with $\phi_{11}(t)$, $\phi_{22}(t)$ being the auto-correlation functions of the signals received by the first (1) and second (2) sensor, respectively.

Equation (8.3) is of interest regarding the position of the maximum of the cross-correlation function that is used for determining the convection velocity. If the impulse response is symmetric, the time coordinates where the impulse response and the cross-correlation function have their maxima will coincide. If the impulse response is asymmetric, as it is the case in turbulent pipe flow, the shape of the auto-correlation function $\phi_{11}(t)$ influences the position of the maximum of the cross-correlation $\phi_{12}(t)$ (see Shu and Tebrake, 1989, and Fig. 8.4). The maxima positions of impulse response and cross-correlation function do not coincide; the difference between the two positions increases with increasing width of the auto-correlation (Fig. 8.4), i.e., with the bandwidth of the signal received by sensor (1).

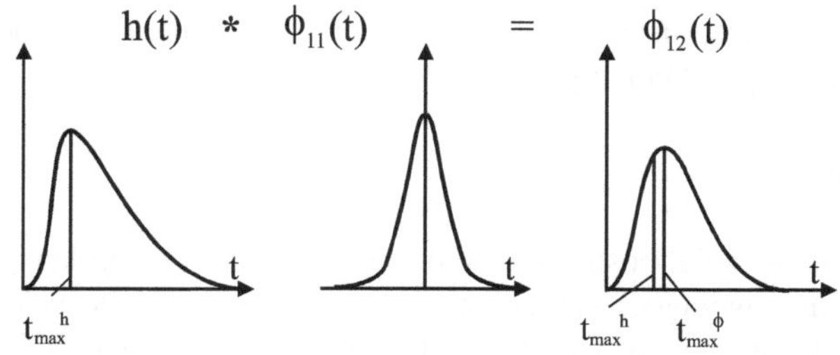

Fig. 8.4. Superposition of auto-correlation ϕ_{11} and an asymmetric impulse response $h(t)$ results in a difference between the position for the maxima of $h(t)$, $t^{\rm h}_{\rm max}$, and cross-correlation ϕ_{12}, $t^{\phi}_{\rm max}$.

In our measurement system the markers are not released instantaneously to the flow. Instead, the markers (turbulent structures) are convected continuously with the turbulent pipe flow, from which one derives that $\phi_{11}(t) = \phi_{22}(t)$. This is in contradiction to (8.4) or, $h(t) * h(-t) = \delta(t)$ (δ = Dirac's function), i.e., the postulated linear model could be applied only to a measuring system with infinitesimally short distance between the two sensors. In order to account for this problem Shu (1987) introduces into (8.3) a noise term that is linked with $\phi_{11}(t) * h(t)$ in a nonlinear way. This noise term also includes the effect of the continuous formation and disappearance of the turbulent stuctures. There are no means for giving an estimate of the magnitude of the noise term, and only from the quantitative results presented in this paper we may conclude that this term that is neglected in our analysis must be very small.

An alternative theoretical model for the considered cross-correlation flow metering systems was reported by Braun (1984). In this model a co-variance matrix is derived that includes a probability density function for describing the transport of the markers. This way, the problem of (8.4) being inconsistent with the real measuring system is not present. However, for determining the maximum of the co-variance matrix function Braun introduces a number of analytical simplifications regarding certain properties of turbulence; e.g., the probability density function (PDF) of the velocities, which in turbulent pipe flow is skewed, is assumed to have a Gaussian distribution, and diffusion as caused by the turbulent fluctuations is totally neglected. We decided to follow Shu's approach because it allows us to account for the skewness of the PDF of the convection velocities and to include the turbulent diffusion processes, as it is outlined in the following sections.

8.3.2 Impulse Response of the Ultrasound System

The impulse response $h(t)$ needed, according to (8.1), (8.2), for determining the cross-correlation function $\phi_{12}(t)$ and its maximum position depends on a number of parameters characterizing the flow and the setup for taking measurements, e.g., velocity profile of pipe flow, turbulence characteristics, properties of the coherent structures, sensor size, distance between sensors. The influence of these parameters on $h(t)$ is studied quantitatively in this section.

Convection Process

According to their radial position r the turbulent structures (markers) are convected at different (time-averaged) velocities $\bar{u}(r)$, where $\bar{u}(r)$ is the velocity profile of the pipe flow for a given Reynolds number. We define a function $C_{\text{convection}}(x, t)$ that describes the spatial and temporal distribution of the markers as caused by the convection process. Figure 8.5 illustrates the technical meaning of $C_{\text{convection}}(x, t)$ under the assumption that the markers are evenly distributed at $x = x_1$ and $t = t_0$. $C_{\text{convection}}(x, t)$ gives information on the arrival times of the markers at the position of the second sensor, $x = x_2 = x_1 + L$.

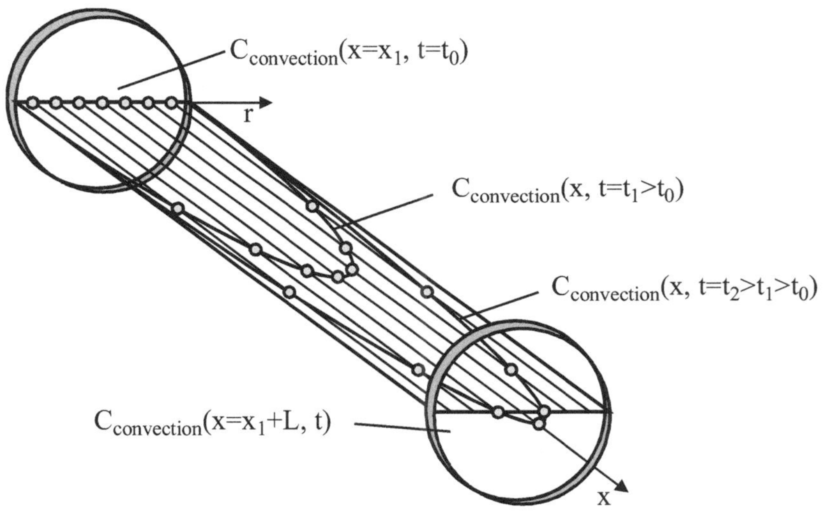

Fig. 8.5. Distribution of the markers, $C_{\text{convection}}(x, t)$, for three different instants of time, t_0, t_1, t_2. The markers are evenly distributed along a pipe diameter at position $x = x_1$ and time $t = t_0$. The markers are convected in axial direction according to the pipe flow velocity profile, $\bar{u}(r)$.

According to the transformation laws for probability functions (Schrüfer, 1990), the function $C_{\text{convection}}(x, t)$ for the convection process is related to the probability density function (PDF) of the velocity, $f(\bar{u}(r))$, by

$$C_{\text{convection}}(x, t) = \left. \frac{f(\bar{u})}{\left| \frac{\mathrm{d}x}{\mathrm{d}\bar{u}} \right|} \right|_{\bar{u}=\frac{x}{t}} \tag{8.5}$$

with $\bar{u} = x/t$ and $0 \leq x \leq \bar{u}_{\max}\, t$ ($\bar{u}_{\max}$ is the maximum velocity of $\bar{u}(r)$).

If one assumes that all structures are detected by sensor (2) with the same local sensitivity and that sensor (2) has an infinitesimally small extension in x-direction, so that its sensitivity can be described with the properties of Dirac's function, $\delta(x-L)$, then the impulse response at sensor (2) can be expressed by

$$h(t) = k \int_{0}^{\infty} C_{\text{convection}}(x,t)\delta(x-L)\,\mathrm{d}x = kC_{\text{convection}}(L,t) \tag{8.6}$$

where k is a constant for normalization, such that $\int_{L/\bar{u}_{\max}}^{\infty} h(t)\,\mathrm{d}t = 1$.

The time for which $h(t)$ has a maximum, $t^{\mathrm{h}}_{\max}$, defines a maximum of similarity between the patterns of a turbulent structure convected from sensor (1) to sensor (2).

For calculating the PDF $f(\bar{u}(r))$ we need an analytical form of the (time averaged) velocity profile $\bar{u}(r)$. The most complete theoretical description of this profile is the asymptotic (Re$\rightarrow\infty$) theory of Gersten and Herwig (1992). Since their formulation of the profile includes several procedures that cannot be expressed analytically, we use, as a simple alternative, the power law for turbulent pipe flow, and we combine it with a parabolic velocity distribution for the core region according to Hinze (1975).

The power law is

$$\frac{\bar{u}(r)}{\bar{u}_{\max}} = \left(1 - \frac{r}{R}\right)^{\frac{1}{n}} \tag{8.7}$$

with R being the pipe radius and $\bar{u}_{\max}$ the velocity on the pipe axis. The index n is determined as a function of the Reynolds number (Nikuradse, 1932):

$$\frac{1}{n} = 0.25 - 0.023\log(Re) \tag{8.8}$$

in the range $4 \times 10^3 \leq Re \leq 3.2 \times 10^6$.

The parabolic velocity distribution according to Hinze makes use of the friction velocity u_τ:

$$\frac{\bar{u}_{\max} - \bar{u}}{u_\tau} = 7.2\left(\frac{r}{R}\right)^2 \tag{8.9}$$

The two profiles are combined without adjusting their slopes, so that a discontinuity results when forming the first derivative.

In order to determine the PDF of $\bar{u}(r)$ according to Schrüfer (1990)

$$|f(\bar{u}(r))| = \left| f(r)\,\frac{1}{\mathrm{d}\bar{u}(r)/\mathrm{d}r} \right| , \tag{8.10}$$

we must know the probability distribution of the turbulent structures $f(r)$.

$f(r)$ depends on the distribution of the structures across the pipe cross section and the area range for which the sensor collects the signal information. If the structures are distributed uniformly in a cross section, and the sensor is sensitive with respect to the whole cross section ("area sensor"), one has $f(r) = 2r/R^2$. For a sensor collecting the information on the structures along a pipe diameter ("line sensor"), one has $f(r) = 1/R$. The ultrasound transducers used in our experiments have a diameter of 10 mm. We assume the detection volume of these sensors to be a 10 mm wideband through the pipe axis. In determining $f(r)$ we take care of the distribution of the markers in this detection volume. Equations (8.10), (8.7) resp. (8.9), and (8.5) allow to determine the impulse response (8.6) in accounting for the pure convection process.

Diffusion Process

The fluctuating motion of turbulent flow causes the turbulent structures to diffuse along their path from sensor (1) to sensor (2). In studying the influence of this diffusion process on the signal formation, a few simplifying assumptions will be made. Diffusion is considered in streamwise direction only, which appears to be justified if the distance L between the sensors is small enough such that, for their convection along L, the structures keep their initial radial position r. Furthermore, we assume as an approximation, that the probability density function (PDF) of the velocities $u(r,t)$ is Gaussian (which we did not assume concerning the convection of the turbulent structures):

$$f(u(r,t)) = \frac{1}{\sqrt{2\pi}\sigma_{\mathrm{u}}} \exp\left[-\frac{(u(r,t)-\bar{u}(r))^2}{2{\sigma_{\mathrm{u}}}^2}\right] . \tag{8.11}$$

σ_{u} is the standard deviation of the velocity $u(r,t)$ in x-direction, and with separating $u(r,t)$ into a time-averaged (mean) value $\bar{u}(r)$ and a fluctuation $u'(r,t)$, $u(r,t) = \bar{u}(r) + u'(r,t)$, one has $\sigma_{\mathrm{u}} = \sqrt{\overline{u'^2}}$ (RMS). For turbulent pipe flow σ_{u} is known to be a function of r and Reynolds number; see, e.g., Laufer (1954), Perry and Abell (1975). An exact analytical function for $\sqrt{\overline{u'^2}}$ is not available, and therefore we use a quadratic approximation that is supported by PIV measurements performed in our laboratory and by data from the literature (Nikuradse, 1932; Laufer, 1954):

$$\frac{\sqrt{\overline{u'^2}}}{u_\tau} = \left(\frac{r}{R}\right)^2 + 0.5\left(\frac{r}{R}\right) + 0.85 \tag{8.12}$$

where the shear velocity $u_\tau = \sqrt{\tau_{\mathrm{w}}/\rho}$ includes the wall shear stress that is Reynolds number dependent.

According to the transformation laws for probability functions (Schrüfer, 1990) the influence of turbulent diffusion on the distribution of the structures is then described by (cf. (8.5))

$$C_{\mathrm{diffusion}}(x,t) = \left.\frac{f(u)}{\left|\frac{\mathrm{d}x}{\mathrm{d}u}\right|}\right|_{u=\frac{x}{t}} = \frac{1}{\sqrt{2\pi}\sigma_{\mathrm{u}}\, t} \exp\left[-\frac{(x-\bar{u}\,t)^2}{2{\sigma_{\mathrm{u}}}^2\, t^2}\right] . \tag{8.13}$$

In contrast to Shu (1987) who takes σ_{u} as a constant we account for the dependence of σ_{u} on r according to (8.12).

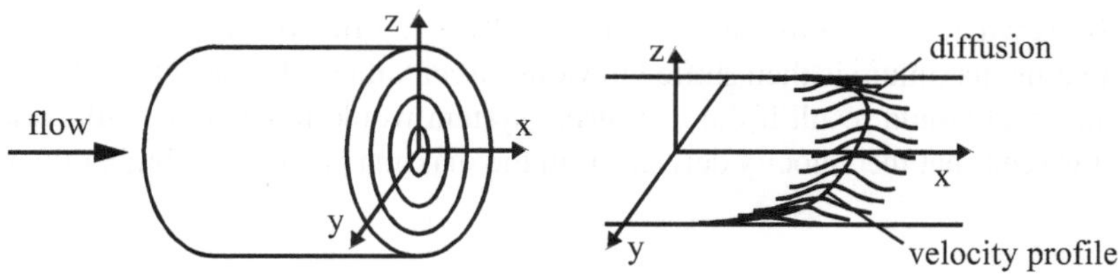

Fig. 8.6. Superposition of axial diffusion to the convection process.

Superposition of Convection and Diffusion

The distribution considering diffusion of the structural patterns as given by (8.13) must be combined with the results obtained for the convection process, i.e., the PDF of the (time-averaged) velocity $\bar{u}(r)$ according to (8.10); (see Fig. 8.6).

As a result one obtains for the distribution of the structures

$$C(x,t) = \int_{0}^{\bar{u}_{\max}} f(\bar{u}(r)) C_{\text{diffusion}}(x,t) \, \mathrm{d}\bar{u} \, . \tag{8.14}$$

The integral must be solved numerically. Figure 8.7 shows the pattern of the impulse response function $h(t)$ when $C(x,t)$ according to (8.14) is used for calculating $h(t)$; the assumed parameter values are given in the figure legend. For comparison, $h(t)$ is also shown if only the convection processes are considered. It is seen that the profile

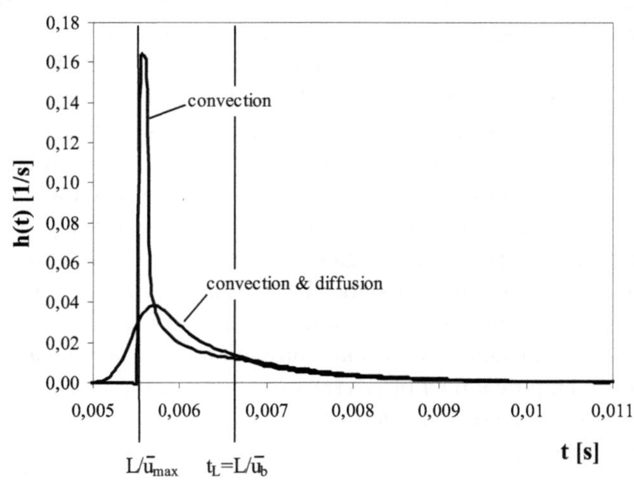

Fig. 8.7. Impulse response $h(t)$ calculated in accounting for pure convection and for convection plus diffusion. This example was calculated with the assumed parameter values $L = D = 100\,\text{mm}$, $Re = 10^5$, and it serves only for demonstrating the characteristic pattern of $h(t)$.

of $h(t)$ is widened towards smaller values of t when diffusion is taken into account, and that its maximum is then shifted towards larger values of time t. t_L in Fig. 8.7 is the time that would result if the measuring system would indicate the bulk velocity $\bar{u}_b$. It is seen that the velocity derived from the maxima positions is higher than $\bar{u}_b$.

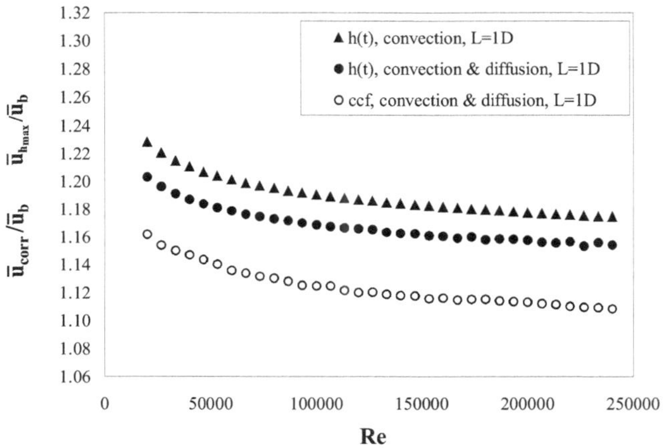

Fig. 8.8. Non-dimensional velocity $\bar{u}_{h_{max}}/\bar{u}_b$ calculated by using the maximum position of the impulse response $h(t)$, and non-dimensional velocity $\bar{u}_{corr}/\bar{u}_b$ calculated by using the maximum position of the cross-correlation function calculated according to (8.3). Assumend distance between sensors $L = D = 100\,\text{mm}$ (ccf=cross-correlation function).

The effect of taking the diffusion process into account is also demonstrated in Fig. 8.8 where the velocity $\bar{u}_{h_{max}} = L/t^{h}_{max}$, determined by using the maximum position of the impulse response $h(t)$, is shown as function of the Reynolds number, again calculated with the assumed values $L = D = 100\,\text{mm}$. $\bar{u}_{h_{max}}$ is made non-dimensional by means of the bulk velocity $\bar{u}_b$. $\bar{u}_{h_{max}}$ is shown for the two cases that only convection and convection with superimposed diffusion are considered. The inclusion of diffusion shifts the calculated velocity towards lower values, as was visible in Fig. 8.7. The asymmetry of the impulse response (cf. Fig. 8.4) causes the calculated velocity values to be too high in comparison to the velocity derived from the maximum position of the cross-correlation function. This is evident from the lower curve in Fig. 8.8 that was determined by making use of $\phi_{12}(t)$ according to (8.3). Since theoretical values of the auto-correlation $\phi_{11}(t)$ are not available, it was necessary to use empirical data of $\phi_{11}(t)$ from our measurements. For each value of Reynolds number we used an auto-correlation function derived from a total of 50 individual measurements.

Influence of the Distribution of the Natural Markers

In the previous sections we have assumed, that the markers are uniformly distributed in a pipe cross section. The PIV measurements of Schneider and Merzkirch (2001) have shown, that this assumption is, in principle, justified. As a difference to artificial markers, natural markers might have properties, that depend on their position in the pipe, e.g., the turbulence intensity and the life time of the coherent structures. These properties can have an influence on the signal generation, if they interact with the ultrasound wave that depends on the selected sensor type. Interaction can occur between the ultrasound and turbulent fluctuations (Chernov, 1960; Tatarski, 1971), and of particular interest here is the fluctuation component in the direction of the ultrasound propagation, v'. We account for this effect by weighting the distribution of the markers at the position of the first sensor, according to the radial distribution of the RMS values $\sqrt{\overline{v'^2}}/u_\tau$ as given by Eggels et al. (1994). This has a small, but visible effect on the shape of the impulse response as reported in more detail in Schneider (2001).

8.4 Comparison of Theoretical and Experimental Results

In this section we compare predictions calculated with the model presented in Sect. 8.3 with experimental results measured in the facility mentioned in Sect. 8.2. The aim is to validate the model, explain the observed difference between measured velocity and bulk velocity, and study the influence of a number of experimental parameters on the signal. A set of measured cross-correlation functions and the cross-correlation calculated with the model is shown in Fig. 8.9 for the experimental conditions specified in the figure legend. The cross-correlation function is shown here in the form ("cross-correlation coefficient")

$$\phi_{12}(t) = \frac{\phi_{12}(t)}{\phi_{11}(0)\phi_{22}(0)} \,. \tag{8.15}$$

The asymmetry that was explained in Sect. 8.3.1 and Fig. 8.3 is visible, and it can also be seen that the positions of the maxima of the measured functions are, to a small extent, at higher values of time than that for the theoretical curve, which is in agreement with the results shown in Fig. 8.10.

The non-dimensional velocity determined by using the maximum position of the cross-correlation function, $\bar{u}_{\mathrm{corr}}/\bar{u}_{\mathrm{b}}$, is shown in Fig. 8.10 as function of the (pipe) Reynolds number. Here, the results of the theoretical predictions are presented together with the results of the measurements (cf. Sect. 8.2). The vertical bars indicate the scatter of the experimental data for a set of approximately 10 measurements for each Reynolds number. For comparing the dependence on Reynolds number we also show in this figure the non-dimensional maximum velocity $\bar{u}_{\mathrm{max}}/\bar{u}_{\mathrm{b}}$ calculated with Gersten and Herwig's (1992) theory. The difference between predicted and measured velocities is in the order of 3 to 4%, and this can be taken as proof of the validity of the theoretical model. In the previous sections we mentioned a number of phenomena of minor importance that are not included in the theoretical model and possibly the reason for the 3 to 4% deviation. The position of the maximum of the cross-correlation function, t^{ϕ}_{max} depends on the position of the maximum of the impulse

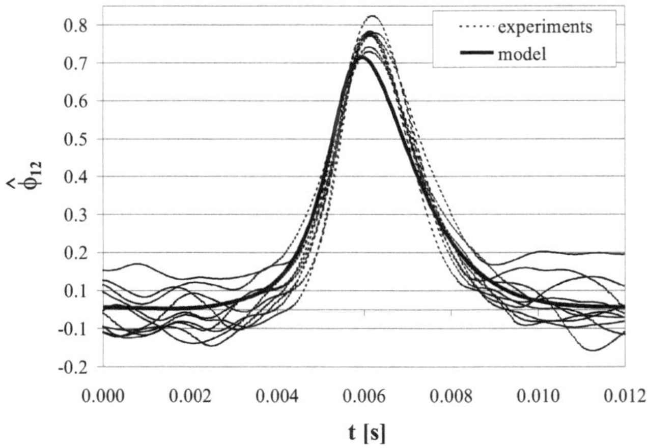

Fig. 8.9. Non-dimensional cross-correlation function $\phi_{12}(t)$: - - - - experiments, —— theoretical model. Parameter valus: $Re = 10^5$, $L = D = 100\,\mathrm{mm}$.

response function, $t^{\mathrm{h}}_{\mathrm{max}}$, as shown in Fig. 8.4. $t^{\mathrm{h}}_{\mathrm{max}}$ is the time at which a maximum of markers arrive at sensor (2) after they had been released at sensor (1) and time $t = 0$, i.e., $L/t^{\mathrm{h}}_{\mathrm{max}}$ indicates a velocity whose presence in the velocity profile has the highest probability ("modal value"), expressed by (8.10). Therefore, one has always $\bar{u}_{\mathrm{corr}}/\bar{u}_{\mathrm{b}} > 1$ as seen in Fig. 8.10.

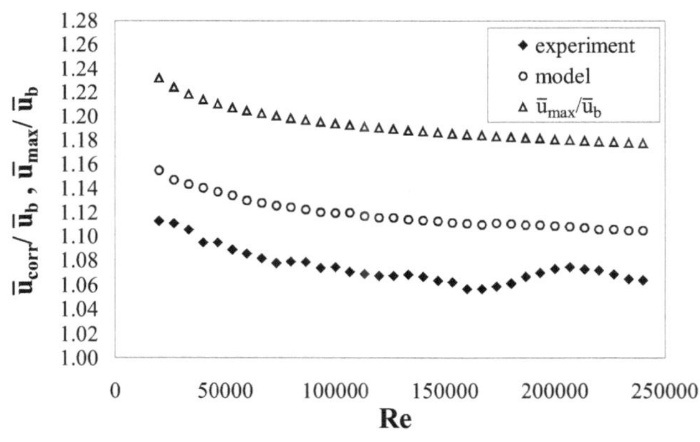

Fig. 8.10. Theoretical and experimental results of the velocity, found by cross-correlation, as function of Reynolds number. Values of the maximum velocity are given for comparison. Distance between sensors $L = D = 100\,\mathrm{mm}$.

A parameter that can be varied in the experimental setup is the axial distance between the two sensors, L (cf. Fig. 8.1). Measured and calculated values of the non-dimensional velocity derived by cross-correlation, $\bar{u}_{\mathrm{corr}}/\bar{u}_{\mathrm{b}}$, are presented in

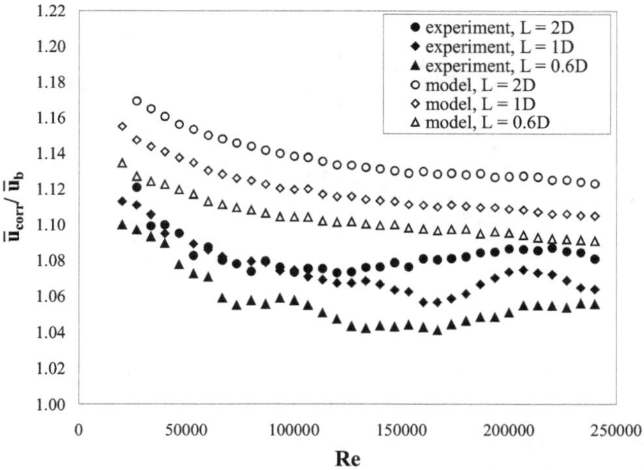

Fig. 8.11. Influence of the distance between sensors, L, on the velocity determined by cross-correlation (experiment and theoretical model) for different Reynolds numbers.

Fig. 8.11 as function of the Reynolds number. The measured curves exhibit certain irregular oscillations, that might be caused by instabilities in the electronic signal processing system. Not much attention should be given to these irregularities at this time. The interesting result in Fig. 8.11 is the dependence of $\bar{u}_{\mathrm{corr}}/\bar{u}_{\mathrm{b}}$ on the distance L, both in the experiment and the theoretical prediction. This indicates that a calibration of a setup for taking measurements of the volumetric flow rate is only valid for a specified sensor distance L. The shown dependence of $\bar{u}_{\mathrm{corr}}/\bar{u}_{\mathrm{b}}$ on L can be explained with the changing distribution of the markers when they are convected in axial direction (cf. Fig. 8.5), which makes the impulse response function in the theoretical model to depend on the axial position x, thereby affecting the maximum position of the cross-correlation function.

The absolute value of the maximum of the cross-correlation is an indication of the degree of correlation. The maximum value of the cross-correlation coefficient, both for experiment and theory, is presented in Fig. 8.12 as function of the Reynolds number and for three values of the distance L between the sensors. The results shown confirm the high degree of agreement between experiment and theoretical model. The slight increase of the maxima with Reynolds number can be explained with the velocity profile that becomes more flat with increasing Reynolds number; this causes more markers to be convected with almost the same velocity, thus increasing the degree of similarity of the received signals. The similarity of the signals decreases with increasing L due to the changing distribution of the markers in axial direction, as it was mentioned before and illustrated in Fig. 8.5. Figure 8.12 also indicates that, for the range of values of L varied in these experiments, the highest degree of correlation is reached here for $L = 0.6$. If one decreases L further below 0.6, the degree of correlation will decrease again due to an undesired mutual interaction

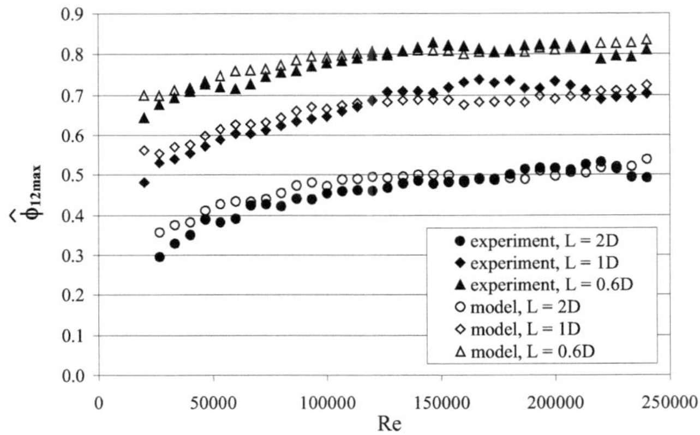

Fig. 8.12. Absolute values of the maxima of the cross-correlation coefficient as function of Reynolds number for three different distances between sensors, L.

(cross-talk) of the ultrasound transducers at positions (1) and (2). We did not attempt to determine the optimum value of L regarding a maximum of correlation.

Experiments were also performed with a setup in which the receiver at the second sensor position was not perpendicularly opposite to the ultrasound emitter. The used geometry is shown in Fig. 8.13. These experiments served only for exploring the capabilities of the theoretical model. The calculated cross-correlation coefficient $\phi_{12}(t)$ shown in Fig. 8.14 for a specific set of experimental parameters ($Re = 10^5$, $D = 100\,\text{mm}$) has two maxima, from which one can conclude that this setup is not useful for practical measurements of the volumetric flow rate. The occurance of the two maxima can be explained with the theoretical model. We have derived that the velocity determined by cross-correlation is nearly the "modal value" of the velocity profile $\bar{u}(r)$. In Fig. 8.13 it is seen that the ultrasound "ray" of the oblique sensor (2) interacts twice with the "modal value", but for different axial distances between the sensor "rays" (1) and (2), thus causing the occurance of two maxima in the cross-correlation function. With (8.11) one derives that, for $Re = 10^5$, the most

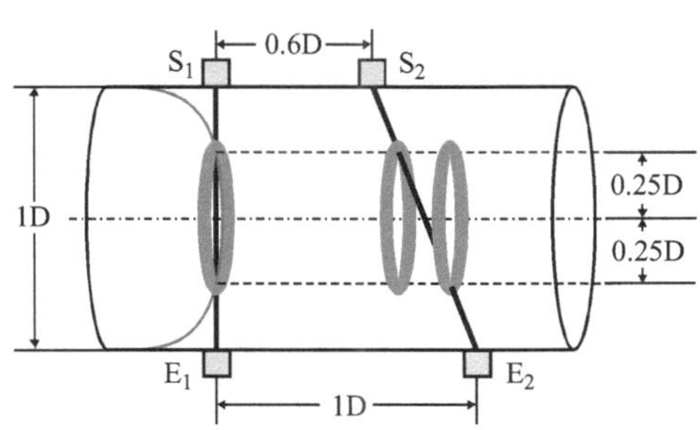

Fig. 8.13. Setup with oblique ultrasound "ray" for sensor (2).

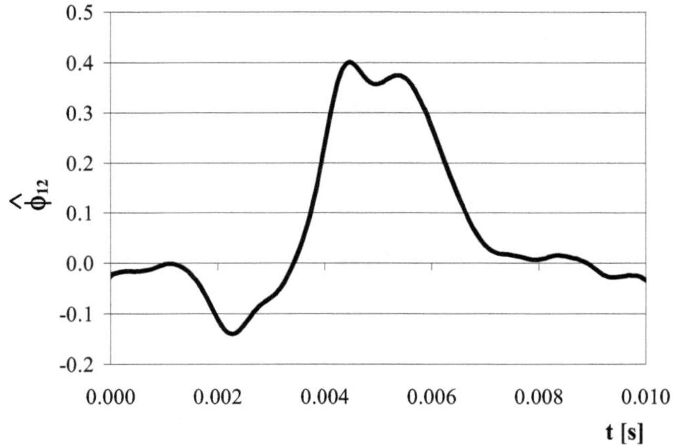

Fig. 8.14. Cross-correlation coefficient obtaind with the setup shown in Fig. 8.13 for $Re = 10^5$, $L(S_1 - S_2) = 0.6D$, $D = 100\,\text{mm}$.

frequently occuring velocity ("modal value") is in the range from $\bar{u}_{\text{corr}} \approx 1.06\bar{u}_{\text{b}}$ to $1.08\bar{u}_{\text{b}}$, and in the turbulent velocity profile $\bar{u}(r)$ this value is found at the position $r \approx 0.25D$. Using these data and the geometry shown in Fig. 8.13 one verifies the temporal positions of the two maxima of the cross-correlation coefficient shown in Fig. 8.14.

In the beginning it was mentioned that the interaction of the ultrasound wave with the turbulent flow can also result in an amplitude modulation of the wave. The form of the cross-correlation function depends on whether the phase modulation or the amplitude modulation is used for the signal generation (Manook, 1981; Skwarek,

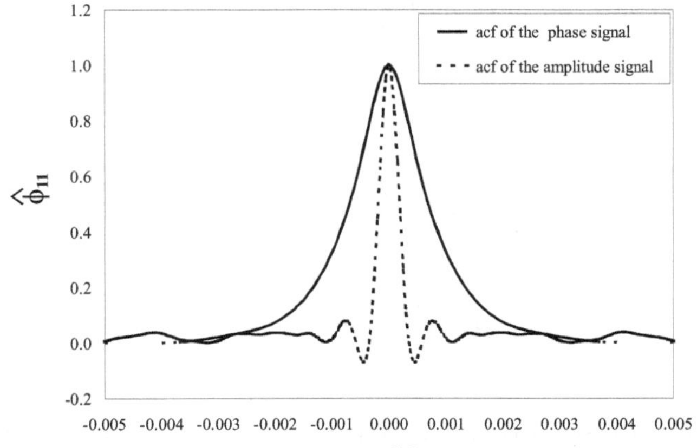

Fig. 8.15. Auto-correlation coefficient derived from phase-modulated and amplitude-modulated signals ($Re = 10^5$) (acf=auto-correlation function).

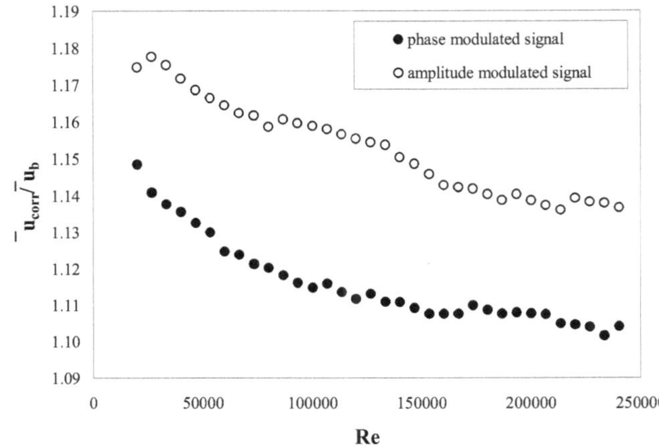

Fig. 8.16. Theoretical results of the non-dimensional velocity determined by cross-correlation by using the phase-modulated and amplitude-modulated signals ($L = D = 100\,\text{mm}$).

2000). The auto correlation $\phi_{11}(t)$ obtained from the amplitude-modulated signals is narrower than $\phi_{11}(t)$ derived from phase-modulated signals as shown in Fig. 8.15 for a special set of parameters ($Re = 10^5$). From Fig. 8.4 it follows that the position of the maximum of the cross-correlation function is then closer to the maximum position of the impulse reponse $h(t)$, i.e., at lower values of time, and the resulting velocity must be higher than the values determined by using the phase-modulated signals. This is in agreement with the experimental findings of Skwarek (2000) who reports, that the velocity values $\bar{u}_{\text{corr}}/\bar{u}_{\text{b}}$ are about 3% higher than those found by using the phase-modulated signals. This observation is validated quantitatively with our theoretical model, as it can be seen in Fig. 8.16. A reason for not using the amplitude-modulated signal is that these signals are more distorted than those obtained from phase modulation (see Fig. 8.17, and Schneider, 2001; Skwarek, 2000).

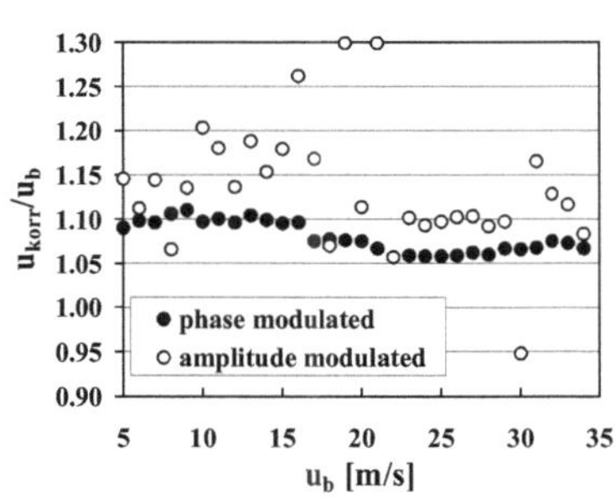

Fig. 8.17. Non-dimensional velocity determined by cross-correlation in turbulent pipe flow using either the phase-modulated or the amplitude-modulated ultrasound signals; $\bar{u}_{\text{b}}$ = bulk velocity from the reference measurements; data from Skwarek (2000).

8.5 Conclusions

The theoretical model based on system analysis gives a complete physical description of the ultrasound cross-correlation method for determining the volumetric flow rate in a pipe. We have extended the model whose principles were described by Shu (1987) and Shu and Tebrake (1989) by introducing additional terms characterizing the flow and by making use of characteristic data of turbulent pipe flow. With the theoretical model it was possible to derive quantitative data that could be compared with experimental results. The theoretically and experimentally determined results for the velocity resulting from cross-correlation of the phase-modulated ultrasound signals deviate by 3 to 4%. The difference must be attributed to the approximation made in (8.2) regarding an additional noise term and to phenomena that, at this time, are difficult to account for in the theoretical model. Because of this difference it is impossible to present here a theoretical estimate of the measurement accuracy of the method.

From the theoretical model it follows that the velocity obtained by cross-correlation is nearly the "modal value" of the time-averaged turbulent velocity profile in the pipe. With (8.10) this "modal value" can be determined if the velocity profile is known, as is the case for fully developed flow. The impulse response depends on the distance of the sensors L, see (8.5) and (8.6), i.e., a calibration of the measurement system is only valid for the value of L used at calibration. Any change of L in a measurement setup would require a new calibration. Also, the calibration only applies to the state of flow present during the calibration process. A system using a multiple-path arrangement is probably not an improvement of the situation, since the signal received is not the result of an integration of the velocity values from wall to wall.

With the theoretical model it was also possible to explain a number of phenomena that have been observed in experiments with setups that turned out to be not appropriate for quantitative measurements of the volumetric flow rate. Finally, an answer could be given to the question whether it is more useful to derive the cross-correlation results from the phase-modulated or the amplitude-modulated ultrasound signals. The answer is clearly that the phase-modulated signals are the better choice.

References

Beck MS (1987) Cross correlation flowmeters: their design and application. IOP Publishers, Bristol

Braun H (1984) Statistik der Signale berührungsloser Strömungssensoren. Dissertation, Universität Karlsruhe (TH)

Chernov L (1960) Wave Propagation in a Random Medium. McGraw-Hill, New York

Eggels JG, Unger F, Weiss MH, Westerweel J, Adrian RJ, Friedrich R, Nieuwstadt FTM (1994) Fully developed turbulent pipe flow: a comparison between direct numerical simulation and experiment. J Fluid Mech 268:175–209

Gersten K, Herwig H (1992) Strömungsmechanik - Grundlagen der Impuls-, Wärme- und Stoffübertragung aus asymptotischer Sicht. Vieweg-Verlag, Braunschweig

Hinze JO (1975) Turbulence, 2nd edition. McGraw-Hill, New York
Kipphan H, Mesch F (1978) Flow measurement systems using transit time correlation. In: Flow Measurement of Fluids, eds. Dijstelbergen HH, Spencer EA, 409–415, North-Holland Publishing Company
Kronmüller H (1980) Durchflußmessung mittels Markierungsverfahren. VDI-Berichte Nr. 375:47–54
Laufer J (1954) The structure of turbulence in fully developed pipe flow. NACA Report 1174
Manook BA (1981) A high resolution cross-correlator for industrial flow measurement. Dissertation, University of Bradford
Mesch F (1982) Geschwindigkeits- und Durchflussmessung mit Korrelationsverfahren. Regelungstechn Praxis 24:73–82
Nikuradse J (1932) Gesetzmäßigkeiten der turbulenten Strömung in glatten Rohren. VDI-Forschungsheft 356
Perry AE, Abell CJ (1975) Scaling laws for pipe-flow turbulence. J Fluid Mech 67:257–271
Poppen G (1997) Durchflußmessung auf Basis kreuzkorrelierter Ultraschallsignale. Dissertation, Universität Essen
Rettich T (1999) Korrelative Ultraschall-Durchflussmessung auf Basis turbulenter Strukturen. Dissertation, Universität Essen (also published as VDI-Fortschr.-Ber., Reihe 7, Nr. 359, VDI-Verlag Düsseldorf)
Schneider F (2001) Eine Analyse der Entstehung der Messsignale bei der korrelativen Ultraschall-Durchflussmessung in turbulenter Strömung. Dissertation, Universität Essen
Schneider F, Merzkirch W (2001) Distribution of coherent structures in turbulent pipe flow. J Flow Visual Image Processing 8:253–261
Schrüfer E (1990) Signalverarbeitung. Hanser Verlag, München
Shu W (1987) Durchflussmessung in Rohren mit Hilfe von künstlichen und natürlichen Markierungen. Dissertation, Universität Karlsruhe
Shu W, Tebrake G (1989) Durchflußmessung in Rohren mit Hilfe von künstlichen und natürlichen Markierungen. Techn Messen 56:58–65
Skwarek V (2000) Verarbeitung modulierter Ultraschallsignale in Ein- und Mehrpfadanordnungen bei der korrelativen Durchflußmessung. Dissertation, Universität Essen
Tatarski VI (1971) The effects of turbulent atmosphere on wave propagation. IPST Keter Press, Jerusalem
Westerweel J, Draad AA, van der Hoeven JGT, van Oord J (1996) Measurement of fully developed turbulent pipe flow with digital particle image velocimetry. Exp Fluids 20:165–177
Worch A (1998) A clamp-on ultrasonic cross correlation flow meter for one-phase flow. Meas Sci Technol 9:622–630

9

Effect of Area Changes in Swirling Flow

Venkatesa Vasanta Ram

Institut für Thermo- und Fluiddynamik, Ruhr-Universität Bochum, 44780 Bochum, Germany

This chapter addresses the question of the effect of changes in the flow-passage area on the radial distribution of the velocity vector, in other words on the profiles of the axial and the azimuthal velocity components, in axisymmetric incompressible swirling flow. The problem is approached by examining the dynamics of the flow through Euler's equations of motion. The method of examination makes explicit use of a well-established physical property of invariants in this flow. This is that the total pressure and the moment of azimuthal velocity remain constant along a (cylindrical) stream surface. The study illuminates the nature of the change in the profiles of the axial and azimuthal velocity distribution, which involve changes of both scale and shape.

Notation

x, r, ϕ	co-ordinate system: axial, radial and azimuthal directions respectively
ρ	density of the flowing fluid
u_x, u_r, u_ϕ	velocity components in x, r and ϕ directions respectively
H	total head, (9.5)
C	moment of azimuthal velocity, (9.6)
$R(x)$	radius of flow passage of circular cross section
R_1, R_2	radius of flow passage at inlet and outlet sections
Q	volume flow rate through the passage
U_{ref}	bulk velocity, reference velocity, defined through $Q = \dfrac{\pi R^2}{U_{\text{ref}}}$
Ω	angular velocity
ψ	$\psi = constant$ defines stream surface
Ψ	value of ψ at outer boundary
ψ^*, r^*	defined in (9.26)
ϵ_{in}	parameter characterizing inhomogeneity, (9.28), (9.29)

9.1 Introduction

Many engineering applications involve metering of the flow in a pipe. A problem often faced in this task is that, for some reason or other, there often exists some swirl in the flow to be metered, and swirl introduces an error in metering. The swirl present is often traceable to factors of installation that generate a swirling motion of the fluid, which is a kind of motion to be defined shortly. Some common examples of such installational elements are: an elbow or a succession of bends in the pipe, an axial compressor installed upstream for boosting purposes, and so on. Swirling motion, once it arises, has the property of decaying only rather slowly, and installational constraints do not often permit a sufficient length of piping for the flow to settle down sufficiently to a state mandatory for correct metering. While it is true of any metering device that it can only "read" the flow it is "exposed" to, there are some peculiarities in the operation of a turbine flow meter that call for a closer examination of the effect of area changes on the flow characteristics. The peculiarities just referred to are derivable from the physical principles underlying conversion of the flow rate into a signal in the turbine flow meter, and these are the subject of Chap. 10. Here, it suffices to note that, after the swirling motion is originated, the flow will have undergone some modification on passing through sections of changing effective area prior to being "perceived" by the flow meter. For this reason the effects of area changes on the flow warrants closer study, and this may be identified as the first of two distinct steps in the engineer's task of assessing the response of the flow meter. Its significance for a correct estimation of the error in metering caused by swirl in the flow cannot be overemphasised, neither can it be overlooked that this step involves a deeper understanding of fluid flow phenomena. The relevance to practical applications is better appreciated on noting that in many applications there is a relatively short section of a pipe introduced between the swirl generating element of the installation, such as the elbow or the booster, and the flow meter. This section of the pipe, while being insufficient to condition the flow to the required standards for metering, has boundary layers growing on its walls that constrict the effective area available for the flow passage. The second and subsequent step of answering the question of the signal response of the flow meter to the flow it "perceives", which is at the exit section of this pipe, is conceptually different, hence meaningful to be treated separately. Here in the present chapter we address the question in the first step, viz. of the flow response to area changes.

Flows with swirl have a long tradition in fluid mechanics research, in the course of which many of the diverse physical phenomena herein have been uncovered, addressed and explained. Section 9.2 titled *Physical Background* attempts to provide a proper perspective for the present work against this background. The kinematic and dynamic features of a flow with swirl that are relevant to metering are summarised for preparatory purposes, and the scope of the present work defined. Section 9.3 that follows carrying the title *Mathematical Formulation of the Problem* is the main part of the work. It sets out the equations and boundary conditions for the problem, introduces the approximation on which the rest of the work is based and outlines the method of solution. The subsequent section contains sample results that illustrate the applicability of the method in the context of flow metering.

9.2 Physical Background

The term "swirl" has been used here to denote a certain outstanding kinematic feature of the flow. It is that in a polar coordinate system placed with its axis coinciding with the pipe axis which is the natural choice for a pipe flow, the azimuthal component of the velocity vector is not zero. The salient characteristic feature of a flow with swirl is therefore defined through the topology of its streamlines. The non-zero azimuthal velocity component in conjunction with the axial velocity component that is decisive for the flow rate results in helically wound streamlines. Dynamically, swirl in the flow is associated with a radial pressure gradient of a strength that depends upon the square of the magnitude of the azimuthal velocity component and the curvature of the projection of the streamlines on to a plane perpendicular to the pipe axis.

9.2.1 Kinematic and Dynamic Characteristics

Swirling motion in a pipe may occur in widely differing patterns. For an initial classification of the same one may adopt a kinematical standpoint according to which a suitable criteron would be the change of sign, if any, of the (non-zero) azimuthal velocity component along the circumference. No change of sign of this quantity is tantamount to the region of swirling motion extending over the entire cross section with one center identifiable. Alternatively, the swirling motion can also occur in cells in which case more than one center is discernible around which the fluid particles describe a "swirling motion". This is a kind of motion in which the fluid particles apparently describe closed circuits; apparent, since it appears so only in the projection and the streamlines are not in fact closed on themselves. It is however important to note that, regardless of the means of swirl generation, the flow is essentially three-dimensional, and the shape of the streamlines is closer to helices than to straight lines.

The actual pattern of the swirling motion assumed by the flow in a pipe depends on various factors, with geometrical features and flow parameters both contributing likewise to its formation. Helical streamlines characteristic of swirling motion also imply that the streamline curvature is spatial as opposed to planar. Given the numerous possibilities of swirl generation, it is obvious that a wide variety of flow patterns is permissible with which the flow can enter the pipe section upstream of the flow meter. The attendant characteristics of the velocity and pressure distributions are also therefore multifarious. The velocity distribution established at the inlet section of this pipe then evolves to the state at its end that, in turn, is perceived by the flow meter at its inlet. The evolution of the flow in the pipe from one section to the other is governed by the physical laws of fluid motion, and these are required to be properly taken into account in analyzing the effect of swirl on metering.

From the standpoint of the dynamics of the flow, the swirling flow in a pipe may be considered as a special case of a more general class of flows characterized by their topological feature of helically wound streamlines. Flows with helically wound streamlines are of inherent scientific interest due to the phenomena they exhibit, of which vortex breakdown holds a key position. They are also of relevance to understanding flow phenomena both in nature and in a wide variety of engineering applications. It is therefore only very natural to find flows with helical streamlines being the subject of extensive studies in fluid mechanics. Literature on this subject is accordingly voluminuous, and we confine ourselves here to citing here only a short selection from recent publications that are representative of the questions addressed, Brown and Lopez (1990), Buntine and Saffman (1995), Escudier (1988), Faler and Leibovich (1978), Gallaire and Chomaz (2003), Rusak et al. (1998), Wang and Rusak (1997). Fundamental properties of these flows that are of significance for the objective pursued here have been established in these studies, and these provide the foundation and a convenient starting point for our analysis.

Studies hitherto, see eg. Chap. 4 in this book, have focussed attention on to several features exhibited by flows with swirl in a pipe that are of importance in the context of flow metering. Perhaps the first and foremost among these, is that swirl, when it arises, decays only rather slowly, see also Chap. 2. It may be noted in this context that swirl in a flow, when understood to be characterized through helical winding of streamlines, does not necessarily imply rotational symmetry. The reader is referred to Chaps. 2, 14 and 15 in this book for examples of rotational asymmetry arising in the flow. However, the special case of the rotationally symmetric flow with swirl is of special significance for various reasons. Firstly, as already stated earlier, the axisymmetric mode is the one with the lowest rate of decay, hence likely to persist longest in the flow. Secondly, being simpler, the axisymmetric case permits analysis to be taken further than the others, thus enabling deeper physical insight to be gained into the mechanisms at work, of course in axisymmetric flow. Strictly speaking, the effect of the area change on profiles of both the mean and the fluctuating velocity components should be sought since the metering device, depending upon the physical principle on which it is based, could be sensitive to one or both. But in the present work we restrict our attention to the mean profiles alone. Again, even with this restriction, we confine ourselves to the inviscid flow. It is appropriate to note at this juncture that a complete picture of the effects of area changes on the flow can of course be obtained only after supplementing the insight gained from the study of the inviscid flow through a study of the boundary layer. Since knowledge of the inviscid flow is a prerequisite for studying the boundary layer, the present study should be regarded as but a first step in the study of the effects of area changes on the flow. Effects of viscosity in this flow are the subject of other chapters, Chaps. 1, 2.

9.2.2 Scope of the Present Work

Euler's equations of motion for the steady axisymmetric flow of a fluid with constant density are therefore at the foundation of our study. Our aim is to extract, based on this foundation, salient characteristics of the flow at the outlet section for a given distribution of the axial and azimuthal velocity components at the inlet when there is a change in area/radius of the flow passage. Literature covering studies of this problem is indeed very extensive, particularly when the flow is irrotational. The flow under study here is, although inviscid, not irrotational, and many methods have been developed to solve the Euler's equations for this class of problems too. These methods are invariably numerical, and results of numerical methods inherently have the character of being solutions to approximations in some sense. Therefore, in order to judge the validity of the results, it is highly desirable that they be subjected to one or more tests. One such test is outlined in the following and will be discussed in more detail later: It can be derived from the differential equations of motion that two flow quantities of significance, viz. the total head and the moment of the azimuthal velocity, remain constant on streamsurfaces. Since this physical constraint follows from the differential equations themselves, compatibility of the numerical results with this condition of invariance is a test for the validity of the approximations inherently present. Since, to the author's knowledge, none of the methods incorporate such a check, it appeared meaningful to develop one in which such a test is an integral part of the method, and this is the course charted out for the present work.

9.3 Mathematical Formulation of the Problem

A sketch of the flow problem addressed is shown in Fig. 9.1. It is the internal flow in a passage of circular cross section with radius $R(x)$. R_1 and R_2 denote the radii at the inlet and outlet sections respectively. The analysis to follow does not place any restriction on the manner of change of the cross-sectional area as long as its character of rotational symmetry is maintained and the rate of change of the radius is only gradual. Contraction and expansion of the passage area are both permissible

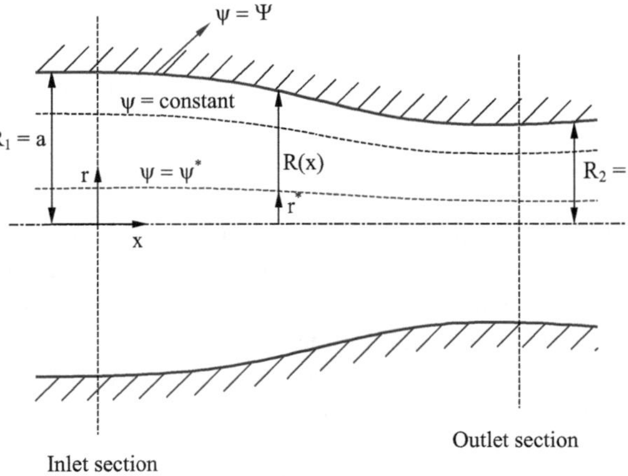

Fig. 9.1. Geometric configuration.

in principle for our analysis, but our interest lies primarily in the case of contraction, $R_1 > R_2$, since this is the case most frequently met with in flow metering practice. The profiles of the axial and azimuthal velocity components at the inlet cross section may be specified arbitrarily, again with the restriction that they be axisymmetric. The flow quantities sought are the axial and the azimuthal velocity profiles at the exit cross section. These shall be obtained as a solution of the Euler's equations of motion satisfying the boundary condition that the specified geometry $R(x)$ be a stream surface.

With the x-axis chosen to coincide with the axis which is the main direction of the flow, the equation of continuity and the Euler's equations of motion governing the steady axisymmetric flow of a fluid at constant density are as follows, see eg. Batchelor (1967), Schlichting and Gersten (2000):

$$\frac{\partial(ru_x)}{\partial x} + \frac{\partial(ru_r)}{\partial r} = 0 , \tag{9.1}$$

$$u_x\frac{\partial u_x}{\partial x} + u_r\frac{\partial u_x}{\partial r} = -\frac{1}{\rho}\frac{\partial p}{\partial x} , \tag{9.2}$$

$$u_x\frac{\partial u_r}{\partial x} + u_r\frac{\partial u_r}{\partial r} - \frac{u_\phi^2}{r} = -\frac{1}{\rho}\frac{\partial p}{\partial r} . \tag{9.3}$$

The equation of continuity, (9.1), may be satisfied by introducing a stream function ψ defined through (9.4) below:

$$u_x = \frac{1}{r}\frac{\partial\psi}{\partial r}; \qquad u_r = -\frac{1}{r}\frac{\partial\psi}{\partial x} . \tag{9.4}$$

An important result that is derivable from the above (9.1), (9.2), (9.3), (9.4), is that the two quantities, the total head H and the moment of azimuthal velocity u_ϕ, defined as,

$$H = \frac{1}{2}\left(u_x^2 + u_r^2 + u_\phi^2\right) + \frac{p}{\rho} = \frac{1}{2}u_x^2 + \int \frac{C}{r^2}\frac{\mathrm{d}C}{\mathrm{d}r}\mathrm{d}r , \tag{9.5}$$

and

$$C = ru_\phi , \tag{9.6}$$

remain constant along streamsurfaces ψ = constant, i.e., $H = H(\psi)$ and $C = C(\psi)$. For a derivation of this result the reader is referred to standard literature on fluid dynamics, eg. Batchelor (1967). Pressure may be eliminated from the equations, (9.2) and (9.3) to yield an equation for the azimuthal component of the vorticity, in which the definition of ψ may be used to obtain a single governing partial differential equation for ψ. This is as follows (see Eq. 7.5.11 in Batchelor (1967)):

$$\frac{\partial^2\psi}{\partial x^2} + \frac{\partial^2\psi}{\partial r^2} - \frac{1}{r}\frac{\partial\psi}{\partial r} = r^2\frac{\mathrm{d}H(\psi)}{\mathrm{d}\psi} - C(\psi)\frac{\mathrm{d}C(\psi)}{\mathrm{d}\psi} . \tag{9.7}$$

The velocity field satisfying the equations (9.1), (9.2), (9.3) is required to fulfill only the kinematic boundary condition of the velocity component normal to the bounding surface being zero. For (9.7) this may be written as follows:

$$r = R(x) : \psi = \Psi \, . \tag{9.8}$$

When the rate of change of the cross section radius is small, the derivative with respect to x can be neglected compared to the other terms and we get the following ordinary differential equation for $\psi(r)$, (9.9), as an approximation. This approximation then captures the salient features of the flow at an arbitrary station x where the cross-sectional radius is $R(x)$. The distribution of H and C at the initial station of cross-sectional radius R_1 is to be regarded as given, and hence known. The approximation then obeys the following equation:

$$\frac{\mathrm{d}^2\psi}{\mathrm{d}r^2} - \frac{1}{r}\frac{\mathrm{d}\psi}{\mathrm{d}r} = r^2\frac{\mathrm{d}H(\psi)}{\mathrm{d}\psi} - C(\psi)\frac{\mathrm{d}C(\psi)}{\mathrm{d}\psi} \, , \tag{9.9}$$

which is the same as for the problem of *cylindrical flows* dealing with transition from one long cylindrical section to another of different radius. It has been treated in Batchelor (1967), Chap.7.5. If the above equation is understood as an approximation for the flow at any value of x in a passage of circular cross section with a gradual change in radius, the x-dependence of the flow is visible only in a parametric form through the outer boundary condition which remains the same as in (9.8).

The difficulties of obtaining a solution of (9.9) for arbitrary distributions of H and C at the inlet cross section are well-known, see, e.g., Batchelor (1967). However, the particular case of a constant axial velocity U and solid body rotation (constant angular velocity Ω) at the inlet section turns out to be relatively simple and the solution for this problem is given in Batchelor (1967). In view of the importance of this special case for the more general problem to follow, we summarise in the subsection below the outlines of this special case.

9.3.1 Solution for the Special Case of Homogeneous Axial Velocity and Solid-Body Rotation at Inlet

When the flow conditions at inlet are such that the axial velocity profile is homogeneous with a value U and the swirl is a solid-body rotation with angular velocity Ω, H and C may be written as linear functions of ψ, viz.

$$C = \frac{2\Omega}{U}\psi \, , \tag{9.10}$$

and

$$H = \frac{1}{2}U^2 + \frac{2\Omega^2}{U}\psi \, , \tag{9.11}$$

and the solution to (9.9) is expressible through Bessel's functions. We need this solution for later reference and summarise its essentials here for the reader's convenience. In our present notation the solution may be written as follows:

$$\psi = \frac{1}{2}Ur^2 + rF(r) , \tag{9.12}$$

where $F(r)$ is a solution of the Bessel's equation of order unity given by

$$\frac{\mathrm{d}^2 F}{\mathrm{d}r^2} + \frac{1}{r}\frac{\mathrm{d}F}{\mathrm{d}r} + \left(k^2 - \frac{1}{r^2}\right) F = 0 , \tag{9.13}$$

where

$$k = \frac{2\Omega}{U} . \tag{9.14}$$

The solution for F satisfying the boundary conditions of the problem may be written as

$$F = AJ_1(kr) + BY_1(kr) , \tag{9.15}$$

where J_1 and Y_1 denote Bessel functions of the first and second kind respectively, and A and B are constants related to the radii at the inlet and outlet sections. This solution corresponds to the axial and the azimuthal velocity distribution at the outlet section given by the following expressions:

$$u_x = \frac{1}{r}\frac{\partial \psi}{\partial r} = U + AkJ_0(kr) + BkY_0(kr) \tag{9.16}$$

and

$$u_\phi = \frac{C}{r} = \frac{2\Omega\psi}{Ur} = \Omega r + kAJ_1(kr) + kBY_1(kr) . \tag{9.17}$$

The solution in (9.16), (9.17) is applicable to the more general geometrical configuration of the swirling flow in the annulus between concentric passages of circular cross section. In the absence of the inner passage the constant B equals zero. We complete the summary of this simple case by giving the expressions for the axial and azimuthal velocity components u_x and u_ϕ which will be needed for extension to the more general case to be dealt with shortly. Evaluating the constants A and B in the expressions (9.16), (9.17) for the flow geometry of transition from a radius a upstream to b downstream, and a homogeneous axial velocity at the value U and a homogeneous angular velocity (solid body rotation) of value Ω, we get:

$$\frac{u_x}{U} = 1 + \left(\frac{a^2}{b^2} - 1\right) \frac{kbJ_0(kr)}{2J_1(kb)} , \tag{9.18}$$

and

$$\frac{u_\phi}{\Omega r} = 1 + \left(\frac{a^2}{b^2} - 1\right) \frac{bJ_1(kr)}{rJ_1(kb)} . \tag{9.19}$$

An important property of this result, which will be of use later, is that u_ϕ at outlet scales with Ω, the angular velocity of the swirling motion at inlet.

9.3.2 The General Case of Arbitrary Profiles of Axial and Azimuthal Velocity Components at Inlet

The main source of difficulty in solving the equations of motion for the boundary conditions specified of the present problem is traceable to the vorticity at the inlet section. This difficulty is more readily visible in the vorticity equation that does not contain the pressure, viz. (9.7) than in the original form with the "primitive variables" (9.1), (9.2), (9.3). The left-hand side of the vorticity equation (9.7) is the Laplace operator for the stream function which would have been equal to zero if the flow had been irrotational. We may recall that, in this case, the coupling between the kinematic and the dynamic problems is only in one way, i.e., the kinematic and the dynamic problems may be solved successively in this order, since the kinematic problem of determining the velocity field is not coupled with the dynamic problem. Pressure is then determinable from the Bernoulli equation after the velocity field is obtained. The rotational nature of the flow destroys this simplicity and the kinematic and dynamic problems then require to be solved simultaneously. The right-hand side of (9.7) illuminates the nature of the coupling involved. The mutual coupling between the kinematic and the dynamic problems is present even in the absence of swirl ($\Omega = 0$) due to the total head H being a function of ψ, but the presence of swirl complicates the problem further. The difficulty is not removed if attention is restricted to gradual changes of the cross section alone, i.e., if the approximation (9.9) is regarded as the governing equation instead of (9.7).

For arbitrary distributions of the axial and azimuthal velocity components, u_x,u_ϕ, at the inlet cross section, it will not be possible to express H and C in terms of ψ in an analytical form. As a matter of fact it may not be feasible to find convenient analytical expressions for describing the profiles of u_x and u_ϕ at the inlet section as functions of r, and even then only with difficulty. Keeping the desirability of application of this study for questions arising in metering in mind, it is more reasonable to work on the hypothesis that u_x and u_ϕ will be available only in tabulated form. The table can be thought of as originating from experimental data. The data in tabular form may also be extracted from another kind of source, say, from post-processed numerical solutions of the equations of motion for the swirl-generating element of the installation that preceeds the pipe, eg. the elbow or the compressor. It may therefore be expected that one has to resort to numerical methods for obtaining the solution to the problem on hand.

In the present state of algorithm development, methods to solve the equations (9.1), (9.2), (9.3) or (9.7) are indeed available, see eg. Chap. 11. Many of these methods are available in the form of published algorithms paving the way for the reader to choose one that suits his or her purpose best. However, by their very nature, numerical methods involve some kind of approximation, a matter of fact that, in the author's opinion, should not be lost sight of when judging the validity of the results. It is therefore desirable to keep a check on the validity of the results whenever this is feasible. Since the numerical results are solutions of approximations to the differential equations, a suitable check for their validity would be if they satisfy the invariance constraints that is derivable from the differential equations directly, see eg. Ferziger and Peric (2002). In devising such a test, it is meaningful to critically examine the underlying principles on which the numerical method intended for use rests. We observe in this context that there is a common principle at the roots of most of the numerical methods in current use for which algorithms are readily available. It is that they work with x and r as the independent variables, or r alone (in (9.9)), as the case may be. Choices for the dependent variable are several in number, and so are the approximations for the dependent variable underlying the discretisation process. But regardless of these steps, it is inherent in this principle of approach that, if the flow quantities (u_x, u_ϕ, p) or ψ are obtained as functions of r at some x, a check on their compatibility with the solutions that would be obtainable from the differential equations would involve examining if H and C remain invariant along streamsurfaces $\psi = const.$. A test for this invariance is, to the author's knowledge, not incorporated as a routine in any of the algorithms hitherto published, neither can such a test be added as a simple and straightforward extension to the algorithms already available. For application to the class of problems in question, preference for an algorithm that incorporates the invariance test is evident, and the revised approach presented herein is to be understood as one such.

The principle underlying the present approach is that instead of trying to solve for ψ as a function of r and checking if it satisfies the physical constraints imposed, we interchange the dependent and independent variables and look for r as a function of ψ. In this approach, if the radius r is obtained for a certain streamsurface ψ, the azimuthal velocity u_ϕ follows from C, which, being constant on the streamsurface, has the same value as at the initial station. Further, since H also stays constant on the streamsurface at its initial value, u_x follows from Bernoulli's equation in the form (9.5). In a method based on this approach the invariace property of H and C remaining constant on streamsurfaces is therefore integrated into the algorithm.

The starting point in this approach is therefore (9.9), with the dependent and independent variables interchanged. This step leads to:

$$r\frac{\mathrm{d}^2 r}{\mathrm{d}\psi^2} + \left(\frac{\mathrm{d}r}{\mathrm{d}\psi}\right)^2 + r^3\left(\frac{\mathrm{d}r}{\mathrm{d}\psi}\right)^3 \frac{\mathrm{d}H(\psi)}{\mathrm{d}\psi} - r\left(\frac{\mathrm{d}r}{\mathrm{d}\psi}\right)^3 C(\psi)\frac{\mathrm{d}C(\psi)}{\mathrm{d}\psi} = 0 \qquad (9.20)$$

The wall boundary condition for (9.20) is the same as before, viz. (9.8), of course with the understanding that r is the dependent variable and ψ the independent variable. A numerical method suited to solve this boundary-value problem is the shooting method. However some care is necessary in handling the boundary condition on the axis when r is regarded as the dependent variable for the reason to be outlined below, and the method of solution is required to pay due regard to this aspect of the problem.

The Boundary Conditions on the Axis and at the Wall

For the case of present interest, where the axial velocity u_x and the azimuthal velocity u_ϕ are permitted to have arbitrary distributions along the radius at the initial station, it is not overly restrictive to hypothesize that they are regular and so can be approximated in the following way in the region around the axis $r = 0$:

$$u_x = U_0 + Pr^2 + o\left(r^2\right) , \tag{9.21}$$

and

$$u_\phi = \Omega_0 r + o(r) . \tag{9.22}$$

The expressions (9.21) and (9.22) convey that in a small neighbourhood of the axis the axial velocity is describable by a parabola and the azimuthal velocity corresponds to solid-body rotation with the angular velocity Ω_0. Equation (9.21) implies that the streamfunction ψ according to (9.4) and its derivative may be written around the axis as follows:

$$\psi = \frac{U_0 r^2}{2} + \frac{Pr^4}{4} + o\left(r^4\right) , \tag{9.23}$$

and

$$\frac{\mathrm{d}\psi}{\mathrm{d}r} = U_0 r + Pr^3 + o\left(r^3\right) . \tag{9.24}$$

On interchanging the dependent and independent variables, the expression for $\mathrm{d}\psi/\mathrm{d}r$ according to (9.24) corresponds to the following expression for $\mathrm{d}r/\mathrm{d}\psi$ in the region around the axis:

$$\frac{\mathrm{d}r}{\mathrm{d}\psi} = \frac{1}{U_0 r} - \frac{Pr}{U_0^2} + o(r) , \tag{9.25}$$

which clearly shows the singular nature of this derivative on the axis. This is only a different manifestation of the singularity on the axis that is known to arise when polar coordinates are employed for computation. It necessitates exclusion of the origin $r = 0$ itself from numerical computations.

Since the governing equations for our work are the Euler's, and not the Navier–Stokes, equations, only the kinematic boundary condition can be imposed at the wall. In terms of the stream function, this means that only the value of ψ at the wall can be prescribed. It is in order to note in this context that, when the shooting method is employed, and integration of (9.20) is to be carried out from the wall towards the axis, the starting value for u_x at the wall ($r > 0$) has to be prescribed to be different from zero, since otherwise the derivative $\mathrm{d}r/\mathrm{d}\psi = 1/(ru_x)$ becomes infinite. Since (9.20) is to be satisfied at the inlet section too, the axial and azimuthal velocity profiles at the inlet section, and through them H and C, may be prescribed arbitrarily but u_x at the wall should not be zero. This is tantamount to admitting slip at the wall which is permissible when working with Euler's equations.

Method of Solution

A suitable method of solution to the problem on hand is the shooting method. It is straightforward in principle, and reduces solving the boundary-value problem to two simple more elementary steps, viz. solving an initial-value problem for a range of guessed initial values, followed by finding the correct initial value through finding the zero of a suitably defined function quantifying the error. Accordingly, the first

step is carrying out integration of (9.20) from the wall towards the axis, say with a Runge–Kutta method, with the value $\psi = \Psi$ given at the boundary, for several guessed values of $\mathrm{d}r/\mathrm{d}\psi = 1/(ru_x)$ at the wall. It is in order to recall here that u_x at the wall cannot be prescribed to be zero. This step yields values of the unknown variable $r = r^*$ at a certain value of the independent variable $\psi = \psi^*$, which is chosen to lie close to the axis but be different from zero. We thus have r^* as a function of the derivative $\mathrm{d}r/\mathrm{d}\psi$ at the wall. The second step is finding the zero of an error defined through the difference between the value of this (discrete) function and the "correct" value of r belonging to ψ^*. The "correct" value of r that belongs to ψ^* may be obtained in the following way:
Since the streamsurface ψ^* lies close to the axis, the axial velocity profile *at the inlet section* may be regarded as uniform with the value U_0 in (9.21), and the azimuthal velocity profile *at this section* a linear function of the radius corresponding to solid-body rotation with an angular velocity Ω_0 in (9.22). With this approximation, the axial and azimuthal velocity components at any downstream section x are distributed according to (9.18) and (9.19) respectively and the relation between ψ^* and r^* may be written as follows, see. (9.26):

$$\psi^* = \int_0^{r^*} ru_x \mathrm{d}r = \frac{U_0 r^{*2}}{2} + \int_0^{r^*} r\left(\frac{r_0^{*2}}{r^{*2}} - 1\right)\frac{\Omega_0 r^* J_0(2\Omega_0 r)}{J_1(2\Omega_0 r^*)}\mathrm{d}r \tag{9.26}$$

The above expression is not very convenient for numerical evaluation of r^* at a given value of ψ^*. But, if the arbitrarily specified distribution of the axial velocity component at the inlet section departs only mildly from the constant value U_0, i.e., if the inhomogeneity is small, it is straightforward to obtain r^* for a chosen ψ^* in terms of the local radius of the circular passage $R(x)$ through integration of (9.18). This is as follows:

$$\psi^* = U_0\left(\frac{r^{*2}}{2} + \left(\frac{R_1^2}{R^2(x)} - 1\right)\frac{kR}{4J_1(kR)}r^{*2}\right) \tag{9.27}$$

9.4 Results and Discussion

The equation (9.20) has been solved by the method described for a number of cases, and we present in the following only an illustrative sample. However, at first we wish to subject our method to a test and examine if it recovers the known results for the standard case of homogeneous distributions of both the axial and the angular velocity at the inlet section.

9.4.1 Verification of the Method

The solution of (9.20) for u and w/r when conditions of a homogeneous axial velocity and solid body rotation are prescribed at the inlet are compared with Batchelor's

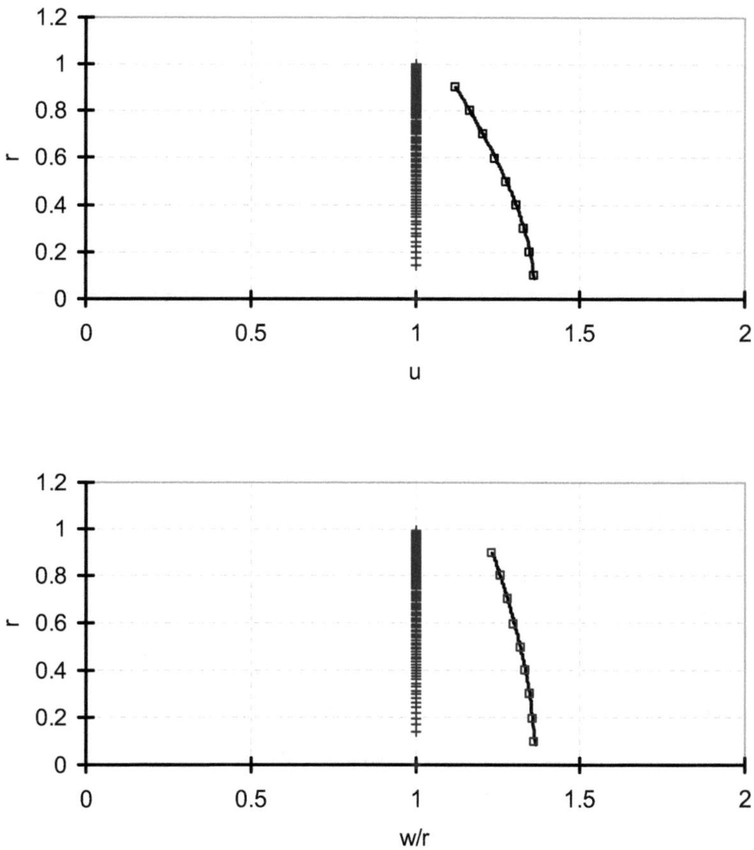

Fig. 9.2. Effect of contraction for homogeneous inlet conditions: Contraction ratio $R_2/R_1 = b/a = 0.9$. Profiles at inlet: (+ + +) $u \equiv 1$, $w/r \equiv 1$. Profiles at outlet: (—) analytical solution according to (9.18), (9.19); (□ □ □) numerical solution of (9.20).

solution for this problem, viz. (9.18), (9.19), for a contraction ratio $R_2/R_1 = 0.9$ in Fig. 9.2. The comparison shows that our method captures the results of the standard case satisfactorily. This relatively simple example already points at an important qualitative character of the effect of an area change on the flow, viz. that u and w/r do not any longer remain constant along a radius downstream of a constriction.

9.4.2 The Effects of Passage Contraction

Figure 9.3 shows profiles of u and w/r caused by a passage contraction when u departs from a constant but w/r remains a constant at the inlet section. The nonhomogeneous axial velocity profile at inlet in this example is prescribed to be "mildly parabolic", i.e., a parabola of mild curvature superimposed on a homogeneous axial

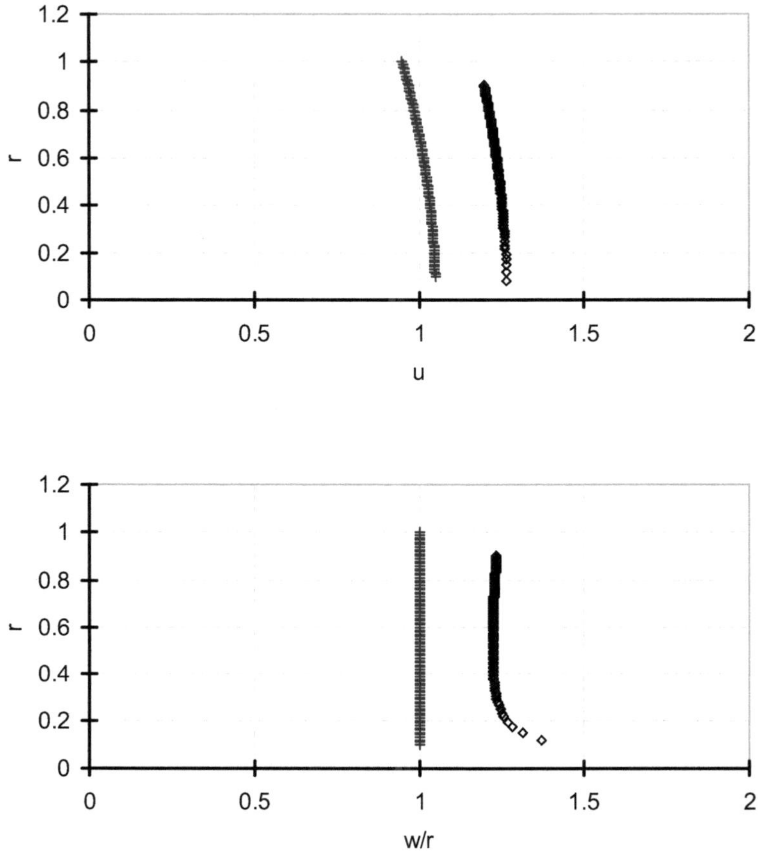

Fig. 9.3. Effect of contraction for nonhomogeneous inlet conditions: Contraction ratio $R_2/R_1 = 0.9$. Profiles at inlet: (+ + +) $u = 0.95 + 0.1(1 - r^2)$, $w/r = 1$. Profiles at outlet: (⋄ ⋄ ⋄) numerical solution of (9.20).

velocity profile. The change in character of the distributons of u and w/r at outlet are evident here too.

A cursory inspection of the plots of u and w/r in Figs. 9.2 and 9.3 already draws attention to an outstanding qualitative feature of the effect of a constriction on the distribution of the flow quantities. It is that the shape of distribution of the flow quantities is no longer preserved. The character of the change is more pronounced in the distribution of the angular velocity w/r than of the axial velocity profile u. Not only is the homogeneous distribution of w/r at inlet, which is solid-body rotation, destroyed as such by the area constriction, but also the nature of the change when the axial velocity profile at inlet is nonhomogeneous departs from that when u is homogeneous. The signs of the curvature of the profiles of w/r at the outlet section are different in the two cases.

The above examples are indicative of the sensitive dependence of the axial and the azimuthal velocity profiles at the exit section to the profiles at the inlet section. Consequently, knowledge of the profile details at the exit section presupposes knowledge of these details at the inlet section which, in most flow metering applications, is rarely, if ever, available. Therefore, conclusive statements of a general nature on the velocity profiles perceived by the metering device do not seem to be within reach at present. However, when the inhomogeneity in the profiles of both the axial velocity and the angular velocity at the inlet section is small, analysis based on the dynamics of fluid flow can be taken a step further along lines to be described below. The more convenient starting point for such an analysis is (9.9).

We introduce for this purpose a small parameter characterizing the inhomogeneity, ϵ_{in} which, for instance may be the maximum departure of the actual profiles from the homogeneous case. We may then write the axial and azimuthal velocity profiles at the inlet section as follows:

$$u_x = U_{\text{H}}(1 + \epsilon_{\text{in}} F_{\text{u}}(r)) \,, \tag{9.28}$$

and

$$u_\phi = \Omega_{\text{H}} r(1 + \epsilon_{\text{in}} F_\phi(r)) \,. \tag{9.29}$$

For the case of solid-body rotation, $F_\phi(r)$ is a constant, and ϵ_{in} may be set equal to 0. It is then straightforward to verify that $\psi, H, C, \mathrm{d}H/\mathrm{d}\psi$ and $\mathrm{d}C/\mathrm{d}\psi$ all assume the same form to the order $O(\epsilon_{\text{in}})$ at inlet. Solutions for (9.9) may then sought in the form of asymptotic expansions in terms of the parameter ϵ_{in}, from which it follows that the velocity profiles at exit are also representable through asymptotic expansions of the same form. The conclusion that may be drawn herefrom is that, to a leading approximation in ϵ_{in}, the effect of an area change on the velocity distribution scales in the same mannor as in the flow with homogeneious inlet conditions.

References

Batchelor GK (1967) An Introduction to Fluid Dynamics. Cambridge University Press, Cambridge UK

Brown GL, Lopez JM (1990) Axisymmetric vortex breakdown, Part 2: Physical mechanisms. J Fluid Mech 221:553–576

Buntine JD, Saffman PG (1995) Inviscid swirling flow and vortex breakdown. Proc. Royal Soc. London A 449:139–153

Escudier M (1988) Vortex breakdown: Observations and explanations. Prog. Aerospace Sc. 25:189–229

Faler JH, Leibovich S (1978) An experimental map of the internal structure of a vortex breakdown. J Fluid Mech. 86:313–335

Ferziger JH, Peric M (2002) Computational Methods for Fluid Dynamics. Springer, Heidelberg

Gallaire F, Chomaz J-M (2003) Instability mnechanisms in swirling flows. Physics of Fluids 15:2622–2639

Marshall JS (2001) Inviscid Incompressible Flow. John Wiley, New York

Rusak Z, Wang S, Whiting CH (1998) The evolution of a perturbed vortex in a pipe to axisymmetric vortex breakdown. J Fluid Mech. 366:211–237

Schlichting H, Gersten K (2000) Boundary Layer Theory. Springer, Heidelberg

Wang S, Rusak Z (1997) The dynamics of a swirling flow in a pipe and transition to axisymmetric vortex breakdown. J Fluid Mech 340:177–223

10

Errors of Turbine Meters Due to Swirl

Venkatesa Vasanta Ram

Institut für Thermo- und Fluiddynamik, Ruhr-Universität Bochum, 44780 Bochum, Germany

This chapter addresses the question of the error in metering of a flow with a turbine flow meter caused by swirl in the oncoming flow. The problem is approached by analyzing the signal-generation mechanism of a turbine flow meter from a fundamental fluid mechanics point of view when the oncoming flow shows departures from its calibration state. The fundamental physical mechanism in question is the driving moment experienced by the rotor of the turbine flow meter due to the aerodynamic forces on the rotor blading. Swirl in the oncoming flow belongs in this scheme to one class of possible disturbances to the calibration flow. The study brings out the salient parameters that govern the error in metering due to swirl, and the pattern of error dependance on swirl strength.

Notation

x, r, ϕ	co-ordinate system: axial, radial and azimuthal directions respectively
ρ, ν	density and kinematic viscosity of the flowing fluid
u_{x}, u_{ϕ}	x and ϕ velocity components at inlet section to turbine flow meter, section A-A in Fig. 10.1
v_{x}, v_{ϕ}	x and ϕ velocity components at inlet section to rotor blading, section B-B in Fig. 10.1
R, D	pipe radius and diameter respectively at inlet section to flow meter, Fig. 10.1, $D = 2R$
$R_{\mathrm{i}}, R_{\mathrm{o}}$	hub and tip radii of rotor
$q(r)$	defined in (10.8), $q = \dfrac{r\Omega_{\mathrm{Rc}}}{v_{\mathrm{xc}}}$
Q	volume flow rate through the flow meter
$\hat{Q}(r)$	defined through (10.14)
U_{ref}	bulk velocity, reference velocity, defined through $U_{\mathrm{ref}} = \dfrac{Q}{\pi R^2}$
Ω_{R}	angular velocity of rotor
Re	Reynolds number, $Re = \dfrac{U_{\mathrm{ref}} D}{\nu}$

Ro	Rotation number, $Ro = \dfrac{\omega_S R}{U_{ref}}$
ϵ_S	parameter characterizing swirl strength, defined in(10.21)
ϵ_Q	error in metering of the flow, defined in (10.39)
$\alpha_{bl}(r)$	blade angle with respect to rotor axis at radius r, Fig. 10.2
$L_{bl}(r)$	chord of blade at radius r, Fig. 10.2
T_{br}	braking torque on rotor of turbine flow meter
T_{aero}	driving (aerodynamic) torque on rotor of turbine flow meter
F_{meter1}, F_{meter2}	functions defined through (10.15), (10.16)
F_T	function defined through (10.20)
B_{br}	defined in (10.34)

Subscripts

c	denotes calibration state, e.g., $v_{xc}, \Omega_{Rc}, \alpha_c, T_{brc}, T_{aeroc}$
1	denotes departure from calibration state, e.g., $\Omega_{R1}, Q_1, \alpha_1, T_{br1}, T_{aero1}$
bl	refers to rotor blade, e.g., α_{bl}, L_{bl}
R	refers to flow meter rotor, e.g., Ω_R
S	refers to swirl or swirl generator, e.g., ϵ_S, ω_S
ref	denotes reference, e.g., Q_{ref}, U_{ref}
meas	denotes measurement in Q_{meas}

10.1 Introduction

The turbine flow meter is a type of flow metering device very commonly adopted in engineering practice. This type of flow meter is often preferred to alternatives in many engineering applications, the reason for which is that it offers the advantages of relative simplicity in installation and operation combined with ruggedness and acceptable accuracy for the purpose of the application in question, see e.g., Baker (2000). The turbine flow meter shares a feature common to metering devices in general in that it needs calibration. Calibration is generally done in the fully developed pipe flow that is free of swirl. Swirl in the flow of measurement causes relevant details of the flow to depart from the calibration state, and this in turn leads to errors in metering. The task of estimating this error calls for an analysis of the physical mechanisms involved in the conversion of the flow rate into the metering signal under measuring conditions. A closer examination of this mechanism for a turbine flow meter reveals what may be regarded as a salient characteristic of this metering device not shared in common by most of the other types. It is that the error is sensitive to the direction of swirl in the oncoming flow. The purpose of the present section is to present an analysis of the characteristics of this error together with a comparison of the same with experiments.

A cursory examination of the operating principle of the turbine flow meter is sufficient to hint at the strong similarities that exist between the questions arising in the context of the turbine flow meter and those in propellers and/or ducted fans. Propellers and ducted fans belong to a classical branch of study in fluid mechanics in which extensive body of literature has been accumulated, see e.g., Leishman (2000), and it is only too natural to look for ways of carrying over results already established and proven, to the flow meter problem. However, a closer examination of the two problems shows that, despite the similarities that undoubtedly exist, there is a significant difference between the two that stands in the way of results from the classical studies being taken over in a straightforward manner to the flow meter. The main difference is that the flow meter is exposed, both in calibration and measurement, to rotational flow, whereas most of the classical studies of propellers and ducted fans are centered around irrotational flow. Thus, although the basic ideas and approach in the classical studies can be carried over to the flow meter problem, differences in results are to be expected, warranting a separate study of the fluid mechanics of the flow meter. Carrying out such a study is the object of the present work.

The standard work of Baker (1993) is a survey of techniques for flow measurement in general, and the reader is referred to this work for a thorough review of the theoretical foundations and operating principles of the turbine flow meter too. The physical principle on which the turbine flow meter rests for its operation is the balance of torque on its rotor. A quantitative formulation of this physical principle leads to the result that the angular velocity of the rotor is decided by the balance between the driving torque due to the aerodynamic forces on the rotor blading and the braking torque (or drag torque) exerted on the rotor. Published literature on the turbine flow meter identifies four sources of drag, viz. the bearing drag, the hub disk friction drag, the tip clearance drag and the hub fluid drag, see Baker (2000). It is a reasonable hypothesis to suppose that swirl in the oncoming flow leaves the physical mechanisms behind the drag torque from these four sources unaffected. So it appears justifiable that an analysis of the error of the turbine flow meter caused by swirl be founded, to start with, on an examination of the effect of swirl in the oncoming flow on the aerodynamic torque exerted on the rotor blading, and this is the subject in focus in the present study.

Scope of the Present Work

The starting point for this examination is the expression for the aerodynamic torque on the rotor under measurement conditions when swirl may be present in the oncoming flow. The relation between the actual flow rate and the angular velocity of the flow meter rotor under measurement conditions follows from this expression. This expression defines the calibration curve of the flow meter which is the relation between the flow rate and the output signal when calibration conditions exist, i.e., when the oncoming flow is free of swirl and the axial velocity profile corresponds to the calibration state. The next step is the derivation of an expression describing the departure of this relation between the flow rate and the output signal (rotor angular velocity) when the velocity profile of the oncoming flow is disturbed from its calibration state. Swirl in the oncoming flow is only one among the several possible classes of disturbance, see Chaps. 2 and 3. Suitable parametrization of this specific class of disturbance leads to an expression relating the flow meter reading error and the swirl parameter. We then describe an experiment conducted in a specially designed facility in which a turbine flow meter is exposed to an oncoming flow into which swirl of known strength and distribution is purposely introduced. The experiment yields the error in the flow rate "read" by the turbine flow meter due to swirl in the oncoming flow. The experimental outcome is set against the analysis over a range of the swirl parameter.

The outline of this chapter is as follows: In Sect. 10.2 we derive the expression for the aerodynamic torque on the rotor under conditions when swirl may be present in the flow. In Sect. 10.3 we describe an experiment and present the data therefrom reduced in a form suitable for setting against the theory of Sect. 10.2. The discussion that follows in Sect. 10.4 then leads to the conclusions drawn in Sect. 10.5.

10.2 Physical and Mathematical Background

Figure 10.1 is a sketch showing the main components of a turbine flow meter relevant for the present study. The physical law underlying the relation betwen the flow rate Q and the output signal Ω_{R} of the flow meter is the balance of torque on the rotor of the flow meter when this is exposed to stationary flow conditions. This may be written as follows:

$$T_{\mathrm{br}} = T_{\mathrm{aero}} \, , \tag{10.1}$$

where T_{br} and T_{aero} are the braking and driving torques respectively. Since the flow meter is neither driven by any external power source, nor is it delivering power, the driving torque is only due to aerodynamic forces on the rotor blading, hence the notation T_{aero} for the driving torque.

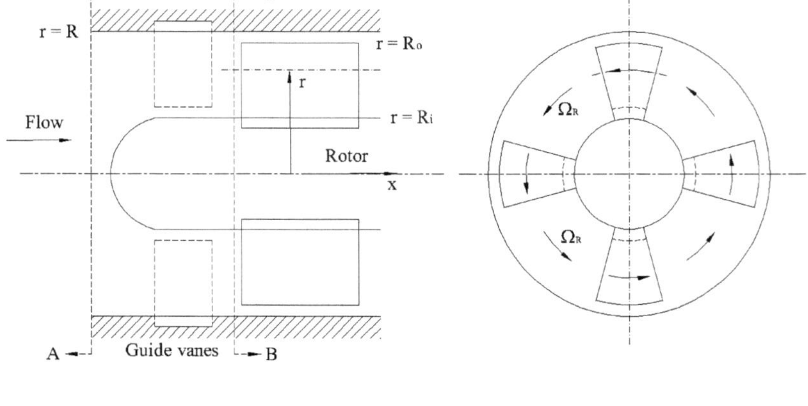

Fig. 10.1. Turbine flow meter components

10.2.1 Aerodynamic Torque on the Turbine Flow Meter Rotor

The velocity and force triangles on a blade element at a radial location r are shown in Fig. 10.2. Figure 10.2a depicts these triangles under calibration conditions whereas Fig. 10.2b refers to conditions when the oncoming flow contains swirl.

The expression for the aerodynamic torque on the rotor may be obtained by integrating the contribution to the torque from the aerodynamic force on a blade element from the hub to the tip. From dimensional considerations the magnitude of the aerodynamic force vector $\mathrm{d}\vec{F}_{\mathrm{aero}}$ on the blade element between the radii r and $r + \mathrm{d}r$ may be written as follows:

$$|\mathrm{d}\vec{F}_{\mathrm{aero}}| = \rho \vec{v}^2 L_{\mathrm{bl}} \mathrm{d}r C \ , \tag{10.2}$$

where ρ is the density of the fluid flowing through the meter, L_{bl} the local chord of the rotor blading, $\vec{v}$ the velocity "perceived" by the blade at the radial location r, see Fig. 10.2, and C a dimensionless coefficient that depends upon the blade form and other geometrical factors entering the design of the meter. The azimuthal component of this force, denoted $\mathrm{d}F_{\mathrm{aero}\phi}$ is then given by

$$\mathrm{d}F_{\mathrm{aero}\phi} = \rho \vec{v}^2 L_{\mathrm{bl}} \mathrm{d}r C \cos\left(\alpha_{\mathrm{bl}} - \alpha\right) \ , \tag{10.3}$$

where α_{bl} is the angle the blade makes with the rotor axis and α the angle of incidence the velocity vector $\vec{v}$ makes with the blade, see Fig. 10.2a, b. The velocity vector $\vec{v}$ is as "perceived" by the blading, i.e., it is in a frame of reference that is rotating about the flow meter axis with the angular velocity of the rotor, Ω_{R}, and α is measured from the direction of zero lift of the blade. It is reasonable to regard this vector $\vec{v}$ as one with negligible radial component even when swirl is present in the oncoming flow. The axial and azimuthal components of $\vec{v}$ are therefore as shown in Fig. 10.2a, b. The torque exerted by the aerodynamic force on the rotor blading may then be obtained by integrating $r\mathrm{d}F_{\mathrm{aero}\phi}$ across the blade from R_{i} to R_{o}. The expression for the aerodynamic torque T_{aero} is then:

$$T_{\mathrm{aero}} = \rho \int_{R_{\mathrm{i}}}^{R_{\mathrm{o}}} r \vec{v}^2 L_{\mathrm{bl}} C \cos\left(\alpha_{\mathrm{bl}} - \alpha\right) \mathrm{d}r \tag{10.4}$$

In the further course of the analysis to follow we work with the hypothesis that the force vector $\mathrm{d}\vec{F}_{\mathrm{aero}}$ is perpendicular to the velocity vector $\vec{v}$, which is tanamount to postulating the aerodynamic torque to be only due to the lifting force on the blade. For a discussion of the meaning and implications of this hypothesis the reader is referred to standard texts on fundamentals of fluid mechanics, e.g., Batchelor (1967), Marshall (2001), Schlichting and Truckenbrodt (1962). The hypothesis is justifiable when the flow meter operates at high Reynolds numbers, which is mostly the case, and boundary layers remain thin and attached to the blading. We also further set the lift to be proportional to the angle "perceived" locally by the blade section, i.e., $C \sim \alpha$. So we may write $C = NC_{\mathrm{L}}\alpha$ where N is the number of blades on the

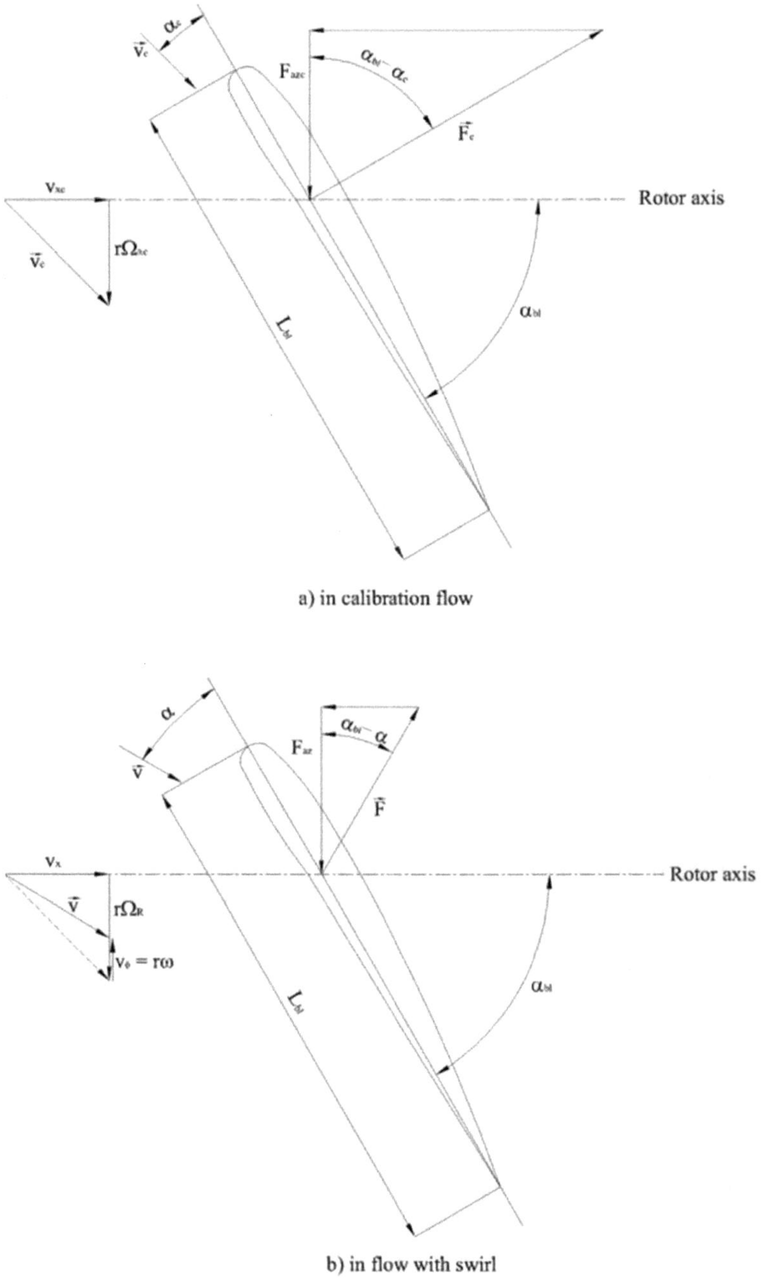

Fig. 10.2. Velocity and force triangles

rotor and C_{L} in (10.5) is the lift coefficient of a single blade, of course in a cascade arrangement. T_{aero} in (10.4) is then given by

$$T_{\mathrm{aero}} = N\rho \int_{R_{\mathrm{i}}}^{R_{\mathrm{o}}} C_{\mathrm{L}} r \vec{v}^{2} L_{\mathrm{bl}} \alpha \cos \alpha_{\mathrm{bl}} \mathrm{d}r. \tag{10.5}$$

The step leading to the expression in (10.5) from (10.4) involves the approximation for small values of α obtainable on use of the standard trigonometric relationship between $\cos(\alpha_{\mathrm{bl}} - \alpha)$ and circular functions of α_{bl} and α. Furthermore, the number of blades on the rotor has been explicitly accounted for through N which is hidden in the dimensionless coefficient C in (10.4). For obvious reasons C_{L} may depend upon the radial location r, hence its position under the integral sign.

It is useful at this stage to introduce the relation between the angle α and flow parameters entering the problem which are the following:

- the axial velocity component $v_{\mathrm{x}}(r)$,
- the azimuthal velocity component in the laboratory coordinate system which may be written in terms of the local angular velocity of the fluid, $\omega(r)$, as $r\omega(r)$, and
- the angular velocity of the flow meter rotor Ω_{R}.

This relation is:

$$\tan(\alpha_{\mathrm{bl}} - \alpha) = \frac{r\left(\Omega_{\mathrm{R}} - \omega\right)}{v_{\mathrm{x}}}, \tag{10.6}$$

which may also be written using standard trigonometric identities as follows:

$$\cos \alpha_{\mathrm{bl}} \cos \alpha + \sin \alpha_{\mathrm{bl}} \sin \alpha = \frac{v_{\mathrm{x}}}{\sqrt{v_{\mathrm{x}}^{2} + r^{2}\left(\Omega_{\mathrm{R}} - \omega\right)^{2}}}, \tag{10.7}$$

It is appropriate at this juncture to pause briefly to reflect upon a certain point in the hypothesis that touches a fundamental aspect of fluid dynamics and is implicit when writing (10.5, 10.6, 10.7). It is that the vector $\vec{v}$ "perceived" by the blade element cannot be set to be just the flow "far upstream", merely seen from an appropriate coordinate system. More properly, it should be the flow far upstream modified by the vortex system in the wake of the rotor, whereby the question of resolving the effective angle of incidence α acquires key importance. The physical reasoning behind this conjecture is essentially the same as in classical wing theory, where the effective angle of incidence that accounts for the lift felt at a certain spanwise section of a wing of large but finite span is given by the flow direction far ahead of the wing corrected by the downwash induced by the vortex system in the wake of the lifting surface. The observation known from classical wing theory, viz. that the downwash depends upon details of the lift distribution along the span of the lifting surface, holds for the problem of current interest too. However, in the present context, which is the task of assessing the effects of swirl on flow measurements with turbine flow meters, carrying over this concept to the problem on hand through accounting for the effect of this vortex system in general terms for turbine flow meters does not seem to be warranted. The reason is that the answer would depend upon such design details of the flow meter that are rarely, if ever, available to the user, hence tending to make the exercise futile. We therefore chart out a course for our present study in which the induced velocity field of the vortex system in the wake of the rotor is divided into two further contributions. These are: 1) the induced velocity field when the flow meter is exposed to the calibration flow, and 2) that when swirl or any other kind of disturbance is present in the flow of measurement. The meaning of this formulation, as it will shortly be seen, is that while enabling the induced velocity field in the calibration flow to be accounted for, that due to swirl in the oncoming flow or another kind of disturbance, is ignored. This approach is analogous to an established method of handling the effects of rotation in the free stream in inviscid airfoil theory, in which it is stipulated that there be no *additional circulation* due to rotation in the free stream e.g., Van Dyke (1975). In the present context the justification for this approach is that the result is applicable more generally to turbine flow meters. A more rigorous treatment of the complete problem would call for an approach through solving the full equations of motion with the appropriate boundary conditions. For an example of this approach for a particular flow meter the reader is referred to Chap. 11.

Calibration Conditions

Under calibration conditions the axial velocity profile at the inlet to the flow meter, this is section A-A in Fig. 10.1, is the fully developed flow in a pipe which we denote as u_{xc}, and the flow is free from swirl. In turbulent flow, the shape of the profile u_{xc} depends upon the Reynolds number, see Chap. 1 for the nature and details of this dependence.

It follows therefore that, for a certain given Reynolds number, there will be, under calibration conditions, a certain velocity distribution u_{xc} at the inlet section of the flow meter (section A-A of Fig. 10.1) to which there will correspond a certain distribution v_{xc} at the inlet section to the rotor blading (section B-B of Fig. 10.1) too. The latter, v_{xc}, depends upon the geometrical details of the meter and hence may differ from design to design. It is reasonable to presume that aspects of meter design, say guide vanes, do not introduce a non-zero azimuthal velocity at the section B-B, so we may work with $v_{\phi\mathrm{c}} = 0$. The velocity triangle for this case is as shown in Fig. 10.2a. The relation sought involves the ratio $(r\Omega_{\mathrm{Rc}})/(v_{\mathrm{xc}})$ for which it is convenient to introduce an abbreviation q in terms of which (10.6, 10.7) then reduce to the following, where the subscript c denotes the calibration state:

$$\tan(\alpha_{\mathrm{bl}} - \alpha_{\mathrm{c}}) = \frac{r\Omega_{\mathrm{Rc}}}{v_{\mathrm{xc}}} = q(r), \tag{10.8}$$

and

$$\cos\alpha_{\mathrm{bl}}\cos\alpha_{\mathrm{c}} + \sin\alpha_{\mathrm{bl}}\sin\alpha_{\mathrm{c}} = \frac{1}{\sqrt{(1+q^2)}} \,. \tag{10.9}$$

In the event of the meter design being such that $v_{\phi\mathrm{c}}$ is not zero on purpose, an extension of the analysis to be presented in the following to account for this design feature is straightforward. However, carrying out the analysis to cover this design feature also would only result in additional algebraic complexity without a corresponding enhancement in transparency, so a treatment in such generalised terms is not warranted at this stage. We will therefore restrict ourselves in the following to $v_{\phi\mathrm{c}} = 0$.

For small values of α_c which may be expected to be the case over the operating range of the turbine flow meter, (10.9) may be written to show explicitly the relationship between the angle α_c, the velocity profile v_xc, and the rotor angular velocity Ω_c. This is:

$$\alpha_\mathrm{c} = \frac{1}{\sin\alpha_\mathrm{bl}\sqrt{(1+q^2)}} - \frac{1}{\tan\alpha_\mathrm{bl}} \,. \tag{10.10}$$

The flow rate Q is related to the velocity profile u_xc through

$$Q = 2\pi\int\limits_0^R r u_\mathrm{xc}\mathrm{d}r = 2\pi\int\limits_{R_\mathrm{i}}^{R_\mathrm{o}} r v_\mathrm{xc}\mathrm{d}r = 2\pi\int\limits_{R_\mathrm{i}}^{R_\mathrm{o}} r|v_\mathrm{c}|\cos\left(\alpha_\mathrm{bl}-\alpha_\mathrm{c}\right)\mathrm{d}r \;, \tag{10.11}$$

which may be approximated for small α_c as

$$Q = 2\pi\int\limits_{R_\mathrm{i}}^{R_\mathrm{o}} r|v_\mathrm{c}|(\cos\alpha_\mathrm{bl} + \sin\alpha_\mathrm{bl}\alpha_\mathrm{c})\mathrm{d}r \;. \tag{10.12}$$

Under calibration conditions in question here, the expression for the aerodynamic torque on the rotor, (10.5), denoted as T_aeroc, may be written as follows:

$$T_\mathrm{aeroc} = N\rho\int\limits_{R_\mathrm{i}}^{R_\mathrm{o}} C_\mathrm{L} r(v_\mathrm{xc}^2 + r^2\Omega_\mathrm{Rc}^2)L_\mathrm{bl}\alpha_\mathrm{c}\cos\alpha_\mathrm{bl}\mathrm{d}r \;, \tag{10.13}$$

where $\vec{v}_\mathrm{c}^2$ has been written as $(v_\mathrm{xc}^2 + r^2\Omega_\mathrm{Rc}^2)$. It it illuminative to rewrite T_aeroc in a form that brings out more explicitly its relation to the overall flow rate through the device, Q, and the other relevant parameters associated with the flow and the flow meter geometry. Introducing for this purpose the notation $\hat{Q}(r)$ defined through

$$\hat{Q}(r) = \int\limits_{R_\mathrm{i}}^{r} r' v_\mathrm{xc}\mathrm{d}r' \;, \tag{10.14}$$

and the abbreviations with the dimensions of a length, viz.

$$F_\mathrm{meter1}(r) = \frac{C_\mathrm{L}L_\mathrm{bl}}{\tan\alpha_\mathrm{bl}} \,, \tag{10.15}$$

and

$$F_\mathrm{meter2}(r) = \frac{C_\mathrm{L}L_\mathrm{bl}\cos\alpha_\mathrm{bl}}{\tan\alpha_\mathrm{bl}} \,, \tag{10.16}$$

the expression (10.13), together with (10.10), may then be written as follows:

$$\frac{T_{\text{aeroc}}}{N\rho} = \left[\left(F_{\text{meter1}} v_{\text{xc}} \sqrt{(1+q^2)} - F_{\text{meter2}} v_{\text{xc}} (1+q^2)\right) \hat{Q}\right]_{R_\text{i}}^{R_\text{o}}$$
$$- \int_{R_\text{i}}^{R_\text{o}} \hat{Q}(r) \frac{\mathrm{d}}{\mathrm{d}r} \left[F_{\text{meter1}} v_{\text{xc}} \sqrt{(1+q^2)} - F_{\text{meter2}} v_{\text{xc}} (1+q^2)\right] \mathrm{d}r \, . \tag{10.17}$$

The expression for T_{aeroc} in (10.13), which is complete within the framework of the physical hypotheses set out earlier, contains some physical features which are not easily visible at first sight but are of significance when the theory derived is to be employed to set its outcome against experimental observations of turbine flow meter performance. We therefore prepare the ground for uncovering these features by deriving approximate forms that render the hidden structure more transparent. Rational approximations for T_{aeroc} may be obtained by examining the limiting cases of (10.13) or (10.17) when the dimensionless quantity $q(r)$ is either large or small. If $q(r)$ assumes large values everywhere within the region $R_\text{i} < r < R_\text{o}$, it is permissible to replace $(1 + q^2)$ by q^2, on which T_{aeroc} according to (10.13) or (10.17) gets reduced to the following form:

$$\frac{T_{\text{aeroc}}}{N\rho} = \Omega_{\text{Rc}} \int_{R_\text{i}}^{R_\text{o}} v_{\text{xc}} r^2 F_{\text{meter1}} \mathrm{d}r - \Omega_{\text{Rc}}^2 \int_{R_\text{i}}^{R_\text{o}} r^3 F_{\text{meter2}} \mathrm{d}r \, . \tag{10.18}$$

The corresponding expression for small values of $q(r)$ is as follows:

$$\frac{T_{\text{aeroc}}}{N\rho} = \int_{R_\text{i}}^{R_\text{o}} v_{\text{xc}}^2 r (F_{\text{meter1}} - F_{\text{meter2}}) \mathrm{d}r$$
$$+ \Omega_{\text{Rc}}^2 \int_{R_\text{i}}^{R_\text{o}} r^3 \left(\frac{F_{\text{meter1}}}{2} - F_{\text{meter2}}\right) \mathrm{d}r \, . \tag{10.19}$$

Two features worthy of special note in the expression for the driving (aerodynamic) torque, (10.13), or (10.17), which are retained in the approximations too are the following:

- T_{aeroc} taken as a whole is not proportional to the flow rate Q. T_{aeroc} may be regarded as a sum of two contributions, only one of which is proportional to the flow rate Q. It may be noted that $\hat{Q}$ evaluated between the limits R_o and R_i is the flow rate through the flow meter, Q. The other contribution is a weighted integral of a function of the axial velocity profile v_{xc} where the weights further depend upon geometrical design details of the flow meter geometry.
- T_{aeroc} is not expressible through Q alone, even when a nonlinear function of Q is permitted. Factors governed by meter geometry enter into the expression of T_{aeroc}.

The latter property in particular, it will shortly be seen, has far reaching implications regarding casting the error of the turbine flow meter due to swirl in a concise and universal form. To state the result in advance, this error will not be expressible in such a general form that is independent of flow meter geometry and hence applicable to all flow meters. Besides, in the approximations for neither of the limiting cases, (10.18) and (10.19), does the coefficient of Ω_{Rc}^2 contain v_{xc}, the axial velocity component determining the flow rate Q, and it may therefore be conjectured that the exact expression retains this feature in the whole range of q too, although perhaps only approximately. Furthermore, for small q, the aerodynamic torque T_{aeroc} does not contain any term linear in Ω_{Rc}. In contrast, for large q there is a term linear in Ω_{Rc} which is linear in v_{xc} too.

Although the aerodynamic torque is not proportional to the flow rate, it is convenient for the purposes of comparison at the end of this section to write T_{aeroc} formally as a product of Q and a quantity F_{T} as follows:

$$\frac{T_{\mathrm{aeroc}}}{\rho} = QF_{\mathrm{T}} \,. \tag{10.20}$$

Written as above, the coefficient F_{T} with the dimensions of a length squared divided by time, besides containing a rest dependence on Q, is influenced by characteristics governed by the flow meter design through F_{meter1} and F_{meter2}. The meaning of F_{T} follows from a straightforward comparison of (10.20) with (10.17).

A sample survey of published catalogue data on commercially availabe flow meters indicates that they are designed to operate with the rotor angular velocity Ω_{R} lying mostly in the range of the dimensionless parameter $(R\Omega_{\mathrm{R}})/(U_{\mathrm{ref}})$ between 0.9 and 1.5. Therefore, although the approximations (10.18, 10.19) do indeed preserve the salient feature of T_{aeroc} not being proportional to Q, their applicability to evaluate meter characteristics theoretically is rather limited. The nature of dependence of the aerodynamic torque on the flow rate is rather complex, in view of which the case for an examination of the aerodynamic torque obtainable from solving the full relevant equations of motions for the problem, say along lines set out in Chap. 11, is indeed strong. Such solutions, even though feasible for a few selected cases only, would, through a judicious comparison with suitable approximations, guide in assessing the regions of validity of the approximations, which would be useful for further analysis. For the present, the approximations provide essentially a qualitative insight into the phenomenon, leaving no option but to resort to (10.13) or (10.17) for quantitative evaluation.

Summing up, given a certain flow rate Q which sets the Reynolds number, there will be a certain profile u_{xc}, and hence v_{xc}, from which the aerodynamic torque T_{aeroc} follows from (10.10, 10.13). Taken together, (10.11-10.13) define the relationship between the flow rate Q and the rotor angular velocity Ω_{Rc} under calibration conditions. This relationship is, of course, obtained experimentally during calibration. Therefore, the effects of the induced velocity field due to the vortex system in the wake of the rotor when the oncoming flow is the calibration flow are accounted for.

Measurement Conditions with Swirl in the Oncoming Flow

The presence of swirl in the oncoming flow implies that the azimuthal velocity component at the inlet section to the flow meter (A-A in Fig. 10.1), u_ϕ is non-zero. A crucial step in our study of the effects of swirl on the turbine flow meter reading is characterization of the swirl in the oncoming flow through a dimensionless parameter ϵ_S. It is convenient, but not necessary, to regard ϵ_S as, say, the value of the absolute maximum of the angular velocity of the fluid in the laboratory coordinate system, referred to the angular velocity of the flow meter rotor, i.e.,

$$\epsilon_\mathrm{S} = \pm||\frac{\omega}{\Omega_\mathrm{R}}|| = \pm||\frac{u_\phi}{r\Omega_\mathrm{R}}|| \,, \tag{10.21}$$

with the sign $+$ or $-$ denoting whether the azimuthal velocity in the oncoming flow is in the same direction as that of rotation of the flow meter rotor or otherwise. A cursory survey through catalogues of commercially available turbine flow meters indicates that they mostly operate at values of Ω_R for which the parameter ϵ_S assumes values in the range $0.025 - -0.25$ in our experiments to be shortly described, which may therefore be regarded as small.

In measurement, flow conditions exist at the inlet section to the flow meter that differ from those at calibration at the same flow rate, and we denote the axial and azimuthal velocity distributions at this section (A-A in Fig. 10.1) as u_x and u_ϕ respectively. The radial distribution of this set of flow quantities undergoes a change from the inlet section of the flow meter to the inlet section to the blading, section A-A to B-B in Fig. 10.1, that depends upon meter geometry and hence differs from meter to meter. However, it is reasonable to hypothesize that the change is adequately described by Euler's equations of motion. If furthermore, the change in flow cross-sectional area from A-A to B-B is also moderate, the parameter ϵ_S is suited to characterize flow perceived the blading also, and we may write the azimuthal velocity at the section B-B, v_ϕ, in terms of the local angular velocity $\omega(r)$ as

$$\frac{v_\phi}{r\Omega_\mathrm{Rc}} = \frac{\omega(r)}{\Omega_\mathrm{Rc}} = \epsilon_\mathrm{S} f_\omega(r) \,. \tag{10.22}$$

The axial velocity prevalent at the section B-B, $v_\mathrm{x}(r)$ may also be written in terms of ϵ_S as

$$\frac{v_\mathrm{x}(r)}{v_\mathrm{xc}(r)} = 1 + \epsilon_\mathrm{S} f_\mathrm{v}(r) \,. \tag{10.23}$$

The condition that axial velocity distribution departs from that at calibration without any change in the flow rate as such imposes an integral constraint on $f_\mathrm{v}(r)$ which is as follows:

$$\int_{R_\mathrm{i}}^{R_\mathrm{o}} r v_\mathrm{xc} f_\mathrm{v} \mathrm{d}r = Q_1 = 0 \,. \tag{10.24}$$

The above constraint shows that $f_\mathrm{v}(r)$ has to change sign at least once within the region $R_\mathrm{i} < r < R_\mathrm{o}$. The precise form of $f_\mathrm{v}(r)$ depends upon the method of swirl generation in the experiment and of course on the design features of the flow meter.

The change experienced by the rotor angular velocity, $\Delta\Omega$, as a consequence of the swirl may be written as $\epsilon_S \Omega_{R1}$, or

$$\frac{\Omega_R}{\Omega_{Rc}} = 1 + \epsilon_S \frac{\Omega_{R1}}{\Omega_{Rc}} + O(\epsilon_S^2) \ . \tag{10.25}$$

Consequently, the azimuthal velocity at the inlet section to the blading (B-B in Fig. 10.1), v_ϕ may also be expected to differ from zero, which in turn modifies the velocity triangle. The modified velocity and force triangles are shown in Fig. 10.2b. Here again, as already set out, we work with the hypothesis that at any radial location r the velocity vector $\vec{v}$ and the force vector $\mathrm{d}\vec{F}_{\mathrm{aero}}$ are perpendicular to each other.

Under these conditions the angle of incidence α in (10.6) or (10.7) may be written as

$$\alpha = \alpha_c + \epsilon_S \alpha_1 + O(\epsilon_S^2) \ , \tag{10.26}$$

where α_c is given by (10.10) and α_1 is as follows:

$$\alpha_1 = -\frac{q^2 \left(\dfrac{\Omega_{R1}}{\Omega_{Rc}} - f_\omega - f_v \right)}{(1+q^2)^{\frac{3}{2}} \sin \alpha_{bl}} \ . \tag{10.27}$$

The functions of the radius r, these are q, f_ω and f_v, are defined through (10.8, 10.22) and (10.23) respectively.

Writing the aerodynamic driving torque under measuring conditions, T_{aero} given by (10.5), as an asymptotic expansion in terms of the small parameter ϵ_S as follows,

$$T_{\mathrm{aero}} = T_{\mathrm{aeroc}} + \epsilon_S T_{\mathrm{aero1}} + O(\epsilon_S^2) \ , \tag{10.28}$$

and expanding the integrand in (10.5) also in terms of ϵ_S, shows that the expression for T_{aero1} is a sum of integrals. T_{aero1} may be written as

$$T_{\mathrm{aero1}} = N\rho \int_{R_i}^{R_o} C_L r (T_\alpha + T_v + T_\Omega + T_\omega) L_{bl} \alpha_c \cos \alpha_{bl} \mathrm{d}r \ , \tag{10.29}$$

where the functions in the integrand T_α, T_v, T_Ω and T_ω are as follows:

$$T_\alpha = \alpha_1 (v_{xc}^2 + r^2 \Omega_{Rc}^2) \ , \tag{10.30}$$

$$T_v = 2 f_v \alpha_c v_{xc}^2 \ , \tag{10.31}$$

$$T_\Omega = 2\Omega_{R1} \alpha_c r^2 \Omega_{Rc} \ , \tag{10.32}$$

and

$$T_\omega = -2 f_\omega \alpha_c r^2 \Omega_{Rc}^2 \ . \tag{10.33}$$

The expression for T_{aero1} in (10.29) shows that to a linear approximation in ϵ_S, the change in the aerodynamic driving torque experienced by the rotor is a weighted sum of the effects due to the change in the angle of incidence α_1, in the axial velocity profile at the inlet to the blading f_{v}, in the rotor angular velocity Ω_{R1} and in the swirl induced azimuthal velocity in the flow f_{ω}. It may also be noted that α_1 itself is a linear weighted sum involving Ω_{R1}, f_{v} and f_{ω}, see (10.27). T_{aero1} is therefore a weighted sum of Ω_{R1}, f_{v} and f_{ω}. The weights themselves, as an inspection of (10.30, 10.31, 10.32, 10.33) shows, are integrals of quantities that depend upon the details of the departure of the oncoming flow from the calibration state and on the geometrical details of the flow meter. The former depends upon the mode of swirl introduction into the flow, whereas the latter is a design feature of the flow meter.

10.2.2 Braking Torque on the Turbine Flow Meter Rotor

In Baker (1993, 2000), the braking torque T_{br} is itself written as a sum of four contributions, which are named as that due to bearing drag, hub disk friction drag, blade tip clearance drag and hub fluid drag. For details of the characteristics of these contributions the reader is referred to Baker (2000) and original literature cited therein. The precise form of these individual contributions, and in particular the parameters affecting them and their dependence on these parameters, is still a matter of continuing investigations. For our present purposes, however, it suffices to note that T_{br} depends upon the rotor angular velocity Ω_{R}. We also wish to add that the relationship between the contribuition to T_{br} and Ω_{R} is linear as long as the specific braking mechanism is adequately describable by classical lubrication theory involving the Stokes' approximation, and bearing drag may be expected to follow this behavior. Under this hypothesis T_{br} depends linearly upon Ω_{R} and may be written as follows:

$$T_{\mathrm{br}} = B_{\mathrm{br}} \Omega_{\mathrm{R}} \, . \tag{10.34}$$

The coefficient B_{br} may then be identified as arising from bearing drag. It has the dimensions of a torque multiplied by time and is dependent upon geometrical particulars of the bearing and lubricant viscosity.

The braking torque under measurement conditions, T_{br}, may also be written as an asymptotic expansion in terms of the parameter ϵ_{S} as follows:

$$T_{\mathrm{br}} = T_{\mathrm{brc}} + \epsilon_{\mathrm{S}} T_{\mathrm{br1}} + O(\epsilon_{\mathrm{S}}^2) \, , \tag{10.35}$$

where T_{brc} is the braking torque under calibration conditions and the dependence of each of the terms, T_{brc} and T_{br1}, on Ω_{R} is itself expressible in the form according to (10.34). The torque balance on the flow meter rotor (10.1) then reduces to

$$T_{\mathrm{brc}} = T_{\mathrm{aeroc}} \, , \tag{10.36}$$

and

$$T_{\mathrm{br1}} = T_{\mathrm{aero1}} \, . \tag{10.37}$$

The relation between the rotor angular velocity under calibration conditions Ω_{Rc} and the flow rate Q which is obtained experimentally during calibration, obeys (10.36), where T_{aeroc} is given by (10.17), or (10.20), and Q by (10.12). This leads to the following relation between the rotor angular velocity, Ω_{Rc}, and the flow rate Q under calibration conditions:

$$\Omega_{\mathrm{Rc}} = \frac{QF_{\mathrm{T}}}{B_{\mathrm{br}}} \tag{10.38}$$

The relation between the K factor commonly quoted for linear flow meters - see e.g., Baker (2000) - and the quantities F_{T} and B_{br} introduced in this work is evident on inspection. We wish to redraw the reader's attention to the point made earlier, viz. that F_{T} cannot be strictly a constant, even for a specific flow meter.

The corresponding relation between the departure from the calibration state and the rotor angular velocity under measuring conditions are contained in (10.29, 10.37). We postpone employing this relation to get an estimate of the error in metering to the next section in this chapter and note the following for the present: Given profiles of the axial and azimuthal velocity components, u_{x} and u_{ϕ} entering the turbine flow meter at section A-A (Fig. 10.1) under measurement conditions, the components v_{x} and v_{ϕ} at the blade inlet section may be calculated if geometrical design details of the turbine flow meter are known, and which, in turn, leads to Ω_{R1} and Q_1, which are the quantities sought.

10.3 Experiments and Data Reduction

Experiments were conducted in a specially designed facility in which the turbine flow meter could be exposed to an oncoming flow with swirl. Swirl over a wide range of the parameter ϵ_{S} could be generated in this facility by a rotating tube bundle, identical in type and dimensions with that described in Rocklage-Marliani et al. (2003). The direction of rotation of the swirl generator could also be reversed, so that both positive and negative values of the parameter ϵ_{S} could be realised. Provisions were made in this facility for an independent measurement of the flow rate. The latter was regarded as the reference measurement for the present purpose, denoted Q_{ref}. For a description of this facility the reader is referred to Chaps. 4, 8. The uncertainty of the flow rate in the reference measurement is $\pm 0.3\%$. The flow rate "read" by the turbine flow meter, Q_{meas}, together with the reference measurement thus enabled a plot of the error "read" by the turbine flow meter vs. the swirl parameter ϵ_{S} to be obtained experimentally.

A schematic diagram of the facility is shown in Fig. 10.3. It is essentially a pipe 100 mm inside diameter and 120 diameters long, at the downstream end of which the turbine flow meter is installed. In this facility, the swirl generator could be mounted at a certain number of chosen locations along the pipe upstream of the flow meter. The evolution characteristics of both the mean and fluctuaing motion in the flow generated by this device over the range of flow parameters (Reynolds number and swirl number) of interest in the present work were ascertained through extensive experiments conducted in a separate facility, see Rocklage-Marliani et al. (2003). The turbine flow meter could thus be exposed to an oncoming flow with swirl whose distribution characteristics along the radius at the flow meter inlet section were known.

For purposes of later reference in Sect. 10.4 we wish to note here a salient property of the distribution characteristics of the azimuthal velocity profiles at different streamwise locations within the measurement regime. It is that the azimuthal velocity profiles essentially scale with the angular velocity of the swirl generator, ω_S, i.e., they can be written as $(u_\phi)/(r\omega_S) = f_\phi(x/R, r/R)$. In the core region with a solid-body like rotation the function f_ϕ is independent of r, however slowly changing with respect to x.

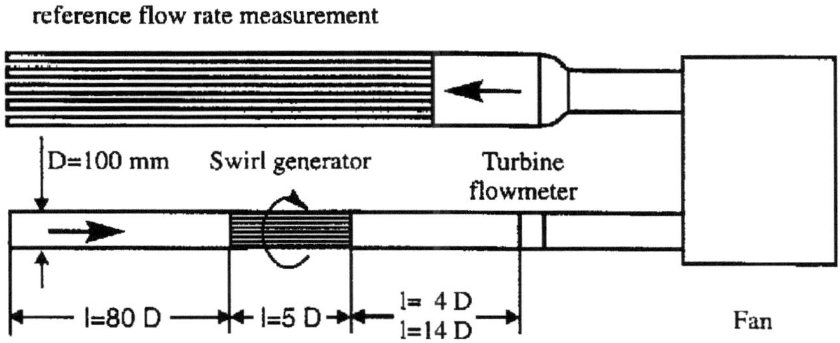

Fig. 10.3. Schematic diagram of experimental facility.

This facility was operated with the flow rate set at a certain level and the angular velocity of the swirl generator varied over a wide range of parameters. The flow rate "read" by the turbine flow meter, Q_{meas}, is the value read off from the calibration curve of the flow meter corresponding to the angular velocity of the flow meter rotor. The reference measurement of the flow rate, Q_{ref}, which is done independently, then enabled the error to be evaluated.

In order to set the experimental outcome against the theory meaningfully, the error evaluation from the measurement data should correspond in spirit to the error definition in the theory. In the present context, satisfying this requirement, which is generally considered too obvious to deserve any special reference to, is not as straightforward as it may seem to be at a first glance. The reason is that the theory deals only with such disturbances traceable to swirl in the oncoming flow, whereas in the experiment the flow meter "perceives" all the disturbances that are present regardless of their origin. The scope of the theory is clear on examining the mathematical formulation in which is the disturbance in both the axial and azimuthal velocity components is characterized through one and the same parameter ϵ_S. In the experiment on the other hand, the departure from the calibration state "perceived" by the flow meter, depends upon the method of disturbance generation. The question of parametrical characterization of the disturbance in the experiment has therefore to be addressed. In particular, characterization through a single parameter may not be feasible. The disturbance caused by a stationary honeycomb as in the experiments reported in Chap. 4 or when the tube-bundle of our present is not set in rotation illustrate this point. In short, it is quite possible to create a departure from the calibration state in the axial velocity component only without a inducing a corresponding azimuthal component, a case clearly not encompassed by the present theory. In order to set the present experimental outcome against theory, it is therefore necessary to evaluate the measurement data according to a procedure that separates out the error traceable to swirl.

In summary, for a proper comparison between experiment and theory, therefore, the following two points need to be addressed:

- Definition of error.
- Data reduction procedure for extraction of the error due to swirl.

10.3.1 Definition of Error

The theory presented earlier relates changes in the flow rate and in the angular velocity of the flow meter rotor to departures of the velocity profiles in the oncoming flow from their calibration state. Within this framework, two distinct approaches of ascertaining the error can be identified.

- The first is the change in angular velocity experienced by the flow meter rotor at a certain flow rate. This corresponds to keeping track of Ω_{R1} under the integral constraint (10.24).
- The second is evaluating the change in the flow rate at $\Omega_{\mathrm{R1}} = 0$. This corresponds to evaluating the integral on the left hand side of (10.24). Defining the error as $(Q_{\mathrm{meas}} - Q_{\mathrm{ref}})$ corresponds to an experimental measurement of this integral.

The second approach is the simpler of the two from the point of view of conducting the experiment in our setup in which there was no provision made for a direct and independent measurement of the rotor angular velocity.

10.3.2 Extraction of Error Due to Swirl

The quantity $(Q_{\mathrm{meas}} - Q_{\mathrm{ref}})$ contains, as stated earlier, the total error due to the disturbance actually present in the experiment. In particular it includes the error caused by the departure of the axial velocity profile from the calibration flow in the experimental configuration in question. Strictly speaking, the departure in this quantity caused by a rotating tube bundle at any particular streamwise location would depend upon the angular velocity of the tube-bundle, and need not necessarily be the same as with the tube bundle stationary. However, the experiments of Rocklage-Marliani et al. (2003) indicate that, in the flow in our experiment the Reynolds-averaged axial velocity profile is relatively insensitive to tube-bundle rotation. The main source of error for the flow meter reading is the azimuthal velocity profile. Therefore an estimate for the contribution to the error due to swirl may be obtained by subtracting out the error measured when the swirl generator is in its place but with the tube-bundle not set in rotation. We denote this quantity by the subscript 0, i.e., as $(Q_{\mathrm{meas}}-Q_{\mathrm{ref}})_0$.

A simple procedure for extracting the error caused by swirl is then forming the difference $(Q_{\mathrm{meas}} - Q_{\mathrm{ref}})$ - $(Q_{\mathrm{meas}} - Q_{\mathrm{ref}})_0$. The error ϵ_Q evaluated as

$$\epsilon_Q = \frac{(Q_{\mathrm{meas}} - Q_{\mathrm{ref}}) - (Q_{\mathrm{meas}} - Q_{\mathrm{ref}})_0}{Q_{\mathrm{ref}}} \tag{10.39}$$

has been plotted against the rotation number, *Ro*, in Fig. 10.4. The straight line drawn is one that satisfies the least square error criterion through the measured data.

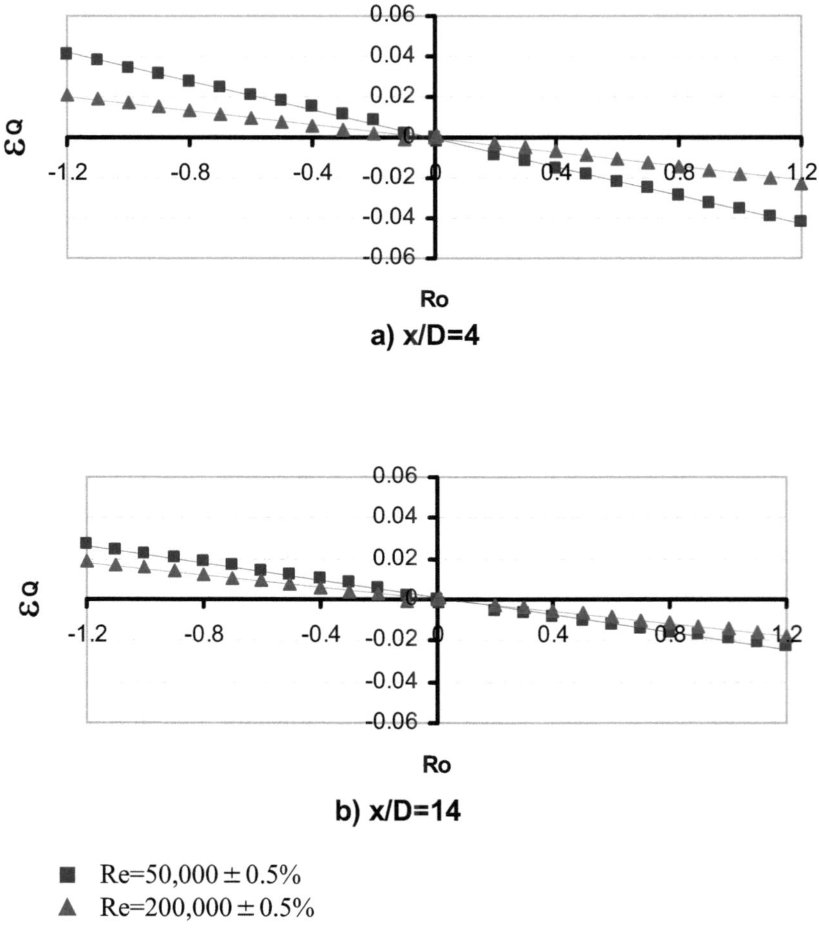

Fig. 10.4. Error in flow metering.

10.4 Discussion

The results of the analysis in Sect. 10.2 show that the output signal of the turbine flow meter, and hence the error in reading, depends not only upon details of the radial distribution of the disturbance but also on geometrical design features peculiar to a particular turbine flow meter, viz. F_{meter1} and F_{meter2}. These quantities also enter the problem not by themselves but as products integrated across the blade height. Furthermore, the design data are rarely, if ever, available in sufficient detail to the user, which makes it a virtually hopeless undertaking to try to formulate error characteristics of turbine flow meters in universal terms.

Against the background of this limitation, the task of comparing the error measured in the experiment with the theory presented can be charted out to proceed along several conceptually different routes. The most immediately obvious choice would be to set the measured quantity $(Q_{\text{meas}} - Q_{\text{ref}})$ against the corresponding quantity evaluated from theory. Such a conceptually simple and straightforward route of comparison, although meriting consideration for the study a particular design as such for which the relevant design data are available, does not entirely fall within the scope of the present work. The reason is that such a study of a single flow meter diverts attention away from the main objective set for the present work, which is to illuminate the gross characteristics of the behavior of swirl-induced error in flow-rate measurement through turbine flow meters. This purpose is better fulfilled by the alternative route of examining the "scaling characteristics" of the error, on which we focus our attention in the following.

For this purpose it is only necessary to note an outstanding point made by the theory that the relation between the error and the strength of the disturbance of a given shape in the oncoming flow is a linear one. The strength of the disturbance used in the theory is characterized by ϵ_{S}, and a suitable measure for the disturbance in the experiment which, incidentally is primarily in the azimuthal velocity, is the rotation number Ro. The experiments of Rocklage-Marliani et al. (2003) indicate that at any particular location downstream of the swirl generator within the region $x/D < 14$, the azimuthal velocity distribution scales with the rotation number Ro. The shape of the azimuthal velocity distribution itself is however dependent on x/D. For this reason, the linear relationship between the error ϵ_{Q} and the swirl parameter ϵ_{S} may be expected to be carried over to the pair ϵ_{Q} and Ro, and the plot in Fig. 10.4 shows this indeed to be the case. The magnitude of the error itself at any particular location is primarily dependent upon the radial distribution of the azimuthal velocity at this location, and the differences in the slope of the ϵ_{Q} vs. Ro plot are a manifestation of this property of the error. The plot also brings out the feature expectable on physical grounds, viz. the change in sign of error with respect to the direction of the swirl in the oncoming flow.

10.5 Conclusions

The main conclusion that can be drawn from the present study regarding the behavior of swirl-induced error in metering by a turbine flow meter is as follows: The error shows a linear relationship with respect to the parameter characterizing the strength of the disturbance in the oncoming flow, but the actual magnitude of the error depends both upon the radial distribution of the disturbance in the oncoming flow and upon geometrical design features of the flow meter.

References

Baker, RC (2000) Flow Measurement Handbook. Cambridge University Press, Cambridge UK

Baker, RC (1991) Turbine and related flowmeters: I. Industrial practice. Flow Meas Instrum 2:147–162

Baker, RC (1993) Turbine flowmeters: II. Theoretical and experimental published information. Flow Meas Instrum 4:123–144

Batchelor GK (1967) An Introduction to Fluid Dynamics. Cambridge University Press, Cambridge UK

Leishman JG (2000) Principle of Helicopter Aerodynamics. Cambridge University Press, Cambridge UK

Marshall JS (2001) Inviscid Incompressible Flow. John Wiley, New York

Rocklage-Marliani G, Schmidts M, Vasanta Ram VI (2003) Three-dimensional laser-Doppler velocimeter measurements in swirling turbulent pipe flow. Flow, Turbulence and Combustion 70:43–67

Schlichting H, Gersten K (2000) Boundary Layer Theory. Springer, Heidelberg

Schlichting H, Truckenbrodt E (1962) Aerodynamik des Flugzeuges, Erster Band. Springer, Heidelberg

Schmidts M, Marliani G, Vasanta Ram VI (1998) A study of the metering error of turbine flow meters caused by swirl in the flow. FLOMEKO 98, Proceedings:399–403

Van Dyke M (1975) Perturbation Methods in Fluid Mechanics, Annotated edition. The Parabolic Press, Stanford California USA

11

Investigation of Unsteady Three-Dimensional Flow Fields in a Turbine Flow Meter

Ernst von Lavante

Institut für Strömungsmaschinen, Universität Essen, 45117 Essen, Germany

Three-dimensional internal flow in a complete stage of a modern turbine flow meter was numerically simulated in order to investigate the flow field in general, and possibilities of improving the accuracy of the meter in particular. The governing equations of the three-dimensional, compressible, viscous ideal gas flow were solved numerically using two simulation programs, the commercially available Fluent and the academic code ACHIEVE. The resulting flow fields, including the details of various flow effects, are discussed. By close consideration of the flow fields, the so-called pressure shift can be explained as an effect depending on the Reynolds number.

11.1 Introduction

The requirements imposed on the measuring systems used in global natural gas market are very high. One flow meter that fulfills the high requirements and is capable of measuring very accurately even under high pressures in field use is the turbine flow meter.

In the past, most of the investigations of turbine flow meters were done analytically and experimentally. Several authors focused on the development of an equation of motion of the meter by relating the driving torques of the fluid to the friction torques due to friction of the fluid on the hub, in the tip gap or in the bearing (Baker, 1993; Lehmann, 1990; Lee and Karlby, 1960). Many investigations (Brümmer, 1997; Lehmann, 1990; Mickan et al., 1996) studied the effect of pulsations or profile deformations on the accuracy of the meter.

In the last two decades research was done on self-correcting and self-checking gas turbine meters. One working principle is based on the coupling of two rotors. A sensor rotor small distance downstream from the main rotor senses and responds to changes in the exit angle of the fluid leaving the main rotor (Lee et al., 1982; Lee and Karlby, 1960).

Another working principle assumes two decoupled rotors, where the second, the so-called reference rotor, works independently of the first stage. The resulting two-

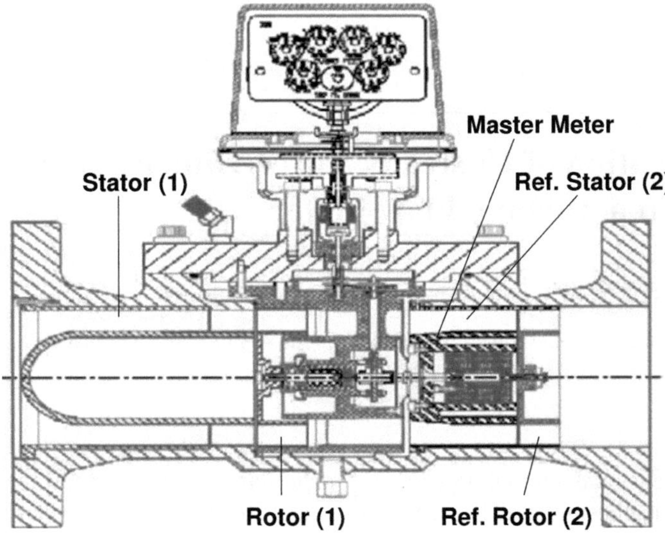

Fig. 11.1. Cutaway Drawing of AccuTest^{TM} Turbine Flow Meter.

stage turbine meter AccuTest^{TM} (Fig. 11.1) has been developed and investigated in recent years experimentally and analytically by Schieber (1998, 2000).

In this work, the flow field in an 6" AccuTest^{TM} turbine meter has been investigated numerically. To this end, the flow in the two-dimensional blade-to-blade plane consisting of both stages has been simulated first, proving the feasibility of the present study. Consequently, the unsteady three-dimensional flow field has been investigated in the first stage of the meter, aimed to describe the main flow effects occuring in the meter and to find promising design modifications to increase the accuracy of the meter. This chapter will concentrate on the three-dimensional results, aiming at a detailed discussion of the flow features in the flow field typical of modern turbine flow meters.

11.2 Theory of Operation

The metering principle of the turbine flow meter is based on a volumetric measurement. Here, parts of the fluid with a certain defined volume given by the geometry of the meter are integrated in time and the transported amount of fluid is displayed (Bonfig, 1997).

The turbine meter is an inferential flow meter meaning that a certain amount of fluid is not directly measured but related to the frequency of a rotor that is impelled by the flow through the meter. The frequency f of the rotor is measured and related to the volume flow rate by the definition of the so-called k factor. The k factor is defined as the ratio of the total number of output pulses to the actual volume of gas that passes through the meter. When considering same time periods, k becomes the ratio of the output frequency over the actual flow rate $\dot{Q}$:

$$k = \frac{f}{\dot{Q}} \tag{11.1}$$

The k factor is a function of the operating condition (pressure, temperature and density of the fluid) and the physical dimensions of the meter. Its value is determined in calibration tests. Figure 11.1 shows a cutaway drawing of the AccuTest^{TM} turbine flow meter that was investigated in this project.

The design of the flow meter consists of two parts. The front part, with the gas entering from the left, is a slight modification of an existing, conventional meter. The second part is a so-called master meter, designed to achieve the self-proving feature of the meter. It is a separate free-running turbine meter cartridge; its blade wheel rotates in opposite direction to the first rotor. Rotor speed is measured with an electronic sensor that detects the rotor blade tips as they pass the proximity of the sensor. The function of the master meter is to evaluate the accuracy of the main meter.

Schieber (1998) derived a working equation of the two-stage turbine meter. He defines the Accuracy of the meter Acc as the ratio of the measured volume flow rate $\dot{Q}$ and the actual volume flow rate $\dot{Q}_{\text{norm}}$. Under the assumptions that the accuracy of the reference rotor is equal one, the bulk flow is one-dimensional, inviscid, incompressible, and steady in time, and the volume flow rate at the main rotor equals the volume flow rate at the reference rotor, one obtains for the working equation:

$$Acc_{\text{SPM}} = \frac{f_{\text{main}}}{f_{\text{ref}}} \frac{k_{\text{ref}}}{k_{\text{main}}} Acc_{\text{cal}} \tag{11.2}$$

where SPM denotes the self proving meter and Acc_{cal} ist the accuracy determined during the calibration. The subscripts main and ref describe the first and second stage of the turbine meter, respectively.

In order to make 11.2 valid, it is assumed that the volume flow rate inferred from the reference rotor is the actual flow rate, the rotors are completely fluiddynamically decoupled, factors influencing the performance of the upstream main rotor will not influence the downstram reference rotor and the master meter performance will not change from initial calibration after extended use in the field.

One major benefit of the two-stage turbine meter over a conventional turbine meter is a reduced need for inspection and maintenance, requiring only one person with a laptop or a so-called electronic flow computer. Furthermore, the meter performance is monitored for immediate detection of accuracy changes and the effects of flow disturbances. Installation effects are weaker due to an improved guide vane design of the first stage.

In order to optimize the design and accuracy of the two-stage turbine meter, the flow field inside the meter was analyzed. Since it is rather difficult to investigate the flow in the meter experimentally, it was decided to simulate the flow numerically using two computer programs for comparison. These two programs are described briefly in the following part 3 of this chapter.

11.3 Numerical Tools: Fluent and ACHIEVE

Nowadays, several computer programs to simulate the 3-dimensional, viscous, unsteady and turbulent flows by solving the Navier–Stokes equations are available. Not all of them are capable of handling unsteady stator/rotor interactions correctly, however. In the present work, the commercial solver Fluent (1998) has been chosen due to its advertized capability to handle the case under investigation. One problem that most commercial solvers have is a high amount of numerical dissipation, which is implemented in order to enhence the robustness of the code. This implies that since the effective Reynolds number could be up to one or more orders of magnitude lower than in the reality, some viscous flow effects could not be modeled accurately. For this reason, the more accurate code ACHIEVE (Algorithm for Solving Chemically Reacting Viscous Flow Problems) (Kallenberg, 1999; Perpeet, 2000; Yao, 1997), also capable of calculating the flow field in the given turbine meter, has been used at the same time. ACHIEVE has been developed at several academic institutions around the world under the guidance of the author, and has, due to its formulation, inherently smaller numerical dissipation than Fluent. Both codes are compared in the following Table 11.1.

Table 11.1. Features of the two solvers.

ACHIEVE	Fluent
Roe's flux difference splitting in finite volume form	segregated/coupled implict solver
two-stage modified Runge–Kutta time stepping	second order implicit method in pseudo-time
k-ω-turbulence model	k-ϵ-turbulence model
compressible flow	compressible flow
ideal gas law	ideal gas law
structured grid	mixed grid(structured in b.l., unstructured in far-field)
multiblock structure; parallel computation on a PC cluster	multiblock structure; computation on a single CPU

11.4 Three-Dimensional Simulations: Geometry and Setup

In the first part of this section, the 3-dimensional model including the corresponding computational grid of the turbine flow meter will be explained. In the next section, the results of the simulation including the stator/rotor interaction, flow in the tip gap and the influence of boundary layers on the angel of incidence β_{i} will be discussed.

In this work, the influence of boundary layer effects and the flow in the tip gap were to be investigated, making a high grid resolution close to the walls essential. In order to achieve a reasonable number of cells in the computational grid it was decided to simulate the flow in the first stage only. Since the first stage closely resembles the convential turbine flow meters, the current analysis of the present results should be of general benefit. Besides, the geometry of the first stage is more complicated than that of the second stage because of the central body (hub) in the inlet section of the guide vanes of the first stage.

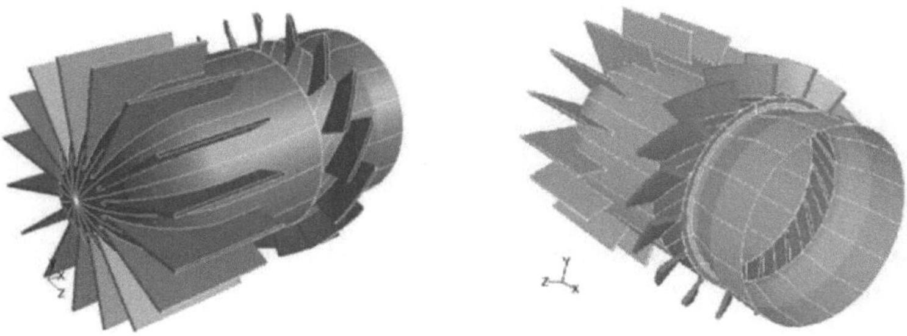

Fig. 11.2. Isometric view of the first stage of the turbine flow meter; *left*: front view looking downstream; *right*: view from the back looking upstream.

Figure 11.2 shows an isometric view of the geometry of the first stage of the turbine meter, as seen from the front (inlet) and from the back (outflow), assumed in the present simulations and plotted without the outer casing of the meter.

The grid had approximately $1,000,000$ cells, covering all relevant boundary layers at hub, casing and the blades. The flow enters the meter axially in the positive x-direction (see Fig. 11.2). It passes at first the central body (hub) with guide vanes of the first stator. Further downstream it flows through the rotor. In the present simulation, the speed of rotation of the rotor is a predetermined input, being initially specified. It was obtained from corresponding experimental data. Only one blade passage of the guide vanes and the rotor have been simulated to save computational time. Therefore, the boundaries between each passage had to be treated as periodic and the incoming velocity profile was assumed to be symmetric fully developed turbulent flow. In the actual meter, there are 16 blades in both the rotor and the stator, making the assumption of periodicity possible. In the display of the results, the computed flow field has been projected to the remaining 15 passages.

Figure 11.3 shows the structured grid in the one-passage model configuration.

In circumferential direction 58 grid points resolved the blade to blade plane. An exponential distribution close to the wall provides very small distances of the first cells to the wall so that at least 10 points normal to the wall were within the boundary layer. In radial direction 51 grid points were distributed over the height of the blade with exponential distributions close to the walls. 15 grid points were equally spaced in the 0.8 mm heigh tip gap.

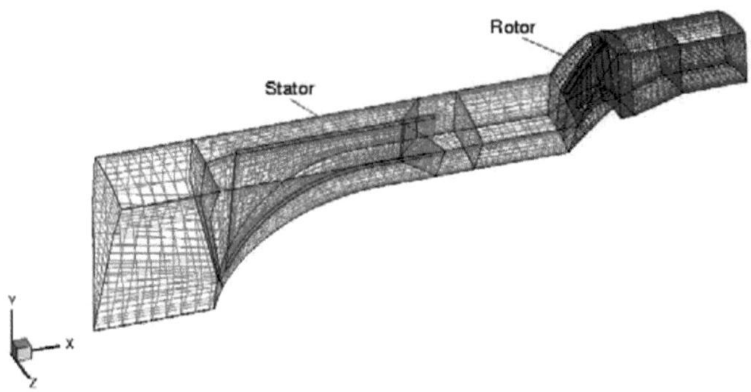

Fig. 11.3. Structured grid of the model configuration.

The three-dimensional Fluent simulations were performed on a 800 MHz Linux-PC with a 768 MB RAM disk. The simulation of one operating point until a time periodic state is reached, took about 3–4 weeks to perform. In addition to the excessive CPU time, the static pressure at the outflow boundary had to be adjusted iteratively in order to obtain the required mass flow.

11.5 Discussion of Results

Four different operating points of the meter were simulated to study the 3-dimensional flow effects. The influence of different flow capacities on the flow field in the meter was investigated by simulations at 10%, 50% and 100% of the maximum flow capacity $\dot{Q}_{\mathrm{max}}$ at an operating pressure of 1 bar. The influence of varying pressure was taken into consideration by the simulation at 10% of maximum capacity and a pressure of 10 bar. The flow at 1 bar and 100% capacity corresponds to the approximately same Reynolds number as the simulation at 10 bar and 10% of maximum capacity. The Reynolds number, calculated based on the incoming velocity of the meter and the diameter of the turbine meter is $Re = 150,000$. Due to the fact that the contour plots at all investigated operating points look very similar, the figures in this chapter concentrate on the simulation at $p = 1$ bar and $\dot{Q} = \dot{Q}_{\mathrm{max}}$. The global range of the computed velocities in the entire first stage was between $12\,\mathrm{m/s}$ and $42\,\mathrm{m/s}$.

As the flow enters the guide vanes the cross-sectional area is reduced by the central body, accelerating the flow as indicated in Fig. 11.4 by the decrease of pressure along the central body. Here, the static pressure on the walls of the entire stage is shown.

A stagnation point develops at the front of the central body. A video animation of the time-dependent static pressure on the walls revealed that there is an influence

Fig. 11.4. Static pressure contours on the walls of the entire stage.

of the rotor-stator interaction on the flow in the guide vanes all the way upstream to the stagnation point. The pressure at the stagnation point changed periodically with the blade-passing frequency. An examination of the flow around the central body showed that no flow separation occured at this part of the turbine flow meter.

As seen in Fig. 11.4 as a dark line on the rotor leading edge, a stagnation point develops at this location. Close to the hub and close to the tip the stagnation pressure decreases due to the reduction of the total pressure. This is due to the influence of the boundary layer that causes lower velocities close to the wall, resulting in smaller values of stagnation pressure. Immediately downstream of the leading edge the flow accelerates, giving rise to a high risk of flow separation in the subsequent deceleration zone. A region of relatively low pressure develops on the suction side close to the trailing edge indicating a risk of flow separation as would be typical for turbine flow.

In Fig. 11.5 the static pressure on the rotor versus the non-dimensionalized length of the rotor is shown in three different radial planes for two time instants. The planes under consideration are described by the percentage height h of the rotor, where zero height is at the hub.

The left graph in Fig. 11.5 shows the static pressure at the time $t = 0\,T_1$ when the rotor leading edge is directly behind a guide vane. The pressure at the stagnation point ($x/L = 0.01$) reaches the greatest values in the 50% h-plane. Here, the boundary layer influences the pressure at the stagnation point since the velocity in the boundary layer approaches zero at the wall. In the 5% h-plane, pressure on the suction side close to the leading edge is greater than on the pressure side. This unfavorable effect is caused by the wake of the stator that reduces the axial velocity thus increasing the incidence angle. At the trailing edge, the pressure in the 50% h-plane reaches again the greatest values.

The pressure distribution in the same planes of constant radius for the time when the leading edge of the rotor is in the middle between two guide vanes of the stator, $t = 1/2\,T_1$, is shown in the right graph in Fig. 11.5. Similar flow effects as at the

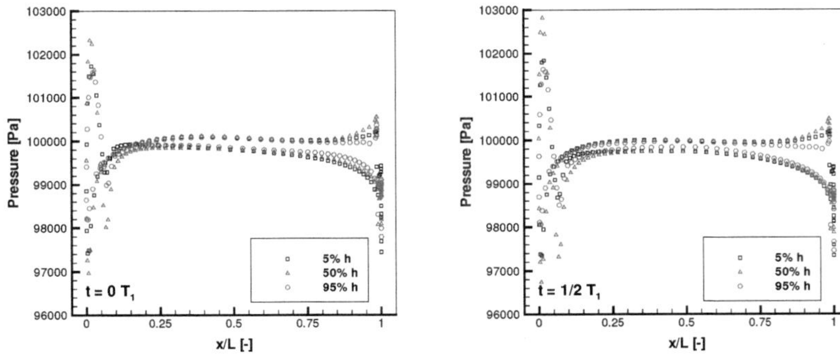

Fig. 11.5. Static pressure on the rotor in different radial planes for two rotor positions relative to the stator.

time $t = 0\,T_1$ can be found for this rotor position. At $t = 1/2\,T_1$, the wake of the stator does not influence the flow at the leading edge of the rotor so that the incoming flow to the rotor is more uniform. The incidence angle does not change significantly between the hub and the tip.

The vortex formation and pressure downstream of the trailing edge vary in radial direction. In the plane close to the tip ($h = 95\%$) of the blade no vortex develops at the top side of the flap. In addition to the boundary layer effects, the influence of the tip gap affects the flow at this location.

A quantitative analysis of the influence of the boundary layer at the hub and the casing on the inlet and outlet angles in the rotor was acomplished by plotting the mass averaged angle $(\beta_1 - \beta_\mathrm{s})$ at the inlet and $(\beta_2 - \beta_\mathrm{s})$ at the outlet of the rotor versus the non-dimensionalized height of the rotor for the two times $t = 0\,T_1$ and $t = 1/2\,T_1$ considered here.

For both times the angle $(\beta_{1,2} - \beta_\mathrm{s})$ is in the range between 3°–5° for a blade height of 30%–70%. Close to hub, $h < 30\%$, and casing, $h > 80\%$, $(\beta_{1,2} - \beta_\mathrm{s})$ increases up to values of approximately 14°. The influence of the boundary layer at the hub is stronger than at the casing because the tip gap reduces the above effects.

In the design of the AccuTest^{TM} turbine flow meter, the radial distribution of the stagger angle β_s is based on the Euler theory. The stagger angle β_s is proportional to the rotational speed, being a function of radius and angular velocity. No friction effects were taken into consideration. In reality, friction influences the flow.

This phenomenon causes the variation of the reduction of the incidence angle in the presence of the boundary layer, shown in Fig. 11.6 (right). The axial velocity in the boundary layer is smaller than in the case without friction. Keeping the rotational speed constant for both axial velocities yields an increased angle of incidence in the boundary layer.

No significant influence of the wake from the stator on $(\beta_{1,2} - \beta_\mathrm{s})$ can be seen at $t = 0\,T_1$ due to the mass averaging. It is easy to realize, however, that smaller

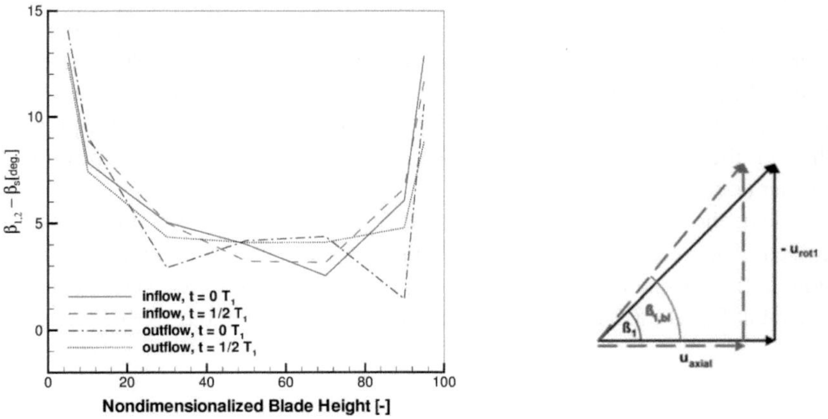

Fig. 11.6. Inflow and outflow angles for two different rotor positions.

axial velocities in the wake of the stator at $t = 0\,T_1$ would increase the local angle $(\beta_{1,2} - \beta_s)$.

The effect of the radially varying incidence angles could be reduced by design of the rotor, but for the given application, this procedure would not be practical.

Concentrating now on the flow in the tip gap, Fig. 11.7 should be consulted, showing density contours and streamlines of absolute velocity in three axial planes ($x = -0.01$, $x = 0.00$ and $x = 0.01$) of the meter.

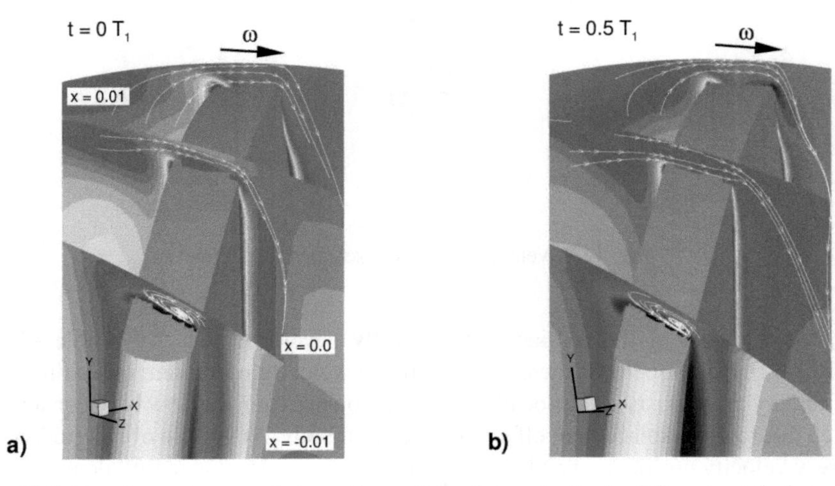

Fig. 11.7. Density contours and streamlines of absolute velocity in different axial planes of the first rotor at two different positions relative to the stator. **(a)** behind the stator, **(b)** in the middle of the channel between two stator vanes.

In the plane close to the leading edge ($x = -0.01$), a vortex rotating counterclockwise originates in the tip gap due to the superimposition of the incoming flow and the flow in circumferential direction imposed by the rotation of the blade. Further downstream, in the middle of the rotor ($x = 0.00$), the pressure difference between the pressure and the suction side of the rotor becomes dominant as the driving force in the blade. Consequently, the streamlines indicate flow from the pressure side through the tip gap to the suction side of the rotor. This behavior does not change further downstream.

Next, the evolution of the axial velocity in the turbine flow meter will be discussed in more detail. In order to obtain an overview of the axial evolution of the flow in the turbine meter, the axial velocity profiles in selected axial planes are displayed in Fig. 11.8. The contours range from a velocity of $0\,\mathrm{m/s}$ to $42\,\mathrm{m/s}$.

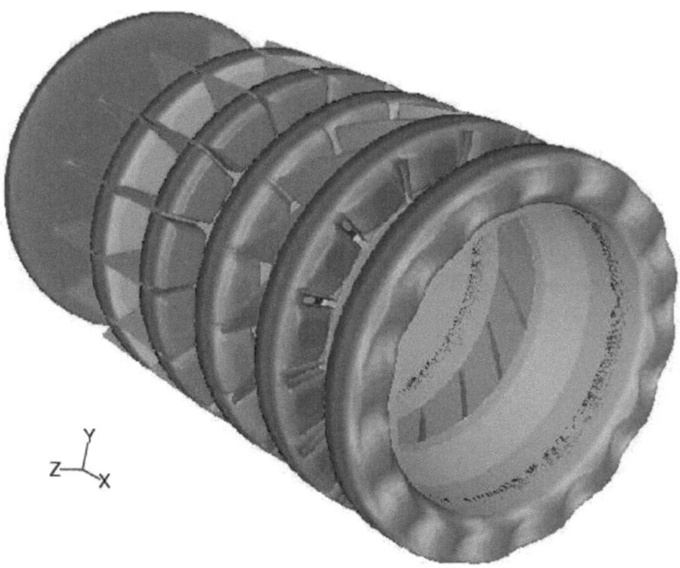

Fig. 11.8. Velocity profiles in selected axial planes.

At the inlet of the turbine meter, a fully developed velocity profile ($Re = 150,000$) can be seen. At the casing, there is a velocity minimum, indicating that a boundary layer exists at this location. Further downstream, the flow has been accelerated due to the displacement effect of the central body. At the end of the guide vanes every velocity profile in one blade section looks like a developed turbulent profile in a rectangular channel flow with boundary layers at the solid surfaces. Downstream of the guide vanes, the wakes behind the guide vanes are being mixed out by natural dissipation, making velocity distribution more uniform.

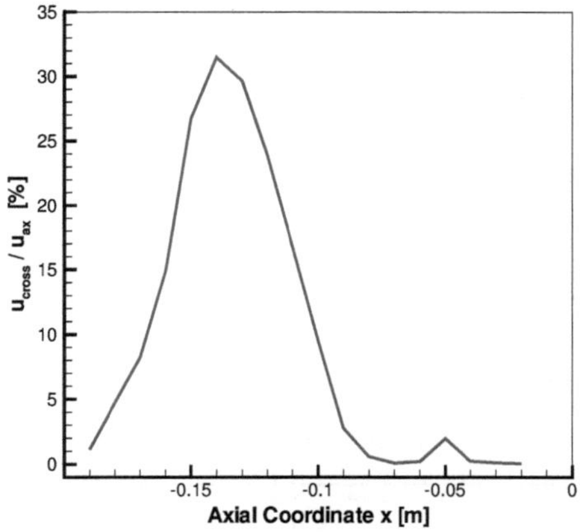

Fig. 11.9. Radial velocity component in the turbine stator.

Just upstream of the rotor, the reduction of the axial velocity in the wakes is smaller but still noticeable. They cause the coefficient of moment c_{m} of the first rotor to vary in time. At the outflow from the rotor, the flow has again been accelerated due to a reduction of the cross-sectional area. A wake forms behind the trailing edge of the rotor. While the flow is convected downstream, the velocity profiles rotate in negative ω-direction because the deflection of the flow in the rotor generates circumferential velocity components. At the outlet of the present configuration, the velocity profile has already become more uniform due to the mixing taking place.

In order to analyze the described flow phenomena quantitatively, Fig. 11.9 shows the radial velocity component, non-dimensionalized by the axial velocity, versus the axial coordinate in the stator.

At the inlet of the meter no radial velocity component can be detected since the flow was assumed to be undisturbed. When the flow passes the central body of the first stator, the flow in the center of the meter follows the geometry of the central body. Therefore, the radial flow component was originated and reached up to 32% of the axial velocity. Downstream of $x = -0.14\,\mathrm{m}$, the radial flow components are reduced by the shape of the guid vanes, so that just at the end of the guide vanes $(x = -0.03\,\mathrm{m})$, no significant radial velocity component can be detected. At $x = -0.05\,\mathrm{m}$, a small radial velocity component is generated by the wake of the stators, but is dissipated later within a short distance.

The wake flow of the stator is analyzed in more detail by plotting the axial velocity in the midplane of the meter versus the circumferential direction of one blade section in 5 axial planes that represent the region between the stator and the rotor.

In all planes the influence of the wake is recognized as a velocity profile deformation. In the plane immediately behind the guide vanes at $x = -0.06\,\mathrm{m}$ the minimum velocity in the wake is only $9\,\mathrm{m/s}$ whereas maximum velocities of up to $43.5\,\mathrm{m/s}$ are reached. Closer to the rotor the minimum velocity in the wake increases due to

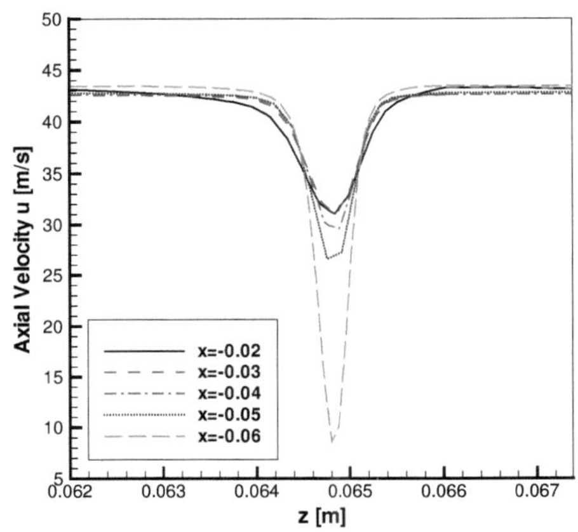

Fig. 11.10. Axial velocity (midplane) in the wake of the stator.

mixing up to 31 m/s. This non-uniform velocity profile is the inlet profile of the rotor and yields a variation of the coefficient of moment in the rotor with time.

11.6 The Pressure Shift

One of the main reasons for the present investigation was the explanation of the so-called pressure shift in the accuracy of the turbine flow meters. This expression denotes an unfavorable change of accuracy with increased pressure of the metered gas.Suspecting that the cause of this effect is a change of the flow in the rotor, it was decided to study the deflection angle as a function of varying flow rate and pressure.

Simulations of the flow were performed between 10% and 90% of the maximum flow capacity at low pressure (1 bar) and high pressure(10 bar). The angle of incidence was chosen to remain constant at $\beta_\mathrm{i} = 2.5°$. It should be obvious that the deflection angle $\Delta\beta = \beta_2 - \beta_1$ must be a nonlinear function of the flow rate $\dot{Q}$ as long as the flow rate is proportional to the axial velocity. For increasing flow capacities $\dot{Q}$ at an operating pressure of 1 bar the deflection angle $\Delta\beta$ decreases until approximately half the maximum flow capacity was reached. For higher flow rates it remained approximately constant. At a pressure of 10 bar, the deflection angle $\Delta\beta$ was approximately constant for every investigated flow capacity.

Plotting the deflection angle as a function of the Reynolds number (Fig. 11.11), a continuous line of a dependence of the deflection angle on the Reynolds number can be found. This is because the maximum flow capacity at 1 bar corresponds to the same Reynolds number of the minimum flow capacity at 10 bar, and the deflection angles are in both cases approximately 5°. A critical Reynolds number Re_crit, describing the smallest Reynolds number for which the deflection angle $\Delta\beta$ remains constant can be given as approximately $Re_\mathrm{crit} \approx 100,000$.

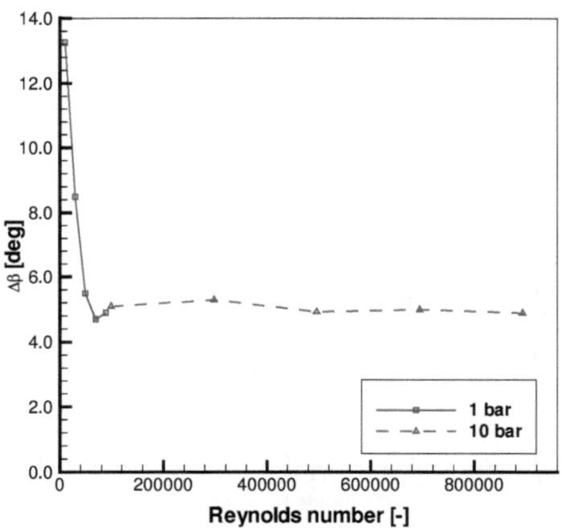

Fig. 11.11. Deflection angle $\Delta\beta$ of the flow through the rotor versus Reynolds number.

The coefficient of moment c_m of the rotor is defined as:

$$c_\mathrm{m} = \frac{M_\mathrm{pr} + M_\mathrm{visc}}{\frac{1}{2}\,\rho_\mathrm{ref}\,c_\mathrm{ref}^2\,A_\mathrm{ref}} \tag{11.3}$$

where M denotes moments due to viscous (visc) and pressure (pr) forces. The subscript ref denotes a reference state defined as the inflow conditions. A_ref is the area of the rotor blade calculated by the product of chord length and height of the blade.

Figure 11.12 shows the dependency of the coefficient c_m on the Reynolds number, displaying a behavior similar to the deflection angle in two dimensions. c_m first

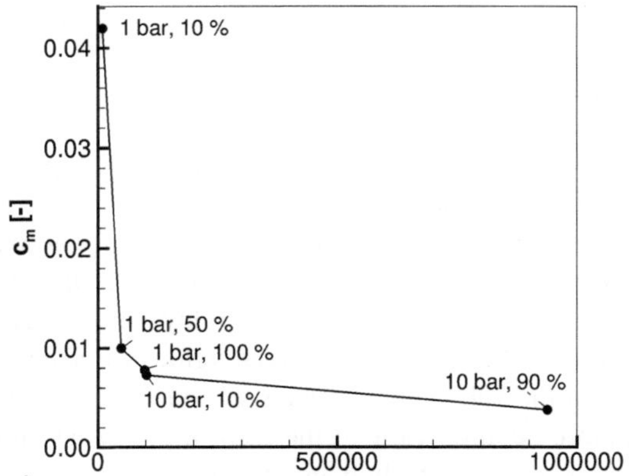

Fig. 11.12. Coefficient of moment from three-dimensional simulation.

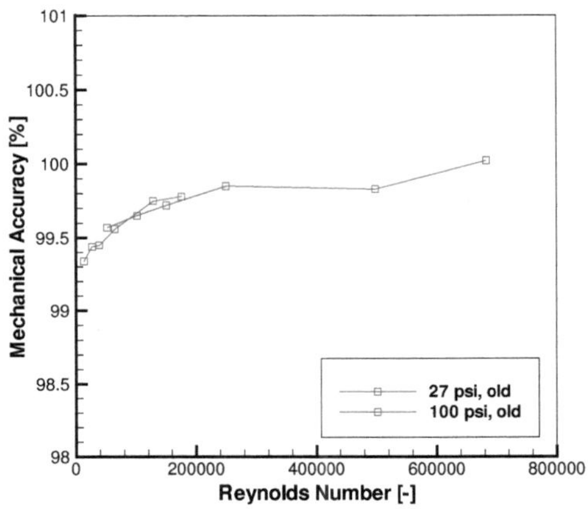

Fig. 11.13. Mechanical Accuracy Acc versus Reynolds number.

decreases for smaller Reynolds numbers, then seems to reach a constant value for Reynolds numbers above a certain critical value. The coefficient of moment determined at 100% flow capacity and 1 bar pressure is very close to that at approximately the same Reynolds number for 10% flow capacity and a pressure of 10 bar.

Recently, experiments were performed to test a design modification of the meter (Schieber, 2000). In Fig. 11.13, the mechanical accuracy Acc of the meter as obtained in experiments is plotted versus the Reynolds number.

The low pressure curve is changing rapidly with increasing Reynolds number, approaching a constant value for higher Reynolds numbers. Starting from a certain critical Reynolds number, in this case about $Re = 150,000$, the high pressure curve remains almost constant. This behavior agrees very well with the results of the two-dimensional study described above.

11.7 Secondary Flow

Due to the local flow acceleration, classical secondary flow patterns develop in the present configuration. The induced secondary velocities in the cross-planes are relatively small, however, as the turbine is extremely lightly loaded. In Fig. 11.14, the contours of axial velocity are shown in a travers plane at the axial position $x = -0.08$ at the end of the stator guide vanes. A boundary layer is clearly visible at all solid surfaces. In the center of the channel, the axial velocity reaches its maximum. Due to the rapid change of the hub diameter, two secondary vortices develop in the lower corners. A much more detailed view of the flow in the lower and upper corners reveals tertiary corner vortices at the hub and the outer casing.

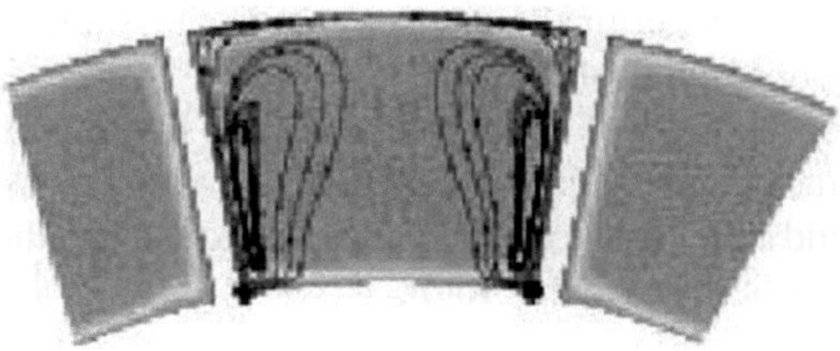

Fig. 11.14. Secondary flow in a travers plane at x=-0.08.

11.8 Summary

The detailed study of the present numerical results indicated that flow separation occured at the leading edge of the pressure side of the rotor for angles of incidence $\beta_\mathrm{i} > 2.5°$ and on the suction side for all investigated angles of incidence $0° < \beta_\mathrm{i} < 5.0°$. Vortices originated at the top and the bottom side of the trailing edge. The vortex formation depends on incoming Reynolds number. Vortex formation at the trailing edge affects the complete flow field around the rotor. The shift of the stagnation point and the change of circulation for different flow capacities $\dot{Q}$ confirm this results. The high pressure shift of the rotor is shown to be attributed to a dependency of the flow character on the Reynolds number. This idea was indicated by the results of two independent simulations and experiments.

The configuration of the three-dimensional model of the first stage of the AccuTest ^{TM} turbine meter, consisting of approximately $1{,}000{,}000$ cells, was presented. In order to simulate the unsteady three-dimensional flow in a geometry including all details of the real meter, e.g., the tip gap, it was decided to simulate only the first stage of the meter. The major results can be briefly summed up as follows:

a) The grid was sufficiently accurate to simulate the unsteady 3-D effects and to investigate the secondary flow effects.
b) Stator wakes cause large amplitude variation of coefficient of moment c_m of first rotor in time.
c) Boundary layers at casing and hub of the rotor increase the angle of incidence in the rotor.
d) Non-axial flow components due to the inlet geometry do not have strong effect on the flow through the rotor.
e) Flow effects that were found in two-dimensional simulations also occured in three dimensions.

In future work there are some phenomena that should be further investigated to improve the accuracy of turbine flow meters. It has been shown that the high pressure shift can be explained as a dependency of the flow character on the Reynolds number. A critical Reynolds number Re_crit, defined as the Reynolds number at which a vortex separation from the trailing edge begins, exists. It is desirable to improve the design of the rotor to reduce this critical Reynolds number to values smaller than those occuring under any operating conditions.

The influence of the wake of the first stator on the first rotor has been shown. In future investigations the axial gap between guide vanes and rotor and the design of the guide vanes could be reconsidered to reduce the extend and the effect of the wake on the incoming flow field in the first rotor.

References

Baker RC (1993) Turbine Flowmeters: II. Theoretical and Experimental published Information. Flow Meas Instrum, 4(3):123–144

Bonfig KW (1977) Technische Durchflußmessung. Vulkan-Verlag, Essen

Brümmer A (1997) Der Einfluß von Druckpulsationen auf die Meßgenauigkeit von Turbinenradzählern. In: Workshop Kolbenverdichter, Rheine, 29.-30.10.

Fluent (1998) Fluent 5, Lebanon, NH, USA

Kallenberg M (1999) Numerische Simulation der chemisch reagierenden Strömung in einer typischen Überschallbrennkammer. Dissertation, Universität Essen

Lee WFZ, Blakeslee DC, White RV (1982) A Self-Correcting and Self-Checking Gas Turbine Meter. Journal of Fluids Engineering, 104:143–149

Lee WFZ, Karlby H (1960) A Study of Viscosity Effect and Its Compensation on Turbine-Type Flowmeters. Journal of Basic Engineering, Transactions ASME, 82:717–728

Lehmann N (1990) Dynamisches Verhalten von Turbinenradgaszählern. Das Gas- und Wasserfach, 131(4):160–167

Mickan B et al. (1996) Die Fehlerverschiebung eines Turbinenradgaszählers in Abhängigkeit vom Anströmprofil. PTB-Mitteilungen, 106(2):113–124

Perpeet S (2000) Numerische Simulation von Strömungsfeldern um Durchfluß-Meßanordnungen. Dissertation, Universität Essen

Schieber W (1998) A Turbine Meter with Built-in Transfer Prover. In Proceedings of the 9th International Conference on Flow Measurement, FLOMEKO '98, pages 509–516, Lund, Sweden, June

Schieber W (2000) United States Patent, No. 6,065,352. American Meter Company

Yao J (1997) Numerische Simulation der Multi-Dimensionalen Strömung in Kolbenmotoren. Dissertation, Universität Essen

12

How to Design a New Flow Meter from Scratch

Franz Peters, Carsten Ruppel

Institut für Strömungslehre, Universität Essen, 45117 Essen, Germany

The idea of this chapter is to base a new pipe flow meter on known fluid mechanics and mechanics from the beginning rather than to explore the underlying physics after the event. Our new method relies on the well-known drag force on a cylindrical body and the force measurement by a load cell (Drag Force Flow Meter: DFFM). By means of an upstream screen the meter is prepared to axial velocity profile distortions and, by the way the force is measured, swirl is invalidated. Tests with various pipe installations demonstrate the main features and the error bar to be expected.

12.1 Introduction

Most of this book deals with the fluid mechanics and signal processing of existing flow metering techniques. In this chapter we want to turn the perspective around. We want to design a new meter from scratch. The meter is to be based on known fluid mechanics and known force measurement and the installation problem is to be included in the design from the beginning. At the end of the design process the meter characteristics are to be clear and available with no need for retrospective research.

Looking for a measuring principle (Ruppel and Peters, 2001) we observed that the force is hardly used among the numerous measured quantities like length (rotameter), time (ultrasound), frequency (vortex meter), pressure (orifice, Venturi) or electrical potential (electromagnetic flow meters). At the same time force measurement is not at all unusual in fluid mechanics, especially aerodynamics, although forces can be small down to the Newton range.

We choose a thin bar (a cylinder in this investigation) mounted radially from wall to wall in a pipe's cross section and record its drag force which is a measure for the velocity. Normal flow past a cylinder is one of the most fundamental flow cases quoted as the drag coefficient vs Reynolds number plot in most fluid mechanics text books (Schlichting, 1960). We have to take into account though that the cylinder collects a velocity profile along the radius rather than a uniform velocity. We also have to consider that in real situations the profile may not be rotationally symmetric and that the flow may be disturbed in the way of swirl. These problems are successfully solved by an upstream screen that flatens the profile and by the way the cylinder is mounted which invalidates the swirl.

Due to expense of time and material the experimental part of the following could only be a case study for gas in a 150 mm i.d. pipe featuring an 8 mm cylinder. Yet, similarity is based on Reynolds number and Dean number which lie in a realistic range. In principle the extension to liquids should be feasible and worthwhile.

12.2 Measuring Principle

We divide the description of the measuring principle into two parts. The first part deals with the problem how the flow exerts a force onto the cylinder and how this force relates to the flow rate. The second part shows how this force is properly measured.

12.2.1 Flow Rate and Drag Force

A body in a uniform stream of velocity u and fluid density ρ experiences a drag force which is proportional to r, the projection area A in flow direction and the square of the velocity. The proportionality coefficient, called drag coefficient c_{D}, depends on the Reynolds number Re

$$c_{\mathrm{D}} = c_{\mathrm{D}}(Re) \tag{12.1}$$

$$Re = \frac{u\, d_{\mathrm{cyl}}}{\nu} \tag{12.2}$$

based on the dynamic viscosity ν and a characteristic length of the body which is the cylinder diameter d_{cyl} in our case. The most frequently quoted data on $c_{\mathrm{D}}(Re)$ of a cylinder are those by Wieselsberger (1921, 1922) reproduced in Fig. 12.1.

When the cylinder is mounted radially across the pipe as shown in Fig. 12.2 flow direction and magnitude at the cylinder will in general depend on the radius r and the

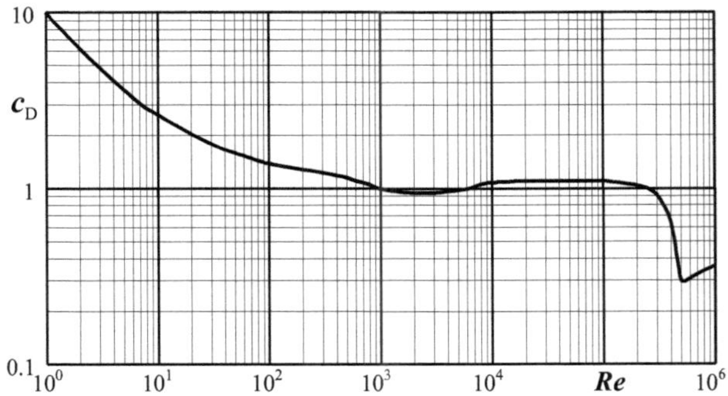

Fig. 12.1. Drag coefficient of a cylinder as function of Reynolds number from Wieselsberger (1921, 1922).

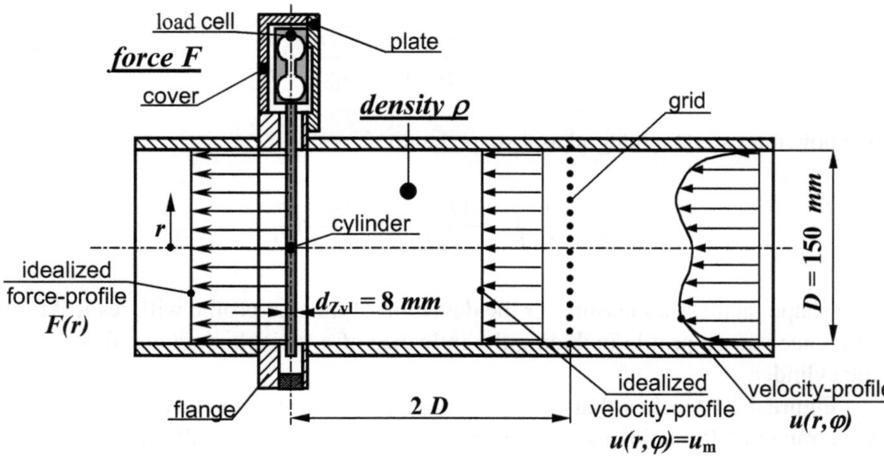

Fig. 12.2. Working principle of the Drag Force Flow Meter (DFFM)

angle φ. In well-defined cases such as developed laminar flow, developed turbulent flow and uniform flow (square profile) the flow remains normal to the cylinder in pipe direction and the velocity is merely a function of the radius. In general, reality differs considerably from these flow patterns, yet they lend themselves for reference purposes. In reality the profiles show distorted shapes and swirl occurs superposed to the main flow not to speak of large scale turbulence which may add time dependence. A key benefit of our design is that we do not need to worry about the swirl component from the beginning. The way the beam is mounted (see below) only the force in pipe direction is measured meaning that only flow in pipe direction is recorded which is exactly the one that contributes to the flow rate.

Velocity and Reynolds number along the cylinder at a fixed angle φ are now functions of the radius

$$Re(r) = \frac{u(r)\, d_{\mathrm{cyl}}}{\nu} \tag{12.3}$$

so that the drag coefficient has to be written in the form

$$c_{\mathrm{D}} = c_{\mathrm{D}}(Re(r)) \tag{12.4}$$

yielding the total force F on the cylinder in flow direction ($D = 2R$)

$$F = \frac{1}{2}\,\rho\, d_{\mathrm{cyl}} \int_{-R}^{R} c_{\mathrm{D}}\,(Re(r))\, u^2(r)\, \mathrm{d}r\ . \tag{12.5}$$

In the simple reference case of the square profile with uniform velocity u_{m} the force reduces to

$$F_{\mathrm{sq}} = \rho\, d_{\mathrm{cyl}}\, c_{\mathrm{D}}\,(Re_{\mathrm{m}})\, u_{\mathrm{m}}^2\, R\ . \tag{12.6}$$

Using this reference force F_{sq} for scaling the general force of (12.5) we have

$$\frac{F}{F_{\mathrm{sq}}} = \frac{1}{2} \int_{-1}^{1} \frac{c_{\mathrm{D}}(Re(r))\, u^2(r)}{c_{\mathrm{D}}\,(Re_{\mathrm{m}})\, u_{\mathrm{m}}^2}\, \mathrm{d}\frac{r}{R}\,. \tag{12.7}$$

We now interpret the integral as a shape factor f

$$f = \frac{1}{2} \int_{-1}^{1} \frac{c_{\mathrm{D}}(Re(r))\, u^2(r)}{c_{\mathrm{D}}\,(Re_{\mathrm{m}})\, u_{\mathrm{m}}^2}\, \mathrm{d}\frac{r}{R}\,. \tag{12.8}$$

The shape factor is a measure for the deviation of the real profile with respect to the reference square profile including the influence of a variable drag coefficient along the cylinder.

Figure 12.1 teaches that the drag coefficient is approximately constant over a wide range of Reynolds numbers which is limited by very small ($Re < 500$) and very large Reynolds numbers ($Re > 2 \times 10^5$). Reynolds numbers $> 2 \times 10^5$ are not envisaged in our design. Reynolds numbers down to zero are present in the wall boundary layer, yet the contribution of this range is very small anyhow. Therefore we approximate

$$c_{\mathrm{D}}(Re(r)) \approx c_{\mathrm{D}}(Re_{\mathrm{m}})\,. \tag{12.9}$$

With this approximation the shape factor can readily be calculated for known profiles. For the parabolic laminar profile we find $f = 2.13$ independent of Reynolds number, which means that the laminar profile more than doubles the force as compared to the square one. This shows the dominant influence of the squared velocity in a peaked profile.

The shape factor of the developed turbulent profile exhibits a slight dependence on pipe Reynolds number Re_{D} as Table 12.1 shows. For the calculation the profile given by Gersten and Herwig (1992) was used.

In order to express the mass flux $\dot{m}$ as function of the force F we eliminate u_{m} from (12.6) and in the resulting equation replace F_{sq} by means of (F/f) from (12.7, 12.8). Then multiplication with the density ρ and the pipe area $R^2\,\pi$ yields

$$\dot{m} = \frac{1}{c_{\mathrm{D}}(Re_{\mathrm{m}})\, f} \sqrt{\frac{R^3\,\pi^2}{d_{\mathrm{cyl}}}}\, \sqrt{F\,\rho}\,. \tag{12.10}$$

Table 12.1. Shape factor f for developed turbulent profiles for various pipe flow Reynolds numbers Re_{D}

$Re_{\mathrm{D}} = \dfrac{u_{\mathrm{m}}\, D}{\nu}$	shape factor f turbulent profile
50,000	1.170
100,000	1.158
150,000	1.151
200,000	1.146
250,000	1.142

Here the force is the measured quantity. The density is either known or measured through temperature and pressure for a gas. Pipe and cylinder diameter are given. The drag coefficient is close to 1 (Fig. 12.1). The crucial parameter is the form factor f. It is only known when the profile is ideal (Table 12.1) or measured which, of course, would save the flow meter. Our answer to the real, distorted profile is a wire screen placed upstream of the cylinder. Such a screen produces relatively small pressure losses and needs little space. It has a smoothing and equilibrating effect on upstream profile distortions such that the downstream profile approaches the square shape. In contrast to the flow straightener it has little impact on the swirl which we are not concerned about anyhow. Squaring the profile the screen pushes the shape factor f closer to one without reaching it, of course. It gets close enough though, so that (12.10) is suitable for the design process of the meter which will normally start with the range of $\dot{m}$ required by the user for a given pipe of radius R. The cylinder ought to be much smaller than the pipe, $d_{\mathrm{cyl}} \ll D$. With the product of c_{D} and f close to one the force F on the cylinder follows for kown density. F then determines the design of the load cell.

After the design has been completed and the load cell has been made calibration of $\dot{m}$ vs $F\,\rho$ according to (12.10) in a reference situation becomes inevitable to accomplish high accuracy. We'll go through these steps further down concluding with an investigation on the sensitivity to installations.

12.2.2 Drag Force and Load Cell

The fluid force on the cylinder is an integrated load with varying load distribution. To measure this load independent of the distribution a load cell as shown in Fig. 12.3 is suitable. Kinematically the four thin portions of the cell constitute the joints of a parallelogram. The front part with the attached cylinder moves under load parallel to the rear part fixed to a rigid base such that the attached cylinder remains in radial direction in the pipe. At the same time the thin portions flex like leaf springs transforming the acting force into deformation. The deformation can be measured by strain gauges or optical displacement sensors (or others).

Our goal is the design of the entire meter. Therefore, as an example, we selected material (aluminum 7075 T7351) and shape of a particular load cell as given in Fig.12.3. All measures but the wall thickness w of the flexible joints were fixed. This given configuration was fed into a FEM program to calculate the displacement

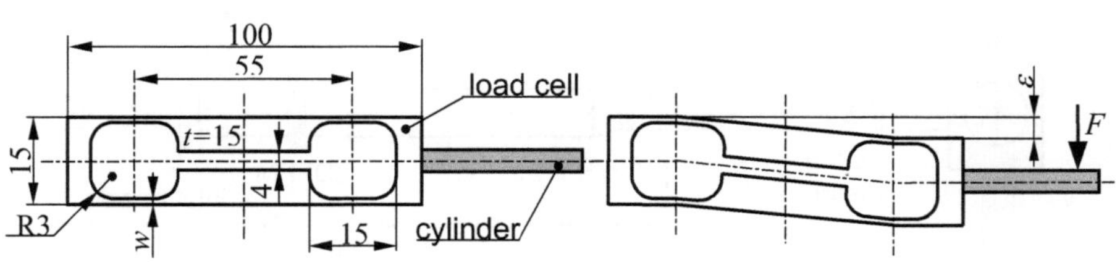

Fig. 12.3. Design of the load cell with (*right*) and without load (*left*)

of the cylinder as function of the load for various wall thicknesses. The calculated results were checked versus a simple measurement in which the load was simulated by weights and the displacement was measured by an optical gauge (optoNCDT by Micro-Epsilon). Excellent agreement within $\pm 2.5\%$ was found. The outcome of this investigation is the following formula

$$\varepsilon\, w^{2.7935} = 0.0982\, F\ . \tag{12.11}$$

It relates the wall thickness w in mm, the displacement ε in mm and the force F in Newton. Most importantly, a linear relationship between force and displacement is achieved. In the design process the maximum force will be known from the flow rate as described above. A corresponding maximum displacement has to be chosen according to the type of length measurement available (strain gauges, optical devices or others). Then (12.11) delivers the thickness w, which of course is bound by upper and lower limits. The lower limit is set by machining and the upper limit by the condition that the flexible joint needs to be thin relative to the side-pieces of the cell. The range $0,5\,\mathrm{mm} < w < 1,5\,\mathrm{mm}$ was found reasonable for the test design.

Please note that this test design was just to show the designability of the load cell. The ranges of measures and forces have to be tuned to the application.

12.3 Setup and Calibration

Figure 12.4 illustrates the basic setup in the calibration mode. An air blower driven by a frequency controlled AC motor circulates room air through a pipe system (precision pipes DIN 2391). Air enters the test pipe of $D = 150\,\mathrm{mm}$ i.d. through a nozzle and a wire screen of $1\,\mathrm{mm}$ wire and $3\,\mathrm{mm}$ mesh size. The pipe is segmented and assembled by flanges (not shown) to allow variable configurations of pipe and installations including the DFFM and the screen which are integrated into flanges themselves. The big pipe reduces to a smaller one of $d = 100\,\mathrm{mm}$ i.d. which incorporates a Venturi meter $20\,d$ upstream of the blower. The Venturi provides the mass flux reference after careful calibration against another reference placed downstream of the blower described elsewhere (Wildemann, 2000). The DFFM is fitted with a $8\,\mathrm{mm}$ cylinder of carbon fibre laminate for best stiffness to mass ratio. The displacement was to be measured by an integrated strain gauge because we were not able to

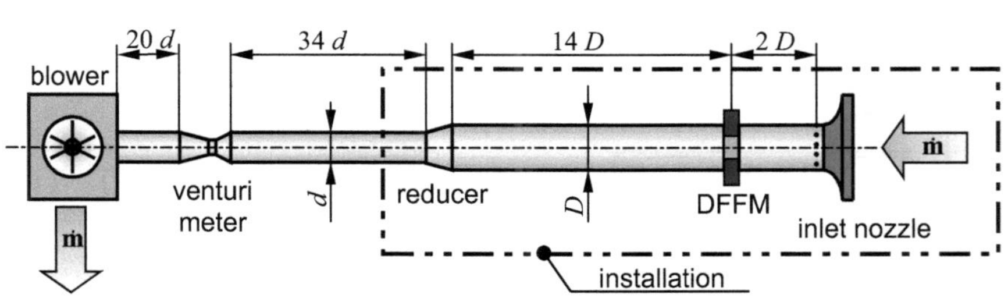

Fig. 12.4. Calibration setup indicating installation range

integrate the optical device mentioned earlier into the flange. A commercial load cell (PW4FC3 by HBM) of the type tested in Sect. 12.2 was purchased in one piece with the strain gauge and an appropriate amplifier (BA 651 by BLH).

Data aquisition is by a National Instruments A/D card "PCI-6024E" in combination with the LabView 6i software. Besides the force temperature and pressure in the pipe are recorded at the sampling rate of $10^4\,\mathrm{s}^{-1}$. Due to an eigen vibration of the bar/cell arrangement of the order of 200 Hz a mean value of the force has to be taken. In order to do so the integration time should not drop below $0.1\,\mathrm{s}$. The LabView program views the recorded data including the reference mass flux from the Venturi.

The calibration procedure is as follows. Starting from $u_{\mathrm{m}} = 3.5\,\mathrm{m/s}$ mean velocity in the big pipe the blower motor frequency is risen stepwise up to the mean pipe velocity of about $32\,\mathrm{m/s}$. For each step ρ is determined from p and T (including relative humidity) and F is measured. The reference mass flux $\dot{m}$ is obtained from the Venturi. This information together with (12.10) yields the calibration diagram Fig.12.5 displaying $c_{\mathrm{D}} f$ versus Re. As c_{D} is close to one (see Fig.12.1 and discussion) the message drawn from this plot is that f in fact comes close to one too meaning that the screen provides an approximately square profile at least in this calibration arrangement right downstream of a rounded inlet nozzle. Of course, the residual Reynolds number dependence must be taken into account. The relationship

$$\begin{aligned} c_{\mathrm{D}}\, f = &\ 0.64 \exp(-64 \times 10^{-5}\, Re) \cos(-535 \times 10^{-6}\, Re - 0.05) \\ &+1.91 \times 10^{-6} Re + 0.997 \end{aligned} \tag{12.12}$$

fits the data quite well constituting the analytical calibration curve of this particular DFFM. With the help of (12.10) it transforms into the mass flux vs force relationship.

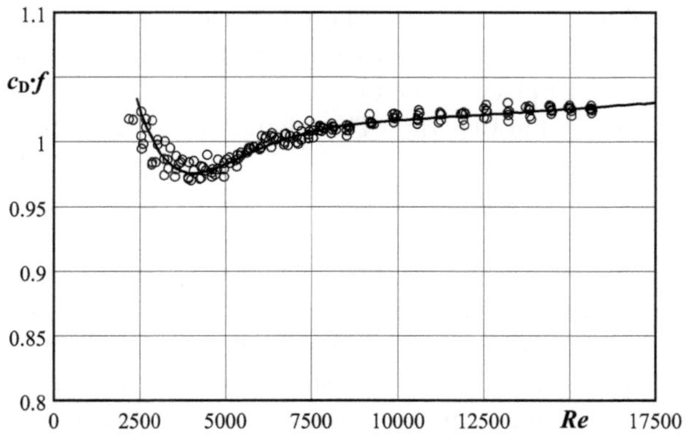

Fig. 12.5. Calibration curve

12.4 Results for Installations

A calibrated DFFM exposed to any square velocity profile in the adequate range will be accurate within a small margin of reproducibility. Obviously, it is not possible to predict the efficiency of the screen for any installed upstream disturbance. As a consequence typical installation situations have to be tested before a reliable error prediction can be made.

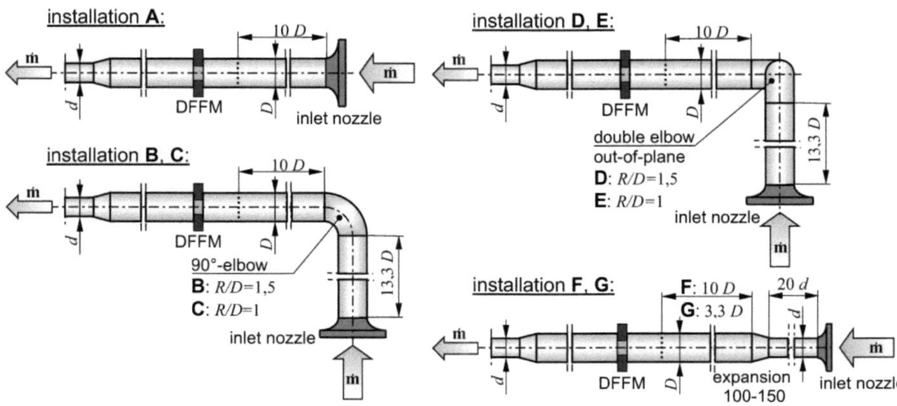

Fig. 12.6. The investigated installations

Figure 12.6 compiles seven variants covering quite common installations and an extreme one. The common ones are the inlet nozzle with 10 D of straight pipe (A), two regular elbow bends (B,C) and two double bends (D,E). A pipe expansion (F,G) opening from 100 to 150 mm i.d. on a short distance of 3.3 D introduces flow separation, an undoubtedly extreme disturbance bound to lead to substantial indication errors of most any meter placed close to it.

Results are compiled in Fig. 12.7. The relative deviation of the DFFM in terms of

$$\Delta\dot{m} = \frac{\dot{m}_{\mathrm{DFFM}} - \dot{m}_{\mathrm{ref}}}{\dot{m}_{\mathrm{ref}}} 100 \qquad (12.13)$$

is plotted versus the reference mass flux. The covered range of mass fluxes corresponds to mean velocities in the big pipe from 5 to 28 $\mathrm{m/s}$.

When $\Delta\dot{m} > 0$ the DFFM overpredicts the flow rate and the other way round for $\Delta\dot{m} < 0$. The first five installations show consistent negative deviations with some scatter. In view of the shape factor discussion the consistency suggests that in all these cases the profile tends to be somewhat “flater” than in the calibration case.

Elbow and double elbow produce strong disturbances which are described by Hilgenstock and Ernst (1996). Generally, the main problem of the disturbance is swirl which is known to decay slowly (Schlüter and Merzkirch, 1996). Here the DFFM is in the advantageous position that it is not affected by radial and tangential velocity

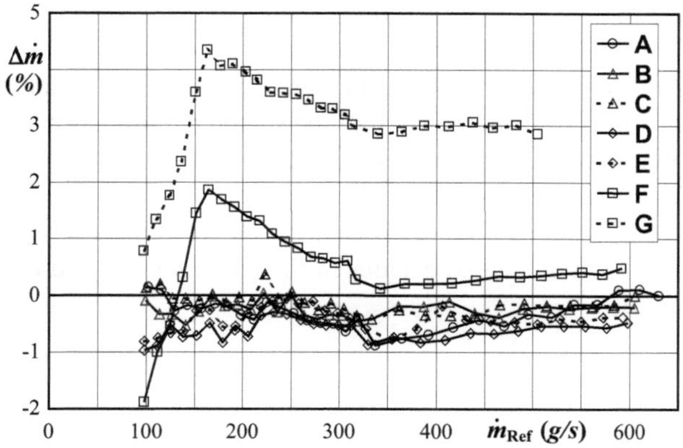

Fig. 12.7. Relative deviation of the mass flux measured by the DFFM versus the reference mass flux for the installations of Fig.12.6.

components as already explained in Sect. 12.2.2. What remains is a moderate dent in the profile which is dealt with by the screen and, to some extent, by the 10 D of pipe length. In sum, the installations lead to deviations from the calibration arrangement, yet, it must be emphasized that the deviation stays within a percent! Furthermore the deviation exhibits a systematic character which could be corrected for.

The pipe expansion of installation F,G produces substantial positive deviations. Certainly, it is not recommendable to place a DFFM (or any other instrument) into such a position. We have included these installations just to show how well the deviations of a DFFM can be interpreted. The diffusor angle of the expansion enforces flow separation with high velocities on the center of the pipe and low ones in the separated region (or even negative). The profile will be extremely peaked entailing large positive deviation up to 4% in case G. In the other case F with the DFFM moved downstream the flow seems to be reattached with a broader profile and correspondingly smaller deviations down to half a percent for the higher flow rates.

12.5 Conclusions

We have shown how to design a flow meter for the mass flux in pipes from scratch. The design is based on fundamental fluid mechanics (flow past a cylinder) and a well understood type of load cell. The load cell determines the force on the cylinder which is an accurate measure of the flow rate when the velocity profile is known. As the profile is normally not known the DFFM generates its own profile by means of an upstream screen. This profile approximates the square one used for calibration. The deviation of the screen profile from the calibration profile causes the uncertainty of the measurement. Tangential and radial disturbances in the velocity distribution are ignored by the meter from the beginning saving further consideration.

In a number of real installation situations different from the calibration situation we have demonstrated that the DFFM works very well. It shows an error bar of less than a percent with a consistent shift that can be rationalized so that accuracy could be even improved by corrections.

The elements of the meter allow refinements which should be looked at in further developments. Especially, the selection of the screen parameters has some potential. Worth mentioning is that two screens, upstream and downstream, would allow to reverse the flow direction.

As the measuring signal, the drag, is based on Reynolds number and as the dimensions of the load cell can be calculated the meter can be laid out for most flow metering situations.

References

Gersten K, Herwig H (1992) Strömungsmechanik - Grundlagen der Impuls-, Wärme- und Stoffübertragung aus asymptotischer Sicht. Vieweg-Verlag, Braunschweig, Wiesbaden

Hilgenstock A, Ernst R (1996) Analysis of installation effects by means of computational fluid dynamics-CFD vs experiments?. Flow Meas.Instrum. 7:161–171

Ruppel C, Peters F (2001) Messung des Massenstroms in Rohrleitungen durch die Widerstandskraft eines quer angeströmten Zylinders (DFFM-Methode). tm-Technisches Messen 68:465–472

Schlichting H (1960) Boundary Layer Theory. 4th ed. McGraw-Hill, New York

Schlüter T, Merzkirch W (1996) PIV measurements of the time-averaged flow velocity downstream of flow conditioners in a pipeline. Flow Meas Instrum 7:173–179

Wieselsberger C (1921) Neuere Feststellungen über die Gesetze des Flüssigkeits- und Luftwiderstandes. Physikalische Zeitschrift 22:321–328

Wieselsberger C (1922) Weitere Feststellungen über die Gesetze des Flüssigkeits- und Luftwiderstandes. Physikalische Zeitschrift 23:219–224

Wildemann C (2000) Ein System zur automatischen Korrektur der Messabweichung von Durchflussmessgeräten bei gestörter Anströmung. Doctor thesis, Universität Essen. See also: Fortschritt-Berichte VDI, Reihe 8, Nr. 868

13

Effects of Disturbed Inflow on Vortex Shedding from a Bluff Body

Ernst von Lavante

Institut für Strömungsmaschinen, Universität Essen, 45117 Essen, Germany

In the present chapter, the problem of accurate determination of mass flows by means of the so-called vortex-shedding flow meter is studied for the case of disturbed inflow. To this end, the flow around bluff bodies of different designs, used in typical vortex-shedding flow meters, was numerically simulated using a solver of the unsteady, compressible Navier–Stokes equations in three dimensions. The computations were carried out for several types of disturbed inflow conditions, including bends, flow with rotation oriented in the axial direction and unsteady flow fluctuations. The results were compared with experimental data obtained in an ongoing investigation. The effects of turbulence were modeled by using large-eddy simulation (LES). The resulting flow fields were analyzed using various methods, including visualization, evaluation of several of the features and DFT of properly chosen variables. Recommendations regarding the operational conditions of the vortex-shedding flow meters are made.

13.1 Introduction

Many aerodynamic problems require a volume- or mass-flow data for its quantitative solution. Therefore, a number of methods for flow rate measurement have been developed. One relatively simple and promissing flow measurement device is the so-called vortex-shedding flow meter, in which the mass flow is determined by observing the relationship between vortex-shedding frequency from a bluff body attached inside a channel, and the corresponding mean velocity about it. The bluff body causes the production of a system of periodic vortices ("vortex street"), whose frequency can be correlated with the mean flow velocity and, therefore, the mass flow.

This procedure assumes a regular and well-defined vortex structure as well as shedding mechanism, resulting mostly in linear dependency of the mass flow on the shedding frequency over a wide range of Reynolds numbers. However, it has been observed that certain types of disturbancies in the inflow result in rather irregular pressure signatures of the vortex system or even shift of its charachteristic

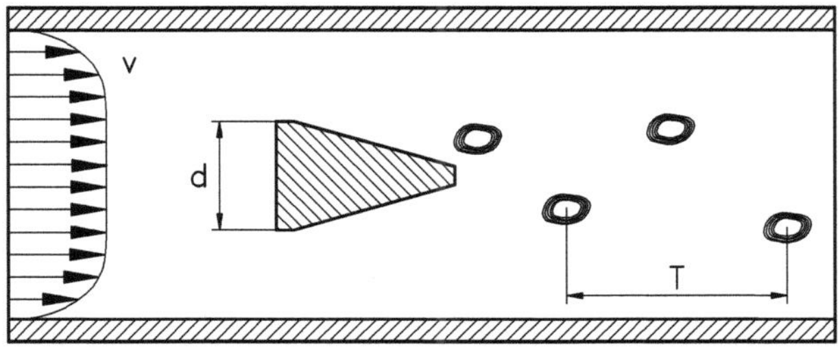

Fig. 13.1. Principle of a vortex-shedding flow meter.

frequencies, leading to unreliable mass flow data. A schematic picture of a typical vortex-shedding flow meter is offered in Fig. 13.1.

Commercial flow meters use a large variety of bluff body shapes, often restricted by the attachment of the pressure sensors or, more likely, patent laws. Previous application of some current bluff body designs lead to fairly irregular pressure signatures, making them unreliable. It was, therefore, decided to investigate not only the known bluff body shapes, but also a few alternate designs, developed empirically by the present authors.

In the past, various shapes of vortex bodies have been tested with regard to their applicability to a simple signal processing. Triangular shapes (Johnson, 1990; Fureby, 1995; Madabhushi et al., 1997; Hans et al., 1998) as well as shapes with truncated tips (e.g., Hans et al., 1997) have been tested. In the preliminary part of this work, tests were performed with T-shaped, rectangular and a new-designed bluff body. Well-defined vortices were generated, giving, after signal processing, excellent measurement results (Hans et al., 1998; Miau et al., 1993). A strong and well-defined dependency of the vortex frequency on the flow velocity and, therefore the Reynolds number, could be obtained.

The theory of operation is treated in more detail in Chaps. 5 and 6. In the work concerning the vortex shedding flow meters included in this book, the method of detection of the vortices that has been employed was always the same. Downstream of the bluff body, an ultrasound barrier was installed, "counting" the vortices. In the numerical simulation, this signal was approximated by an integral of the vertical velocity component over an area closely resembling the ultrasound beam.

In the present work, the flow was investigated using a Navier–Stokes equations solver, capable of handling unsteady, compressible and viscous flows over a wide range of Mach numbers. Several different classes of inflow disturbancies were studied and analyzed in regards to their effects on the flow metering capabilities.

13.2 Numerical Algorithm

The numerical algorithm employed uses the three-dimensional, time-dependent full Navier–Stokes equations describing the conservation of mass, momentum and energy of fluid flows. The divergence form in body-fitted, curvilinear coordinates is:

$$\frac{\partial Q}{\partial \tau} + \frac{\partial (F - F_{\mathrm{v}})}{\partial \xi} + \frac{\partial (G - G_{\mathrm{v}})}{\partial \eta} + \frac{\partial (H - H_{\mathrm{v}})}{\partial \zeta} = 0 \,, \tag{13.1}$$

with $Q = J^{-1}(\rho \; \rho u \; \rho v \; \rho w \; e)^{\mathrm{T}}$ the vector of the conserved variables. J is the Jacobian of the coordinates transformation from physical (x, y, z, t) to computational (ξ, η, ζ, τ) space. The program is based on the finite-volume formulation, using a cell-centered organization of the control-volumes. The spatial discretization is carried out with the help of Roe's Flux Difference Splitting Scheme, a Godunov-typ method providing an approximate solution of the Riemann problem on the cell interfaces. Here, the flux is (Vatsa et al., 1987):

$$F_{i+\frac{1}{2}} = \frac{1}{2}\left\{F(Q)_{\mathrm{L}} + F(Q)_{\mathrm{R}} + \mid \tilde{A}_{i+\frac{1}{2}} \mid (Q_{\mathrm{R}} - Q_{\mathrm{L}})\right\} \,. \tag{13.2}$$

The index $i + \frac{1}{2}$ describes the values on the cell interfaces, R and L the values right and left from it. The Roe-averaged matrix $\tilde{A}$ is given by the differentiation of the local-linearized function $F(Q)$. This scheme is formally central difference plus a damping term. The method has been proved to be very accurate and effective in the simulation of low Mach number viscous flows (von Lavante and Yao, 1993).

Upwind-biased differences are used for the convective terms, central differences for the viscous fluxes. Starting with a constant initialization of the scalar variables and body-fitted velocity components, the integration in time is carried out by a modified explicit Runge–Kutta time stepping as well as, optionally, an implicit Approximate-Factorization method or symmetric Gauss–Seidel scheme (Jameson et al., 1981).

13.2.1 Low Mach Number Modifications

Numerical algorithms for the simulation of compressible flows become inefficient and inaccurate at very low Mach numbers. The difficulties are due to the formulation of the governing equations in their discretized form. The problems have been addressed by many previous investigators, including Shuen et al. (1993), Pletcher and Chen (1993) and Edwards and Roy (1998). The main problem is the stiffness of the governing equations at low Mach numbers. The condition number of the Jacobian matrices (ratio of the maximal to the minimal eigenvalues) increases to infinity as the Mach number approaches zero. The time marching step therefore is restricted because of the large disparity of the eigenvalues of the Jacobian, representing the convective and acoustic signal speeds, respectively. The second problem results from the pressure singularity in the momentum equations at very low Mach numbers. The ratio of the magnitudes among the pressure term p and the convective terms ρu^2 is inversely proportional to the Mach number squared M^2. The large difference in magnitude will yield a large roundoff error.

To enable efficient numerical solutions of the equation system at low Mach number, a pseudo-time term was added to the time-dependent compressible Navier–Stokes equations. The primitive variables were employed as unknowns, rather than the traditional conservative variables. A preconditioning matrix Γ was used in order to eliminate the time-step difference between the convective and acoustic characteristic speeds at low Mach number. The preconditioned equations can be discretized in the delta form:

$$\left[\Gamma + \frac{\Delta\tau}{\Delta t}\hat{A}_{\mathrm{Q}} + \Delta\tau\left(\frac{\partial(\hat{A}_{\mathrm{F}})}{\partial\xi} + \ldots + \frac{\partial(\hat{A}_{\mathrm{H}})}{\partial\zeta}\right)\right]\Delta\hat{Q} = - \Delta\tau\left(\frac{\partial Q}{\partial t} + \frac{\partial F}{\partial\xi} + \ldots + \frac{\partial H}{\partial\zeta}\right), \tag{13.3}$$

where the Jacobian matrices are defined as,

$$\hat{A}_{\mathrm{Q}} = \frac{\partial Q}{\partial\hat{Q}}, \quad \hat{A}_{\mathrm{F}} = \frac{\partial F}{\partial\hat{Q}}, \quad \hat{A}_{\mathrm{G}} = \frac{\partial G}{\partial\hat{Q}}, \quad \hat{A}_{\mathrm{H}} = \frac{\partial H}{\partial\hat{Q}}. \tag{13.4}$$

Here Q is the vector of the conservative variables and $\hat{Q}$ is the vector of the primitive variables $(p, u, v, w, T)^{\mathrm{T}}$. A more detailed description of the matrix Γ can be found in Yao and von Lavante (1997).

To obtain time-accurate solutions, a dual time-stepping integration procedure was used. For each physical time step, the equations were integrated in pseudo-time. The physical time term behaves like a source term during the pseudo-time iterations. The converged solution in pseudo-time corresponds to a time-accurate solution in physical time. For steady flow problems, the physical time terms are neglected. An implicit iterative procedure using symmetric Gauss–Seidel scheme was used for the sub-iterations of the equations in the pseudo-time. The converged equations at each physical time step satisfy the original discretized Navier–Stokes equations.

To circumvent the problem of large difference in magnitude of the convective and pressure terms in the momentum equations, a gradient splitting of the Euler flux into convective terms and a pressure term can be considered:

$$\nabla(\rho u^2 + p) = \nabla(\rho u^2) + \nabla p. \tag{13.5}$$

Both split gradients are of the same order of magnitude, independent of the Mach number. The Liou's Advection Upwind Splitting Method was applied to treat the convective and pressure terms separately. The convective terms were upstream-biased using an appropriately defined advection Mach number at the cell interface, while the pressure term was strictly dealt with by using acoustic waves. One of the advantages of the method is that the upwind effect can be easily achieved with few modifications of the programs.

13.2.2 Turbulence Model

The highly complex geometrical and physical configurations are making the choise of an appropriate turbulence model very difficult. In the vortical, periodically separating flow with strong local rates of acceleration, any isotropic models are not adequate. The present authors decided, therefore, the use of the concept of large eddy simulation (LES).

13.2.3 Verification

The present numerical algorithm was subjected to verification of it's temporal and spatial accuracy and consistancy. The scheme is formally second order accurate in space, since the viscous terms are obtained from second order central differences. The scheme was first verified using the usual grid refinement study for the case of viscous flat plate flow at a free stream Mach number of $M_\infty = 0.5$. Defining the global error as the L_2 - norm of the deviation of the present solution from the Blasius solution, second order accuracy was verified (von Lavante, 1990). Next, the combination of the present solution scheme with the computational grid was investigated for the case of an optimized bluff body shape using three different grids. Figure 13.2 shows the density plot and summarizes the results. It should be noted that the finest grid very closely approaches the Strouhal number of the corresponding experiment.

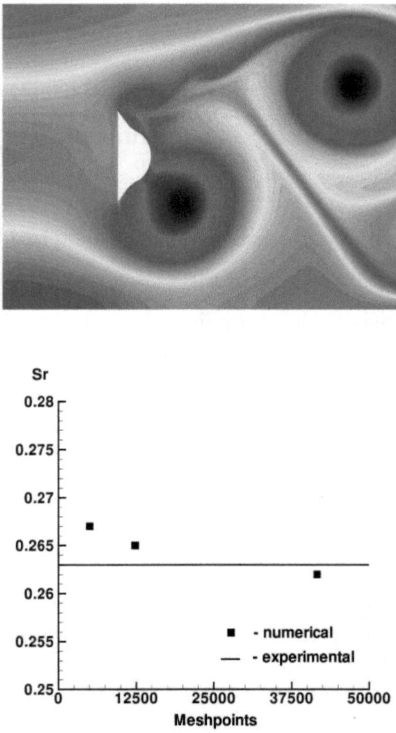

Fig. 13.2. Grid refinement study for an optimized bluff body shape.

The temporal accuracy was tested by simulating the 2-D flow about a cylinder, with free stream Mach number $M_\infty = 0.1$ and a Reynolds number of $Re = 200$. The resulting Strouhal number of the vortex separation is $Sr = 0.196$ and therefore within the range given in literature.

13.3 Results

The bluff body inside the vortex-shedding flow meter causes separation of periodic vortices from the upper and lower edge of a bluff body, similar to the well-known von Kármán vortex street. In the present effort to optimize the bluff body, many different shapes were testet. As a result, an "optimum" shape was found and extensively tested (Hans et al., 1998) in 2-D, 3-D and experimentally. The resulting plot of the frequency of vortex passage through an ultrasound barrier placed downstream of the bluff body is displayed in Fig. 13.3 as a function of the time averaged mean velocity in the pipe. Its highly linear behavior in the given velocity range is obvious. Somewhat surprising is the good agreement between the 2-D and 3-D numerical results. The right side of Fig. 13.3 shows the shape of the optimized bluff body.

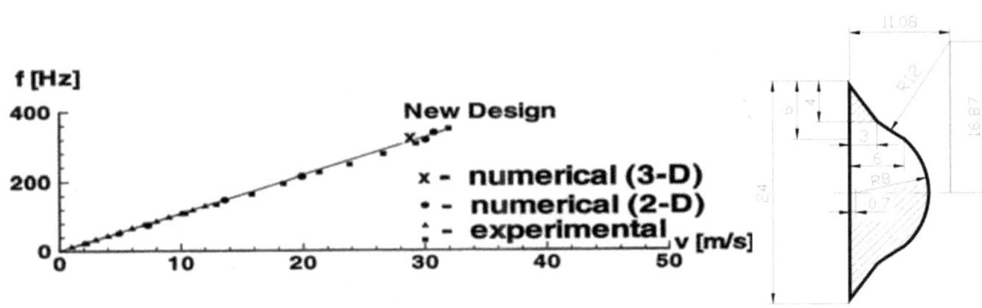

Fig. 13.3. Frequency vs. pipe mean velocity and optimized bluff body shape.

Figure 13.4 shows two basic strategies of mounting the bluff body inside a vortex-shedding flow meter. The left device causes the separation of periodic vortices from the upper and lower edge of a bluff body, similar to the well-known von Kármán vortex street. The right one produces axially-symmetric rings in the wake of a bluff body.

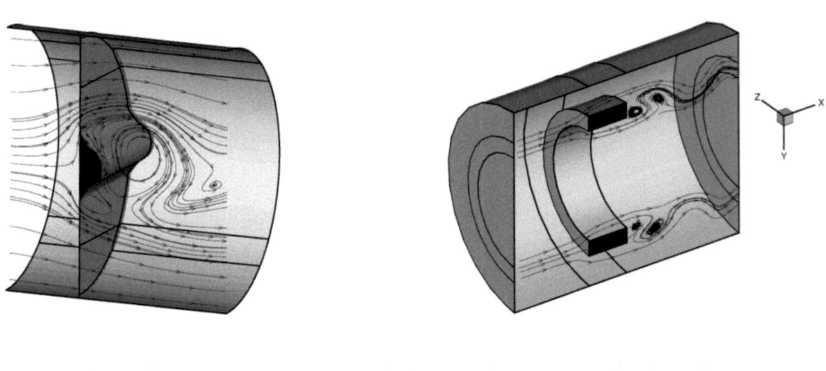

Fig. 13.4. Two different designs of bluff body mounting inside a pipe.

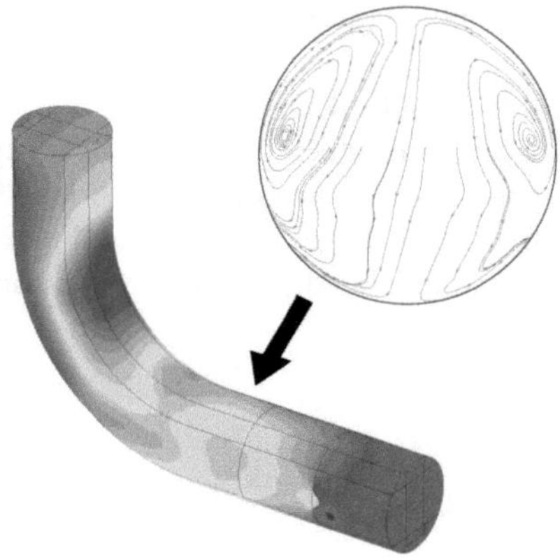

Fig. 13.5. Pressure Contours in a single bend with the vortex flow meter at the lower end.

As long as the inflow remains undisturbed, the shedding frequency in the wake of the bluff body is rather constant for most shapes. There are many studies concerning the optimization of the bluff body shapes, investigated under the condition of undisturbed inflow. We decided to examine the development of the vortex-shedding frequency in a more practical view. Flow meters are normally part of a pipe system and therefore the inflow is often nonuniform or even unsteady. Figure 13.5 shows a section of the pipe, with a single bend, featuring the protruding bluff body downstream of the bend, visible at the lower end of the picture. Figure 13.6 displays the vortex flow meter with the double bend. In the case, the meter is at the upper end of

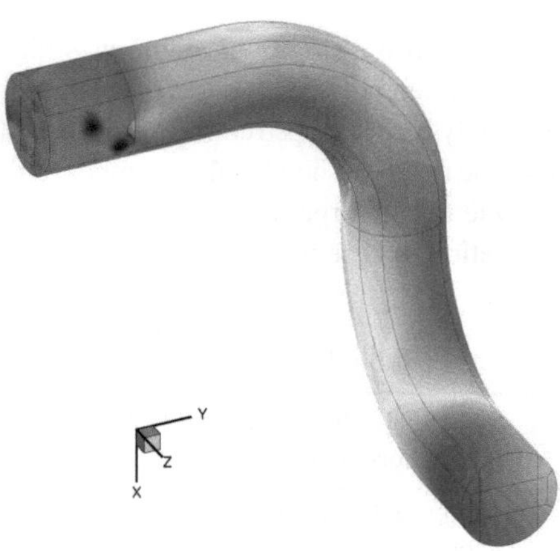

Fig. 13.6. Pressure Contours in a double bend with the vortex flow meter at the upper end.

the pipe. The characteristic double vortex system in the wake of the bend can to be seen.

Figure 13.7 left shows the result of DFT analysis of the flow behind the bluff body. The result is shown after introducing a single bend upstream of the meter, displaying a very strong peak at the meter's natural vortex-shedding frequency. In this case, the dominating frequency remains unchanged. Figure 13.7 right shows the DFT analysis of flow past a double bend. Again, a very strong peak at the meter's frequency is noticeable.

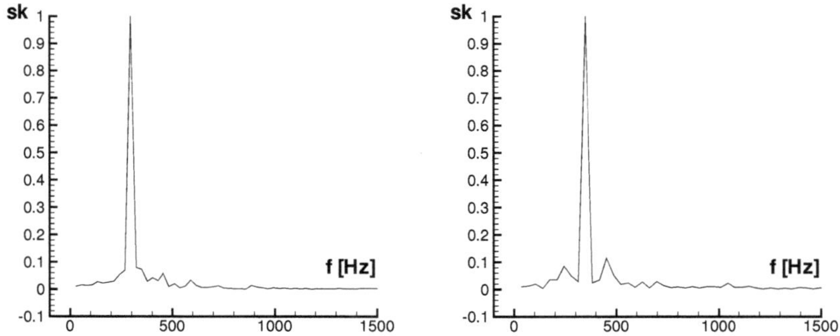

Fig. 13.7. Shedding frequency for flow downstream a single bend (*left*) and with a double bend upstream the meter (*right*).

The signal in Fig. 13.7 was obtained by observing the modulation of the radial velocity component, as given by the present configuration with the ultrasound barrier, described in more detail in Chap. 6.

However, the pressure field in the single 90°-bend was highly unsteady, with a wave that was alternativly reflected between the lower and upper wall while rotating around the pipe axis and slowly moving upstream (Fig. 13.8, to the top).

Table 13.1 summarizes the results. The Strouhal number remains almost constant for all introduced bends. Due to the large tangential velocities in magnitude of the pipe mean velocity the deviation in case of the swirl is noticeable higher.

Table 13.1. Results

	Swirl	90°-Bend horizontal	90°-Bend vertical	Double Bend	Standard Tube	Exp.
Meshpoints	414300	621400	621400	621400	1.07 Mio.	
Reynolds number	183000	180000	173000	213000	191000	175000
Frequency (Hz)	287	300	295	348	307	280
v (m/s)	27.7	27.29	26.18	32.18	28.9	26.48
Strouhal number	0.249	0.264	0.27	0.26	0.255	0.254

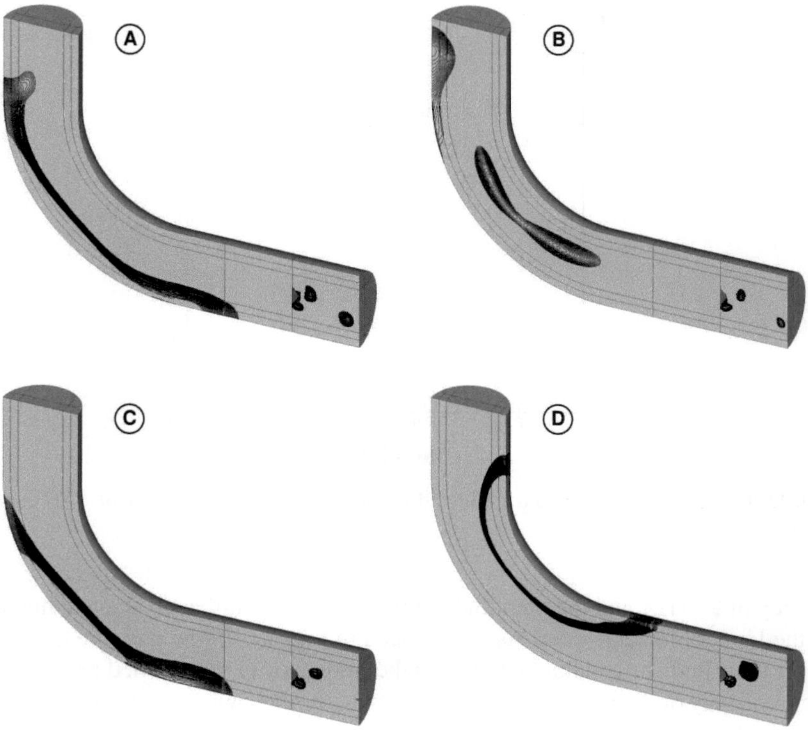

Fig. 13.8. Density waves in the single 90°-bend.

Finally, the effects of pulsating inflow on the accuracy and basic function of the vortex-shedding flow meter were investigated. For this purpose, the two-dimensional computational model was modified by blocking the inflow portion of the pipe far upstream of the bluff body. The complete blockage was executed at a certain frequency for a small percentage of the period T corresponding to the frequency of pulsation. The flow conditions were selected such that the undisturbed frequency of the meter was 250 Hz. The pulses were implemented at 50, 100 and 200 Hz, blocking the flow for between 0.5% of T and 10% of T.

At the lower frequency of pulsation (50 and 100 Hz), the vortex-shedding frequency was basically uneffected. In Fig. 13.9, the static pressure at the bluff body is plotted as a function of the number of iterations in time on the top of the picturefor the case of 100 Hz pulsation with the blockage lasting 0.5% of T. The pulses can be seen as sharp peaks in the pressure history, followed by the same periodic fluctuation as for the undisturbed inflow. The simulated ultrasound signal, shown in the middle of the figure, is practically undisturbed. The mass flow just after the blockage location is displayed at the bottom of Fig. 13.9.

At the highest pulsation frequency of 200 Hz and 10.0% T duration, the situation is completely different. The disturbance frequency is much closer to the natural fre-

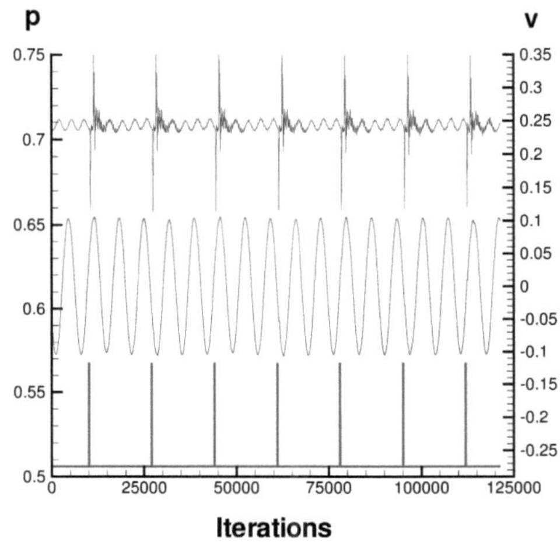

Fig. 13.9. Pressure evolution and simulated signal for pulsation frequency of 100 Hz at 0.5% duration.

quency of 250 Hz, and the amplitude of the pressure disturbance is much higher due to the longer blockage time. As can be seen in Fig. 13.10, the pressure does not recover to its undisturbed fluctuation, and the frequency of vortex shedding "locks" on the frequency of the pulsation. Needless to say, this behavior results in a catastrophic failure of the meter.

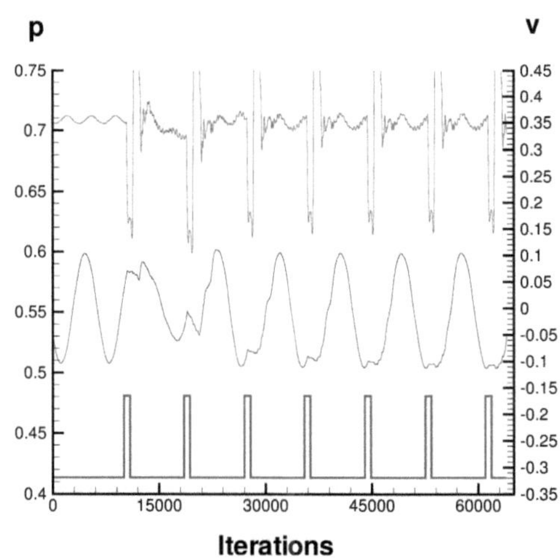

Fig. 13.10. Pressure evolution and simulated signal for pulsation frequency of 200 Hz at 10.0% duration.

The results of the numerical simulation of pulsating inflow are summarized in Fig. 13.11, in which the frequency of the vortex shedding is shown as a function of the iterations in time and, therefore, also as a function of time.

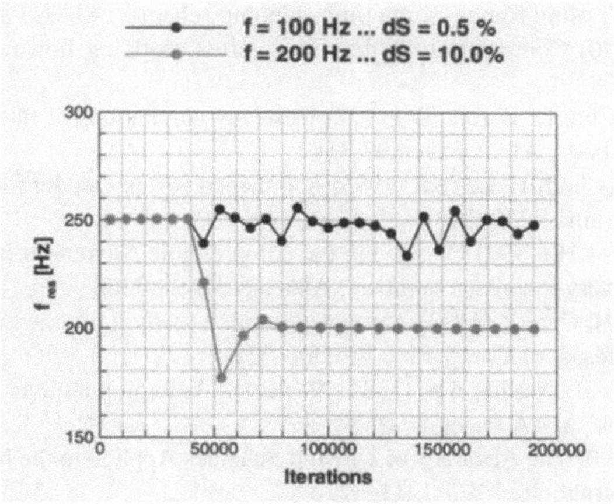

Fig. 13.11. Frequency of vortex shedding with pulsating inflow.

After 50000 iterations, the pulsation was initiated. In the case 100 Hz pulsation signal, shown in Fig. 13.9, the frequency becomes somewhat irregular, but remains basically at 250 Hz. When the pulsation frequency is increased, however, to 200 Hz, and the time of the flow blockage changed to 10% of T, the vortex-shedding frequency jumps to 200 Hz in the apparent "lock-in".

13.4 Conclusions

Several numerical investigations of unsteady and disturbed pipe flows are presented. A verified and validated Navier–Stokes solver for three-dimensional simulations of compressible, viscous and unsteady flow was applied producing results that were in good agreement with the corresponding experimental work. The simulations were useful in determining practical demands concerning the installation of the meter. The DFT analysis showed that the vortex-shedding flow meter will still work properly under disturbed flow conditions.

References

Edwards JR, Roy CJ (1998) Preconditioned multigrid methods for two-dimensional combustion calculations at all speeds. AIAA Journal, 35, No.2:185–192

Fureby C (1995) Large-eddy simulation of turbulent anisochoric flows. AIAA Journal, 33, No.7:1263–1272

Hans V, Poppen G, Lavante E. v., Perpeet S (1997) Interaction between vortices and ultrasonic waves in vortex-shedding flowmeters. FLUCOME '97, Hayama

Hans V, Poppen G, Lavante E. v., Perpeet S (1998) Vortex-shedding flowmeters and ultrasound detection: signal processing and bluff body geometry. Flow Meas Instrum, 9:79–82

Jameson A, Schmidt W, Turkel E (1981) Numerical solutions of the Euler equations by finite volume methods using Runge–Kutta time-stepping schemes. AIAA Paper, 81-1259

Johnson MW (1990) Computation of flow in a vortex-shedding flowmeter. Flow Meas Instrum, 1:201–208

Madabhushi RK, Choi D, Barber TJ (1997) Unsteady simulations of turbulent flow behind a triangular bluff body. AIAA Paper, 97-3182

Miau JJ, Yang CC, Chou JH, Lee KR (1993) A T-shaped vortex shedder for a vortex flowmeter. Flow Meas Instrum, 4:259–267

Pletcher RH, Chen KH (1993) On solving the compressible Navier–Stokes equations for unsteady flows at very low mach numbers. AIAA-paper 93-3368

Shuen JS, Chen KH, Choi Y (1993) A coupled implicit method for chemical non-equilibrium flows at all speeds. J. of Comp. Phy. 106:306–318

Vatsa VN, Thomas JL, Wedan BW (1987) Navier–Stokes computations of prolate spheroids at angle of attack. AIAA Paper, 87-2627

von Lavante E (1990) The Accuracy of Upwind Schemes Applied to the Navier–Stokes Equations. AIAA Journal, 28, No. 7:1211–1212

von Lavante E, Yao J, (1993) Simulation of flow in exhaust manifold of an reciprocating engine. AIAA 24th Fluid Dynamics Conference

Yao J, von Lavante E (1997) Proceedings of 7th International Symposium on CFD:689–694, Beijing, PR China

Hans V, Windorfer H, von Lavante E, Perpeet S (1998) Experimental and numerical optimization of acoustic signals associated with ultrasound measurement of vortex frequencies. FLOMEKO 98, Proceedings:363–367

14

Correction of the Reading of a Flow Meter in Pipe Flow Disturbed by Installation Effects

Carsten Wildemann[1], **Wolfgang Merzkirch**[1], **Klaus Gersten**[2]

[1]Institut für Strömungslehre, Universität Essen, 45117 Essen, Germany
[2]Institut für Thermo- und Fluiddynamik, Ruhr-Universität Bochum, 44780 Bochum, Germany

A method is described that allows the correction of the reading of a flow meter exposed to pipe flow disturbed by installations upstream of the meter. The method is aimed at minimising the distance between installation and meter and is based on characterizing the flow by "fundamental" disturbances, physically interpretable as vortical structures, and on the detection of these disturbances by a measuring device located slightly upstream of the meter. A functional relationship between fundamental disturbances and "error shift" of the meter is postulated, and its existence is demonstrated by a number of experiments. It is concluded that the correction method is applicable to any type of installation and flow meter.

14.1 Introduction

Flow meters are usually calibrated in fully developed pipe flow. For the practical use of a meter, existence of the same state of flow at the position of the meter must be secured, and this requires providing certain lengths of straight pipe upstream and downstream of the meter, as specified in the technical norms. These requirements are often not met in practice; the meter is then exposed to flow that is not fully developed, and measurement errors arise. Deviations from the fully developed flow state in the pipe are caused by installations, e.g., bends, valves, junctions, etc. Two principal approaches are known for using the flow meter with a minimum length of straight pipe between installation and meter, below the lengths prescribed by the norms. One is the use of a flow conditioner that is supposed to accelerate the redevelopment of the flow. The various types of disturbances of the fully developed velocity profile caused by installations disappear at different decay rates downstream of a conditioner; see, e.g., Laws (1990), Xiong et al. (2003), and also Chap. 4 in this book.

The second approach is to calibrate the meter in the presence of the specific installation. This possibility was investigated systematically for orifice meters (Mattingly and Yeh (1994); Reader-Harris et al. (1995)). The velocity distribution downstream of various installations was measured and characterized. The reading of the orifice was then corrected as a function of chararcteristic numbers describing the deviation of the disturbed velocity profile from the fully developed state. While these approaches are restricted to the specific types of installation and meter used in these procedures, we describe here a method of correction that is aimed at being more universal, independent of the knowledge about the specific installations and, in principle, applicable to any kind of flow meter.

Our approach is based on the theoretical results of Gersten and Papenfuss as presented in Chaps. 2 and 3 of this book. It is shown there that any disturbance caused by an installation is composed of a number of "fundamental" disturbances that decay downstream of the installation according to different decay laws. The local (in axial direction) values of the magnitude of the fundamental disturbances characterize the local state of flow, i.e., the deviation from the fully developed state. This is equivalent to describing the state of flow by characteristic numbers, each of them representing one of the fundamental disturbances, see Chaps. 2 and 3 of this book. In contrast to the characteristic numbers as they had been used, e.g., by Mattingly and Yeh (1991) or Mickan et al. (1996), the fundamental disturbances represent specific structures in the flow, i.e., they have a real physical significance; see Chap. 2. The idea is now that the state of the flow, expressed by a set of characteristic numbers, is measured slightly upstream of the flow meter, and that the reading of the meter can be corrected if a relationship between this reading and the set of characteristic numbers is known. Knowledge of this relationship is provided by a calibration: The difference between the discharge coefficient of the meter in fully developed flow, $C_{\mathrm{D}\infty}$ (definition see below), and the discharge coefficient in disturbed flow, C_{D}, at the same value of the volumetric flow rate, is measured for a limited number of disturbances and flow rates. It must be proven that, once the relationship is established from this set of finite data values, the correction can be performed for any kind of disturbances caused by installations.

The relationship between C_{D}, characteristic numbers, and volumetric flow rate (or Reynolds number) must be determined for each type of meter to be used. We apply the procedure to the Venturi and orifice flow meter. The deviation of the flow profile from the fully developed state is determined by measuring the azimuthal distribution of the wall shear stress along a circumference of the inner pipe wall, a short distance ($< 1\,D$) upstream of the flow meter. Using this quantity for characterizing the flow has the advantage that its value, which is known for fully developed flow, can be interpreted in physical terms, and that its measurement is nonintrusive, i.e., not causing an additional pressure drop. In the following we describe the characterization of the flow by the fundamental disturbances as defined in Chap. 2 of this book, the derivation of characteristic numbers by measuring the wall shear stress, and the establishment of the relationship for correcting the reading of the flow meter. The validity of the correction procedure will be demonstrated by a number of experimental applications.

14.2 Characterization of the Disturbed Flow

By means of a theoretical approach Gersten and Papenfuss (Chaps. 2 and 3 in this book) show that the disturbed flow in a pipe of circular cross section can be expressed as the composition of a basic flow pattern and a set of superimposed secondary patterns that are regarded as fundamental disturbances. Each of these disturbances is defined as a specific set of eigenfunctions of the problem and represents a specific flow structure extending in the axial direction, e.g., pure swirl, a pair of two counter-rotating vortices ("secondary flow", see Fig. 14.1a), a quadruple of vortices, etc.. The decay rates of the intensity of these flow structures in the axial direction are different, e.g., the theory confirms the experimentally known fact that the decay rate of pure swirl is the lowest among all fundamental disturbances. Our aim is to characterize the state of the disturbed flow by measuring the wall shear stress along a circumference of the pipe and relating the measured result to the fundamental disturbance structures in Gersten and Papenfuss' theory.

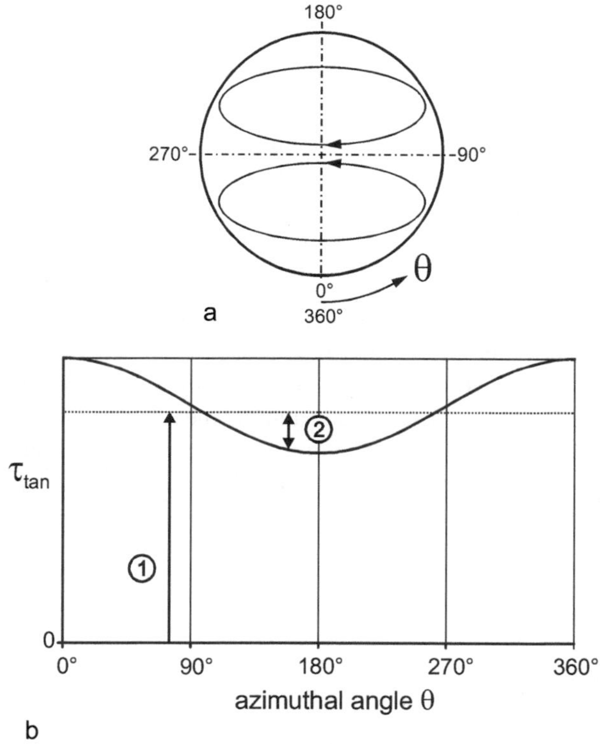

Fig. 14.1. (**a**) Pair of two counter-rotating vortices ("secondary flow") as caused by a 90° single bend and definition of the azimuthal angle with respect to the orientation of the bend. (**b**) Principal distribution of the tangential component of the wall shear stress downstream of the $2 \times 90°$ out-of-plane bend: (**1**) contribution of swirl, (**2**) contribution of superimposed pair of counter-rotating vortices.

The wall shear stress τ has two components, one in the axial and one in the azimuthal or tangential direction, $\tau_{\rm ax}$ and $\tau_{\rm tan}$, that will be measured as functions of the azimuthal angle θ (see Fig. 14.1a). For fully developed flow one has $\tau_{\rm ax} = \tau_\infty = const.$ and $\tau_{\rm tan} = 0$. Figure 14.1b shows the principal distribution of $\tau_{\rm tan}$ for swirl (in the figure designated as 1) plus the superimposed pair of counter-rotating vortices (designated as 2). For the purpose of flow characterization we develop the non-dimensional components of the wall shear stress, T, into Fourier series:

$$T_{\rm ax}(\theta) = \frac{\tau_{\rm ax} - \tau_\infty}{\tau_\infty} = a_{0\,\rm ax} + \sum_{i=1}^{\infty}\left(a_{i\,\rm ax}\cos\left(i\frac{2\pi}{360}\theta\right) + b_{i\,\rm ax}\sin\left(i\frac{2\pi}{360}\theta\right)\right)$$
$$T_{\rm tan}(\theta) = \frac{\tau_{\rm tan}}{\tau_\infty} = a_{0\,\rm tan} + \sum_{i=1}^{\infty}\left(a_{i\,\rm tan}\cos\left(i\frac{2\pi}{360}\theta\right) + b_{i\,\rm tan}\sin\left(i\frac{2\pi}{360}\theta\right)\right) \tag{14.1}$$

It can be shown that the Fourier coefficients a_i, b_i can be associated to the eigen-solutions of Gersten and Papenfuss' theory and the respective fundamental disturbances or flow structures. In particular, the following relationships exist:

Fourier coefficients	flow structure
$a_{0\,\rm ax}$	ring vortex
$a_{0\,\rm tan}$	single axial vortex (pure swirl)
$a_{1\,\rm ax}, b_{1\,\rm ax}$	pair of counter-rotating vortices
$a_{1\,\rm tan}, b_{1\,\rm tan}$	source-sink pair
$a_{2\,\rm ax}, b_{2\,\rm ax}$	quadruple of vortices
$a_{2\,\rm tan}, b_{2\,\rm tan}$	source-sink quadruple

In Chap. 2 of this book Gersten and Papenfuss explain that disturbances associated with higher order coefficients $i > 2$ decay very rapidly and can be disregarded for our further investigations. In the following we shall describe how the first ten Fourier coefficients are determined from the measured distributions $T_{\rm ax}(\theta)$ and $T_{\rm tan}(\theta)$. Important for the envisaged correction procedure is that ten coefficients are in practice sufficient for an efficient correction, i.e., that higher-order coefficients can be neglected.

14.3 Experiments

14.3.1 Flow Facility

The experiments are performed with air flow in a pipe of circular cross section (diameter $D = 100\,\rm mm$) at pipe Reynolds numbers ranging from $Re_{\rm D} = 5 \times 10^4$ to $Re_{\rm D} = 2.5 \times 10^5$. Higher values of the Reynolds number cannot be produced. This flow facility is equipped with a device allowing continuous reference measurements of the volumetric flow rate with an accuracy of $\pm 0.25\%$. This device is based on a one-point measurement of the velocity profile in fully developed flow; for further details of the flow facility see Wildemann (2000) and also Chaps. 4 and 8 in this book.

Three different installations are used for disturbing the flow in the pipe: a 90° single bend (radius of curvature of the centerline 1.5 D), a 2 × 90° out-of-plane double bend (for respective sketches see Chap. 4), and a gate valve, i.e., a circular plate, inserted through a slit from above as a gate, with its plane normal to the pipe axis, thus blocking off a certain percentage of the pipe cross section.

14.3.2 Measurement of Wall Shear Stress

The wall shear stress τ is measured at various axial positions x/D downstream of the installations, ranging from $3 \leq x/D \leq 77$, and in fully developed flow. Sublayer fences (Dengel et al. (1987); Nitsche (1994)) inserted in the inner pipe wall serve for measuring τ. They are calibrated in fully developed flow by making use of the balance between shear force and pressure force. Since it is necessary to determine two components of the wall shear stress (see (14.1)), two fences are used at each

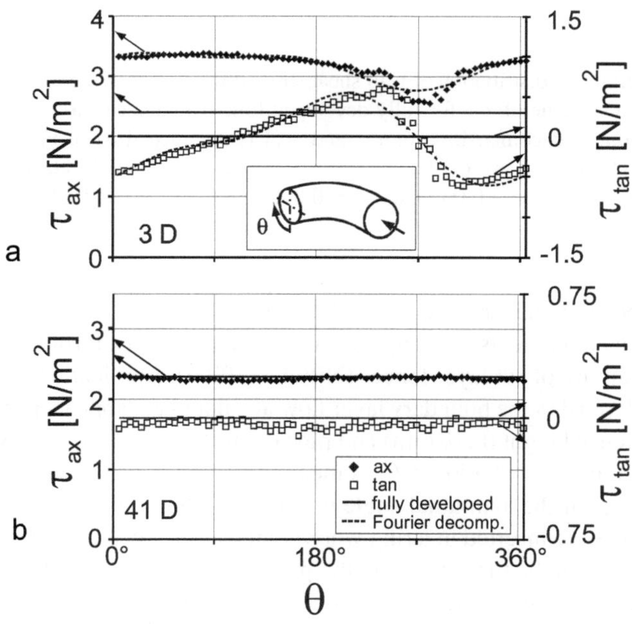

Fig. 14.2. Azimuthal distribution of wall shear stress downstream of the 90° single bend: Coordinate of the axial component, τ_{ax}, is on the left, coordinate of the tangential component, τ_{tan}, on the right. Pipe Reynolds number $Re_{\mathrm{D}} = 2.2 \times 10^5$. The values of τ_{ax} and τ_{tan} for fully developed flow are indicated. (**a**): Axial distance from bend $x/D = 3$, Fourier decomposition of τ_{ax} and τ_{tan} is included; List of Fourier coefficients: $a_{0\,\mathrm{tan}} = -0.0206$, $a_{0\,\mathrm{ax}} = 0.319$, $a_{1\,\mathrm{tan}} = -0.199$, $a_{1\,\mathrm{ax}} = 0.0128$, $b_{1\,\mathrm{tan}} = -0.0080$, $b_{1\,\mathrm{ax}} = 0.112$, $a_{2\,\mathrm{tan}} = -0.0051$, $a_{2\,\mathrm{ax}} = 0.0517$, $b_{2\,\mathrm{tan}} = 0.0781$, $b_{2\,\mathrm{ax}} = -0.0003$. (**b**): $x/D = 41$.

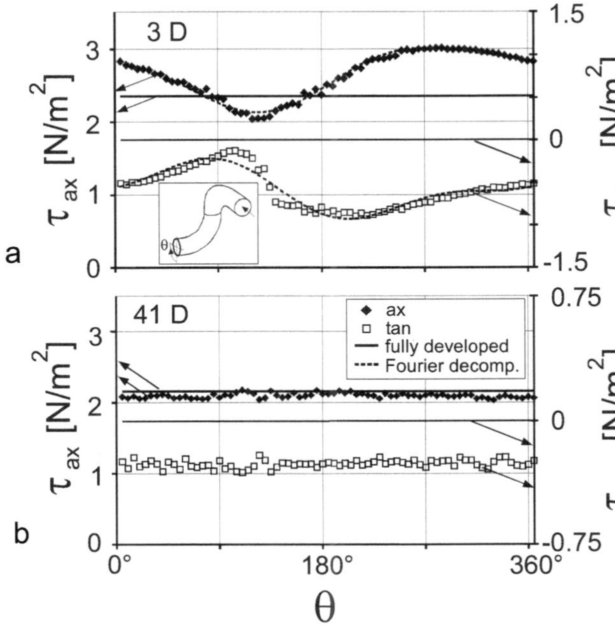

Fig. 14.3. Azimuthal distribution of wall shear stress downstream of the 2 × 90° out-of-plane double bend. The value of τ_{ax} for fully developed flow is indicated. (**a**): $x/D = 3$, $Re_{\mathrm{D}} = 2.2 \times 10^5$. Fourier decomposition of τ_{ax} and τ_{tan} is included; List of Fourier coefficients: $a_{0\,\mathrm{tan}} = -0.246$, $a_{0\,\mathrm{ax}} = 0.122$, $a_{1\,\mathrm{tan}} = 0.0654$, $a_{1\,\mathrm{ax}} = 0.0838$, $b_{1\,\mathrm{tan}} = 0.0974$, $b_{1\,\mathrm{ax}} = -0.157$, $a_{2\,\mathrm{tan}} = -0.0553$, $a_{2\,\mathrm{ax}} = 0.0021$, $b_{2\,\mathrm{tan}} = -0.0145$, $b_{2\,\mathrm{ax}} = 0.0413$. (**b**): $x/D = 41$.

measuring position that are oriented at ±45° with respect to the direction of the pipe axis. The two components, τ_{ax} and τ_{tan}, are calculated with a formula describing the angular sensitivity of sublayer fences (Nitsche (1994)). Application of this formula, which was derived in the boundary layer flow at a flat plate, appears justified because the ratio of fence height (0.15 mm) and pipe diameter (100 mm) is very small.

The azimuthal distribution $\tau(\theta)$ is measured along a circumference of the pipe, i.e., for a range of the azimuthal angle θ from 0° to 360°. For this purpose the two sublayer fences are inserted in a ring whose inner surface is flush with the inner pipe wall and which can be turned around the pipe axis. At stationary flow conditions, τ is measured at angular intervals $\Delta\theta = 5°$, i.e., at 72 equally spaced angular positions θ_i for the whole circumference.

Downstream of the 90° bend and the 2 × 90° out-of-plane double bend, these measurements are performed for the 5 downstream positions $x/D = 3$, 5, 11, 21, and 41, and for the 4 Reynolds numbers $Re_{\mathrm{D}}/10^5 = 0.55$, 1.3, 2.2, and 2.45. From this variety of measured data we show only four typical distributions of $\tau_{\mathrm{ax}}(\theta)$ and $\tau_{\mathrm{tan}}(\theta)$ in Figs. 14.2 and 14.3. Also shown is the Fourier decomposition of τ_{ax} and τ_{tan}; the Fourier coefficients are listed in the figure legend. The different scales for

τ_{ax} on the left and τ_{tan} on the right side of the diagrams should be noted. The distributions of τ_{ax} and τ_{tan} are governed by the existence of the two counter-rotating vortices as shown in Fig. 14.1a,b. Therefore, τ_{tan} changes its sign downstream of the 90° bend (Fig. 14.2a), while this quantity remains always negative downstream of the double bend due to the superimposed negative swirl (Fig. 14.3a). At $x/D = 41$, τ_{ax} approaches closely its value for fully developed flow; this also applies to τ_{tan} downstream of the single bend, i.e., $\tau_{\text{tan}} \to 0$ (Fig. 14.2b), while, for this position, τ_{tan} assumes an almost constant negative value downstream of the double bend (Fig. 14.3b), thus giving evidence of the low decay rate of the swirl generated by the double bend.

Two sketches inserted in Figs. 14.2a and 14.3a, respectively, indicate the spatial orientation of the bends with respect to the flow meter in the straight pipe section downstream of the installations. The four pressure taps for the Venturi and orifice are located at the angular positions $\theta = 0°, 90°, 180°$, and $270°$; the four pressure taps are connected by a hose such that a pressure value averaged along the circumference is measured.

14.3.3 Measurement of Discharge Coefficients

Three different flow meters are exposed to the flow disturbed by the three installations mentioned in Sect. 14.3.1: a Venturi according to ISO 5167-1 with an opening diameter ratio $\beta = 0.7$, and two orifices according to ISO 5167-1 with $\beta = 0.65$ and $\beta = 0.8$. The value $\beta = 0.8$ is beyond the limits set by several norms for practical use of the orifice; this value was chosen here only for demonstrating the method with a meter that is known to have an uncertainty in C_{D} higher than normal. The differential pressure Δp at the flow meters is measured with a transducer (MKS Baratron 698) for which a value of the relative accuracy of $\pm 0.1\%$ is given by the producer, while the air density ρ, necessary for deriving the discharge coefficient, is determined from a measurement of temperature and absolute pressure with a precision of $\pm 0.15\%$. The discharge coefficient C_{D} of the meters, as defined by equation (1) in Mattingly (1983), is then calculated from the measured data Δp, ρ, the volumetric flow rate Q provided by the reference measurement (see Sect. 14.3.1), and the geometry of the meter (β, throat area).

A quantity

$$\Delta C_{\text{D}} = \left(\left(C_{\text{D}} - C_{\text{D}\infty}\right)/C_{\text{D}\infty}\right) 100\% \tag{14.2}$$

in the literature often named “error shift” (Mattingly and Yeh (1991, 1994)), with $C_{\text{D}\infty}$ being the discharge coefficient in fully developed flow, is determined for all configurations of the installations, the indicated distances between flow meter and installation, and the 4 values of the pipe Reynolds number realised in our experiments.

The gate, one of the three installations, was set at three different positions, such that the pipe cross section, which remained free, was 98.7%, 88.2%, and 67.2%, respectively. This is equivalent to the position of the lower edge of the circular plate inserted from above at $0.1D$, $0.3D$, and $0.5D$ (see Fig. 14.4).

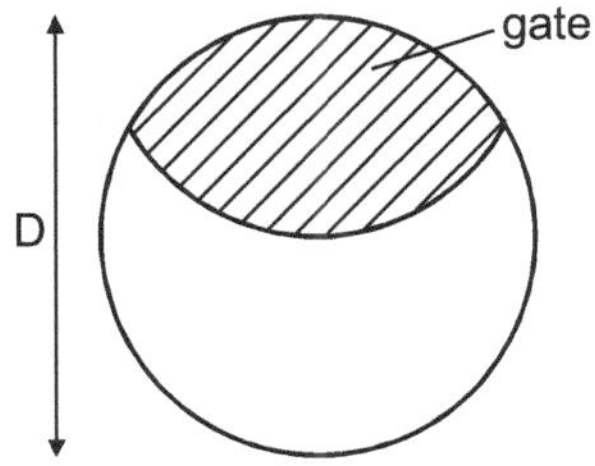

Fig. 14.4. Circular flat plate (gate valve) inserted from above into the pipe of diameter D as an installation.

As typical examples we show the values of ΔC_{D} measured at various axial distances x/D between installation and meter for Venturi (Fig. 14.5) and the $\beta = 0.65$ orifice (Fig. 14.6). Figures 14.5 and 14.6 indicate that the reading of both meters, Venturi and orifice, is less affected by the disturbances caused by the $2 \times 90°$ out-of-plane double bend, i.e., swirl, than by the disturbance due to the presence of a $90°$ single bend, i.e., "secondary flow" due to the counter-rotating vortices. Particularly for the Venturi, the values of ΔC_{D} are of the same order as the uncertainty in the measurement of the volumetric flow rate. It is important to note that the values of

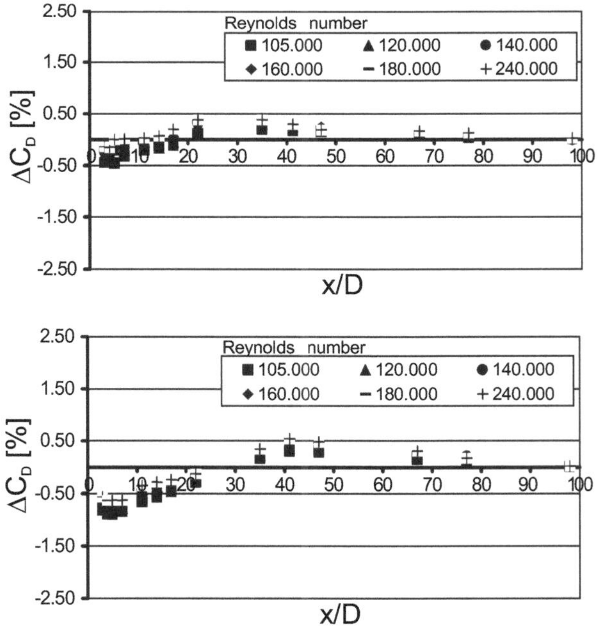

Fig. 14.5. Error shift ΔC_{D} measured for the Venturi meter downstream of the $2 \times 90°$ out-of-plane double bend (*above*) and the $90°$ single bend (*below*) as function of the axial distance x/D between installation and meter (horizontal scale).

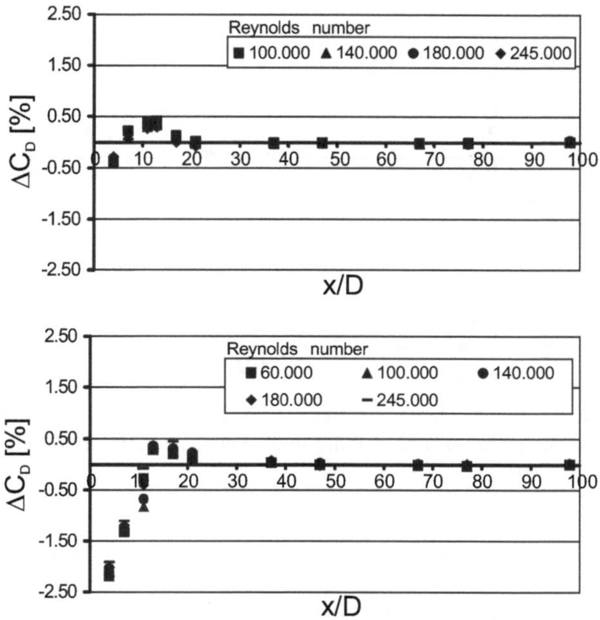

Fig. 14.6. Error shift ΔC_{D} measured for the $\beta = 0.65$ orifice meter downstream of the $2 \times 90°$ out-of-plane double bend (*above*) and the $90°$ single bend (*below*) as function of the axial distance x/D between installation and meter (horizontal scale).

ΔC_{D} shown represent a systematic difference to reproducible values of C_{D}. Here, their absolute numbers are not of any practical relevance, because they just serve to demonstrate the applicability and usefulness of the physical principle of correcting the reading of flow meters.

14.4 Relationship Between Error Shift and Flow Disturbance

According to our earlier assumption, a functional relationship exists, for a given meter, between the error shift ΔC_{D}, measured for the various installations, and the flow profile whose disturbance is characterized by the Fourier coefficients a_i, b_i (14.1). The Fourier coefficients are functions of the distance x/D between installation and meter. This dependence expresses the decay rate of the particular disturbance in axial direction. A further parameter on which a_i, b_i depend is the pipe Reynolds number Re_{D}, in our experiments a measure of the bulk velocity in the pipe or the volumetric flow rate. As an example of the derivation of the Fourier coefficients from the measured distributions $T_{\mathrm{ax}}(\theta)$, $T_{\tan}(\theta)$, Fig. 14.7 shows the dependence of $a_{0\,\tan}$ and the combined coefficient $c_{1\,\tan} = \sqrt{a_{1\,\tan}^2 + b_{1\,\tan}^2}$ on the axial distance x/D for a Reynolds number $Re_{\mathrm{D}} = 1 \times 10^5$; here, $a_{0\,\tan}$ represents the swirl whose low decay rate is evident, while $c_{1\,\tan}$ characterizes the vortex pair as caused by the single

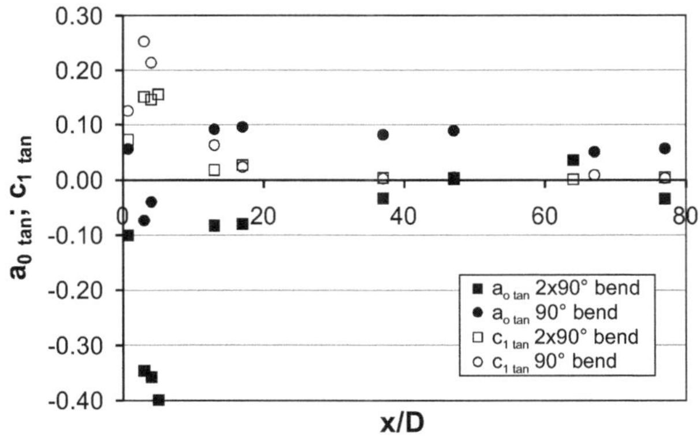

Fig. 14.7. Fourier coefficients a_0, $c_1 = \sqrt{a_1^2 + b_1^2}$ determined at various distances (horizontal scale) downstream of the $2\times90°$ out-of-plane double bend and $90°$ single bend; pipe Reynolds number $Re_{\mathrm{D}} = 1 \times 10^5$.

bend, and it is seen that this disturbance decays much faster in axial direction than the swirl.

The postulated functional relationship between ΔC_{D}, the Fourier coefficients and the Reynolds number is established by means of an artificial neural network. For this purpose we use a "feed forward" type network and the respective "Matlab", version 5.0, software. The network is "trained" according to the Levenberg-Marquardt approximation with a selected, limited set of combinations of Fourier coefficients (up to order 2), related error shifts, and Reynolds number for each of the three meters. These data sets for the training are taken from a certain number of experiments performed with the three flow meters and the two bends, but without using the data measured with the gate valve as an installation. The selection of the limited number of data from the complete data sets as input for the network was performed on the basis of a random choice program. The question for the further investigations is, how accurate can ΔC_{D} be predicted from a measured distribution of $T_{\mathrm{ax}}(\theta)$, $T_{\mathrm{tan}}(\theta)$ for those cases, that were not used in the training of the neural network. For details of the selection and the training of the network, see Wildemann (2000). The neural network is, of course, a substitution for a respective explicit relationship that is not available at this time; the network is easy to implement and it can be used without difficulties.

14.5 Results

The "error shifts" predicted with the artificial neural network, $\Delta C_{\mathrm{D\,ANN}}$, and measured in our experiments, $\Delta C_{\mathrm{D\,exp}}$, are compared in Fig. 14.8 for the case of the Venturi meter. Again $\Delta C_{\mathrm{D\,ANN}}$ is determined via the Fourier coefficients a_i, b_i from the measured distributions $T_{\mathrm{ax}}(\theta)$, $T_{\mathrm{tan}}(\theta)$. The figure evidences which of the data were used for the training of the network and which are actually predicted. The data points shown include all positions x/D and Reynolds numbers as listed above. Ide-

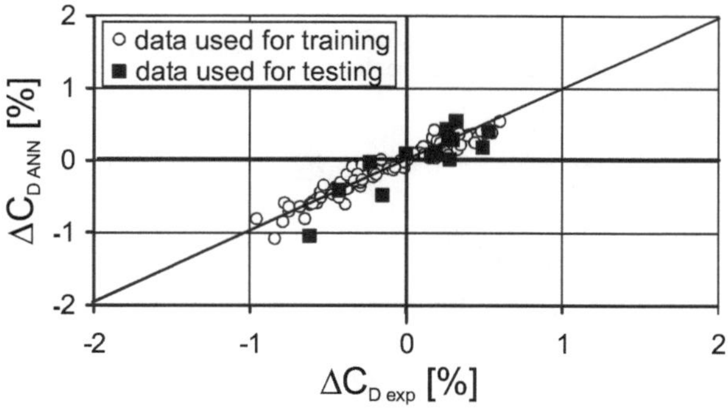

Fig. 14.8. Comparison of measured error shift, $\Delta C_{\mathrm{D\,exp}}$, and error shift predicted by the artificial neural network, $\Delta C_{\mathrm{D\,ANN}}$, for the Venturi meter.

ally, all data points should lie on the oblique straight line through the origin. The scatter which is due to the experimental inaccuracies and the approximations by the neural network indicates the accuracy (or inaccuracy) in the prediction of ΔC_{D}. In order to quantify the precision of the prediciton we form the difference

$$\delta_{\Delta C} = \Delta C_{\mathrm{D\,exp}} - \Delta C_{\mathrm{D\,ANN}} \tag{14.3}$$

whose dimension is %, since ΔC_{D} is expressed in %, too, according to (14.2). Then, a probability density function (PDF) of $\delta_{\Delta C}$ is determined (Fig. 14.9). From the pattern of the PDF and the area it includes one can derive that for 90% of the predictions the difference between the predicted and measured error shift is smaller than $\pm 0.18\%$, while for 100% of the predicitons this difference is smaller than $\pm 0.3\%$. From Figs. 14.8 and 14.9 it follows that the investigated Venturi meter, when being exposed to the pipe flow disturbed by the present installations, measures the volumetric flow rate with an inaccuracy of $-1.0\% \leq \Delta C_{\mathrm{D}} \leq +0.7\%$, and that this uncertainty range is reduced to $-0.3\% \leq \Delta C_{\mathrm{D}} \leq +0.3\%$ with the device for correction described here.

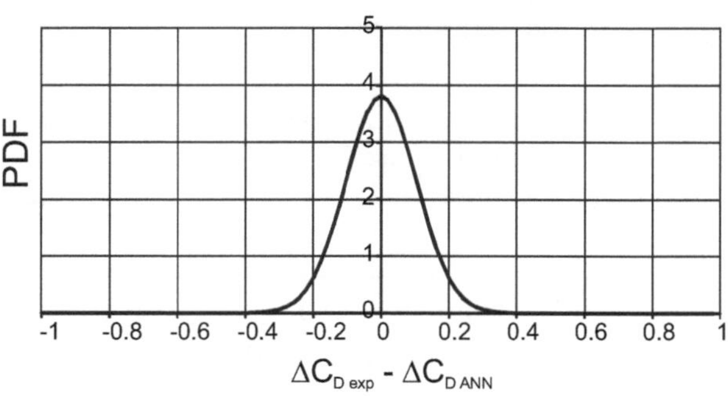

Fig. 14.9. Probability density function (PDF) of the scatter of the data shown in Fig. 14.7.

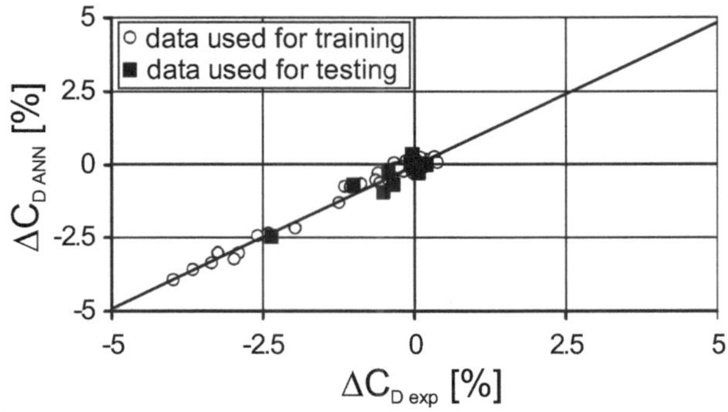

Fig. 14.10. Comparison of measured error shift, $\Delta C_{\mathrm{D\,exp}}$, and error shift predicted by the artificial neural network, $\Delta C_{\mathrm{D\,ANN}}$, for the $\beta = 0.8$ orifice meter.

The respective results for the two orifice meters are presented in Figs. 14.10 and 14.11. The measured error shifts are in most cases negative and their absolute range is larger than that for the Venturi. This confirms the known fact that the Venturi is one of the most robust meters regarding the influence of flow disturbances (see, e.g., Baker (2000)). The orifice with $\beta = 0.8$ that is not defined in the norms has error shifts larger than those for the orifice with $\beta = 0.65$. From the PDF (not shown here) it follows that the uncertainty is reduced by the correction device to $-0.6\% \leq \Delta C_{\mathrm{D}} \leq +0.6\%$, and for the orifice with $\beta = 0.65$ that is designed according to the norms to $-0.35\% \leq \Delta C_{\mathrm{D}} \leq +0.35\%$.

We investigate next how the device and procedure for correction perform when being applied to the flow disturbed by the installation whose data, i.e., the distributions $T_{\mathrm{ax}}(\theta)$ and $T_{\mathrm{tan}}(\theta)$, were not used for "training" the artificial neural network. The installation is the gate sketched in Fig. 14.4 and used with the three different positions of the lower edge of the plate as described above. Also, the distance between installation and meter, x/D, and the Reynolds number are varied as indicated. These

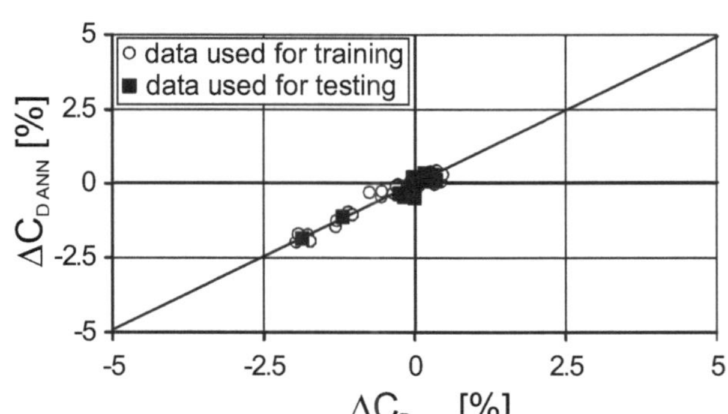

Fig. 14.11. Comparison of measured error shift, $\Delta C_{\mathrm{D\,exp}}$, and error shift predicted by the artificial neural network, $\Delta C_{\mathrm{D\,ANN}}$, for the $\beta = 0.65$ orifice meter.

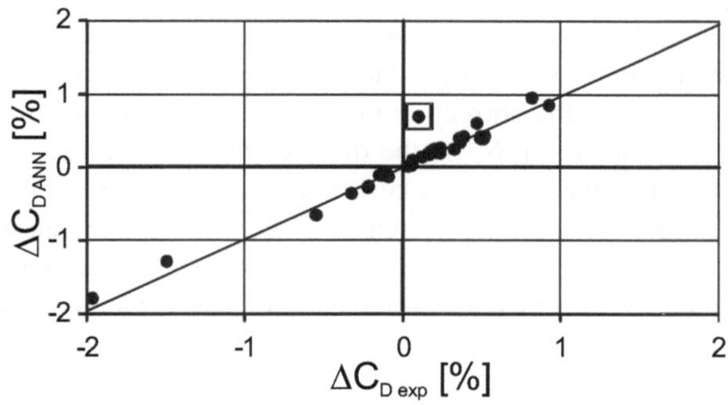

Fig. 14.12. Comparison of measured error shift, $\Delta C_{\mathrm{D\,exp}}$, and error shift predicted by the artificial neural network, $\Delta C_{\mathrm{D\,ANN}}$, for the gate (installation shown in Fig. 14.4). No data measured with this installation were used for the training of the ANN.

experiments are restricted to the Venturi which is, among the three meters tested here, the least sensitive regarding disturbances. That is, the respective ΔC_{D} values for the orifices are expected to be larger than those for the Venturi, thus providing a higher signal amplitude for the correction. We conclude that, if the correction performs well with the Venturi, it should also work with those meters that are more severely affected by the disturbances than the Venturi. The result, presented in Fig. 14.12, is very satisfactory and can be taken as a proof of the applicability of the used principle that any disturbance caused by an installation is composed of a finite number of fundamental disturbances.

In a practical application of this correction method it would be necessary to measure the wall shear stress (or a related quantity) with a set of stationary sensors along a circumference, as done here slightly upstream of the meter, and it is then of interest to minimise the number of sensors. For the results shown above we used the data of 72 measuring positions along the circumference for determining the distributions $T_{\mathrm{ax}}(\theta)$ and $T_{\mathrm{tan}}(\theta)$. In order to provide information regarding the possible minimisa-

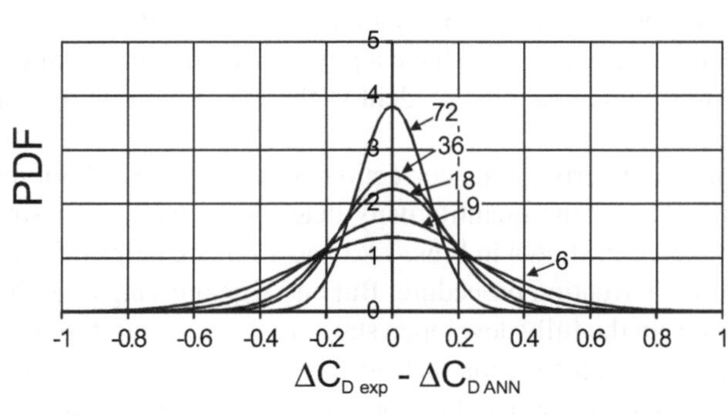

Fig. 14.13. Probability density functions (PDF) as defined and shown in Fig. 14.9 for different numbers of measuring positions used in determining the distributions $T_{\mathrm{ax}}(\theta)$ and $T_{\mathrm{tan}}(\theta)$.

tion of the number of measuring positions or sensors, we have also determined these distributions with data from a reduced number of measuring positions and used the new data for performing the correction procedure. The influence of the number of the measuring positions used, N, on the quality of predicting ΔC_{D} is investigated with the PDFs as they have been explained above. With decreasing number N the amplitude of the PDF's maximum decreases while the width increases, as shown in Fig. 14.13 for the Venturi meter. From this presentation one can easily derive how the accuracy limits within which the corrected error shift ΔC_{D} is predicted vary with N.

14.6 Conclusions

A method is described for correcting the reading of flow meters that are exposed to pipe flow disturbed due to the presence of an installation. The axial distance between installation and flow meter can be minimised to a few pipe diameters; the smallest distance we investigated was 3 pipe diameters. In contrast to flow conditioners that are also used for keeping this distance small, the presented method works without any additional pressure drop.

The principle of the method is based on the theoretically supported assumption that the disturbances caused by installations can be classified into fundamental disturbances, here: specific flow structures, whose linear superposition allows characterization of the disturbed pipe flow (cf. Chaps. 2 and 3). These fundamental disturbances are detected with a device, located slightly upstream of the meter, that, in our case, measures the azimuthal distribution of the two components of the wall shear stress. We postulate that a functional relationship exists between the set of detected characteristic disturbances and the "error shift" of the flow meter. This relationship is established by means of an artificial neural network, and the correction is successfully demonstrated, even for an installation that is, in principle, "unknown" to the network. We have taken this result as proof that the characterization of the disturbed flow by means of the fundamental disturbances is realistic. An advantage of characterizing the disturbed flow velocity profiles in the described way is that the fundamental disturbances used have a physical significance, namely characteristic flow structures, which become evident in the measured distributions of the wall shear stress.

In developing and deriving the correction procedure, we did not make use of the physical significance of the quantitatively determined wall shear stress. The measured distributions of τ, shown in Figs. 14.2 and 14.3, are helpful for the physical interpretation of the correction procedure. But any signal giving quantitative evidence of the deviation from the fully developed state along the circumference could be used for the same purpose. This means that, for a practical application of the correction principle, the sublayer fences can be replaced by other sensors that must only fulfill the condition of having a reproducible sensitivity regarding changes of the wall shear stress. Inexpensive semiconductor sensors of the required type are available that can be considered for a practical realisation. Technical details remain to be investigated for an adaptation of the correction principle to use in practice, e.g., minimisation of the number of sensors along the circumference, optimisation of the neural network and its training, replacing the neural network procedure with an algebraic functional relationship, further testing with a variety of different installations as well as different flow meters. The work presented here can be considered to deliver the physical basis for such developments, see Gersten et al. (2001). Problems of compatibility of the method with rules set by technical norms should only become relevant when investigations of the mentioned technical details have arrived at satisfactory solutions.

References

Baker, RC (2000) Flow Measurement Handbook. Cambridge University Press

Dengel P, Fernholz HH, Hess M (1987) Skin-friction measurements in two- and three-dimensional highly turbulent flows with separation. In: Comte-Bellot G, Mathieu J (Eds): Advances in Turbulence, Springer-Verlag, Berlin/Heidelberg, pp. 470–479

Gersten K, Klika M (1998) The decay of three-dimensional deviations from the fully developed state in laminar pipe flow. In: Rath HJ, Egbers C (Eds.): Advances in Fluid Mechanics and Turbomachinery, Springer-Verlag, Berlin/Heidelberg, pp. 17–28

Gersten K, Merzkirch W, Wildemann C (2001) Verfahren und Vorrichtung zur Korrektur fehlerhafter Messwerte von Durchflussmessgeräten infolge gestörter Zuströmung. Patent No. 197 24 116

Laws EM (1990) Flow conditioning - A new development. Flow Meas Instrum 1:165–170

Mattingly GE (1983) Volume flow measurements. In: Goldstein RJ (Ed.): Fluid Mechanics Measurements, Hemisphere Publ Corp, Washington DC, pp. 245–306

Mattingly GE, Yeh TT (1991) Effects of pipe elbows and tube bundles on selected types of flow meters. Flow Meas Instrum 2:4–13

Mattingly GE, Yeh TT (1994) Pipeflow downstream of a reducer and its effect on flowmeters. Flow Meas Instrum 5:181–187

Mickan B, Wendt G, Kramer R, Dopheide D (1996) Systematic investigation of pipe flows and installation effects using laser Doppler anemometry. Part II: The effect of disturbed flow profiles on turbine gas meters - A describing empirical model. Flow Meas Instrum 7:151–160

Nitsche W (1994) Strömungsmesstechnik. Springer-Verlag, Berlin

Reader-Harris MJ, Sattary JA, Spearman EP (1995) The orifice plate discharge coefficient equation - Further work. Flow Meas Instrum 6:101–114

Wildemann C (2000) Ein System zur automatischen Korrektur der Messabweichung von Durchflussmessgeräten bei gestörter Anströmung. Doctor thesis, Universität Essen. See also: Fortschritt-Berichte VDI, Reihe 8, Nr. 868

Xiong W, Kalkühler K, Merzkirch W (2003) Velocity and turbulence measurements downstream of flow conditioners. Flow Meas Instrum 14:249–260

15

How to Correct the Error Shift of an Ultrasonic Flow Meter Downstream of Installations

Carsten Ruppel, Franz Peters

Institut für Strömungslehre, Universität Essen, 45117 Essen, Germany

A commercial ultrasonic flow meter in combination with a new probe flange is subjected to the disturbed flow downstream of a 90° bend and a double bend, respectively. The probe flange determines the flow structure by swirl-angle measurements. Simultaneously, the error shift of the meter is recorded. A mathematical link is established that relates the flow structure to the error shift. With this link the error shift can be predicted for any position of the meter, which means that the meter can be corrected. It is shown that not only the mean error shift, but also the circumferential error shift of the meter, can be significantly reduced.

15.1 Introduction

The problem of the error shift correction of a flow meter mounted downstream of an installation has been addressed at length in Chap. 14 saving a discussion at this point. However, as the present work builds up on the key results of that chapter we want to recall these as a starting point.

It was shown that the disturbances induced by the installations (secondary flows) can be experimentally identified in terms of the wall shear stress distribution. The modeling of these distributions using Fourier series proved feasible as had been postulated in Chaps. 2 and 3. And considerable corrections of a Venturi and of two orifices could be verified when the distributions were related to the error shifts using a neural network transformation. While these results form the scientific basis of any correction procedure they do not yet concern any application aspects. At present we want to deal with these centered around a commercial ultrasonic flow meter.

The sublayer fence for the wall shear stress measurement as used in Chap. 14 is out of the question for practical application. It is too sensitive to clogging and contact. The encountered low pressure differences require high resolution, i.e., expensive transducers in combination with long calibration times. An array of these sensors along the pipes cirumference, as would be needed for on-line measurements, is unrealistic. In consequence we have developed a new differential pressure probe to meet the following requirements

- rugged design with high signal to noise ratio
- inexpensive pressure transducers
- fast signal generation as a prerequisite for on-line correction
- capability for array measurements
- simple with respect to implementation into real piping systems (flange design)

It no longer measures the wall shear stress but the swirl angle close to the wall, which turns out to be the better choice. Concept and performance of the single sensor has been investigated first (Peters and Ruppel, 2004). Later (Ruppel and Peters, 2004) an array of sensors was integrated into a flange. We are using this flange in combination with a commercial meter with the focus on the correction of the meter. No new technical developement will be reported. As a second accomodation to application we do abandon the neuronal network for two reasons. The inherent mathematical transformation is not transparent such that the influence of parameters cannot be traced. And secondly, for an on-line correction the involved calculation is too complex. A simple expression composed of power functions replaces the neuronal network with success.

At present the pressure probe array is used downstream the installations 90°-bend and double bend in combination with a commercial ultrasonic flow meter. The meter under investigation is of the transit-time type (Baker, 2000) which makes use of the fact that sound is carried by the fluid at an effective speed with respect to the wall which is either the difference or the sum of its own speed and the fluid velocity. The sound is directed across the streaming fluid as a beam such that velocity information is collected along a line (line sensor) in terms of time shift or frequency without being able to reconstruct the velocity profile. Therefore, there is a principle ambiguity in determining the mean transport velocity when the profile is not known. Various designs try to circumvent this problem by multiple paths arrangements which penetrate the stream in more than one direction. Some designs claim (Baker, 2000; Holden and Peters, 1991) to attain measuring uncertainties below 1%. These remarkable achievements together with the crucial benefits of an unblocked cross section and applicability to both gases and liquids have led to a great success with a variety of models on the market. It was important to us to select a widespread commercial meter to be sure that the desired correction is practically relevant. As to the detailled analysis of the meter's error shift the reader is referred to a previous publication (Ruppel and Peters, 2004). The key issue of this work is on-line correction which means that the probe flange is always in place together with the meter. We are neither analyzing the flow nor the meter separately. The combination of both instruments is to be seen as a new instrument.

15.2 Experimental Setup

15.2.1 Test Rig

The experiments were carried out in our multifunctional flow rate test rig a top view of which is sketched in Fig.15.1. A speed controlled blower circulates room

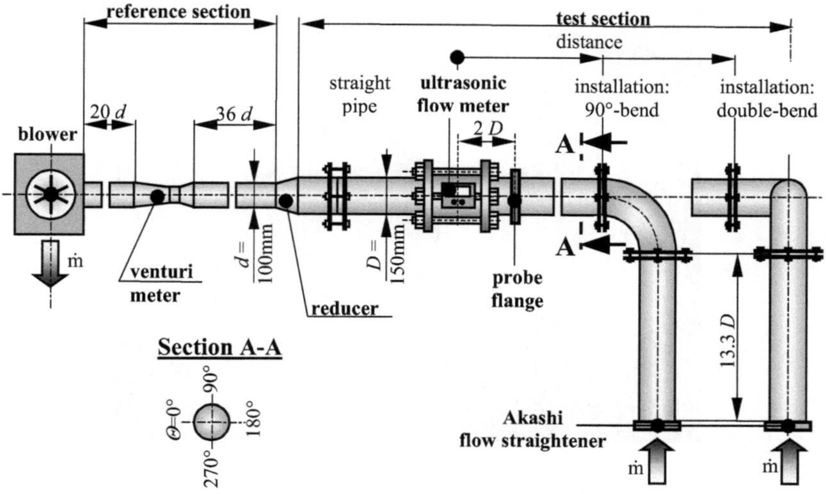

Fig. 15.1. Top view of test rig.

air through the test and reference sections made of precision piping of 150 mm and 100 mm inner diameter, respectively.

The test section is segmented and assembled by flanges to allow variable configurations of pipe lengths and installations including the ultrasonic meter and the pressure probe flange. The reference section incorporates a Venturi at $20d$ upstream of the blower ($d = 100\,\mathrm{mm}$) and $36d$ downstream of a constriction which reduces the wider pipe ($D = 150\,\mathrm{mm}$) to the smaller one. The Venturi was calibrated against another reference based on developed pipe flow which can be placed optionally on the downstream side of the blower (Wildemann et al. , 2002). By means of pressure and temperature measurements in the Venturi the Venturi differential pressure reading is converted into mass flux. The mass flux reproducibility is better than $\pm 0,2\%$ over all tested configurations in the test section upstream of the constriction. This is an important feature because we are looking for error shifts. The absolute measuring uncertainty is less important.

15.2.2 Probe Flange

Figure 15.2 provides a photograhic picture of the probe flange looking downstream. The eight probes protrude radially from the wall into the flow for a few millimeters. They are wedge-shaped with the edge pointing straight upstream. Each probe features two pressure taps symmetrically located on the two sides of the wedge. The two pressures are individually measured against the static pipe pressure amounting to sixteen individual low cost pressure transducers (PXLA 12x5 GN by Sensortechnics). The information drawn from the probe measurement is the flow angle θ close to the wall vs circumferential angle Θ forming a clear picture of the vortex formation behind installations. Eight probes provide sufficient information for the Fourier

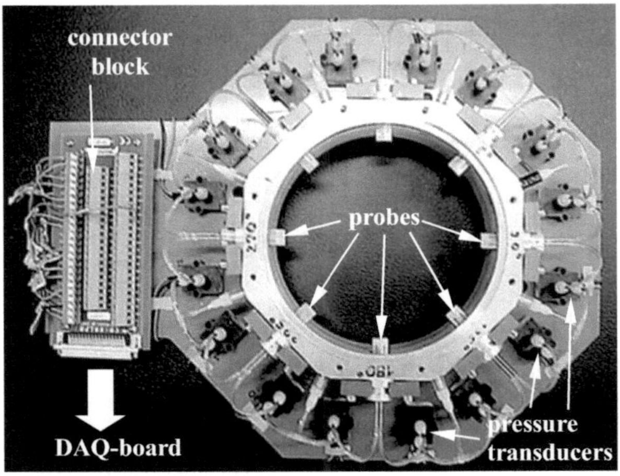

Fig. 15.2. Photography of probe flange.

modeling of the flow angle distribution. All this was shown in Ruppel (2004).The pressure signals were digitized by the A/D board DAQ Pad 6020E of National Instruments at a sampling frequency of $1000\,\mathrm{Hz}$ averaging over $90\,\mathrm{s}$. All data processing was carried out by LabView 6i of National Instruments.

15.2.3 Meter

A commercially available ultrasonic flow meter for gas was used. It is layed out for the pipe diameter of $150\,\mathrm{mm}$ with flange attachements according to DIN 2633. Its design is based on two V paths. The beams are not directed radially and not even in a radial plane. They penetrate a measuring volume in an optimized fashion the background of which is a corporate secret. The meter is sold for the volume flow rate range $14 - -1600\,\mathrm{m}^3/\mathrm{h}$ corresponding to the mass flux range $5 - -533\,\mathrm{g/s}$ under our test conditions. The reading of the meter is in m^3/s. Like with the Venturi we convert to mass flux by measuring pressure and temperature continuously $1D$ downstream of the meter.

The output signal of the meter is a current between $4\,\mathrm{mA}$ to $20\,\mathrm{mA}$ proportional to the flow rate which is processed by a PCI 6024 by National Instruments. Integration time is $90\,\mathrm{s}$ at a sampling frequency of $1000\,\mathrm{Hz}$. Figure 15.3 provides the characteristic curve of the meter with the flange in place in a developed pipe flow($50D$ of straight pipe). A perfect linearity between the output current and the flow rate is observed. From three different test series (see legend) it was calculated that the flow rate can be predicted from the measured current with an uncertainty of less than $\pm 0.15\%$.

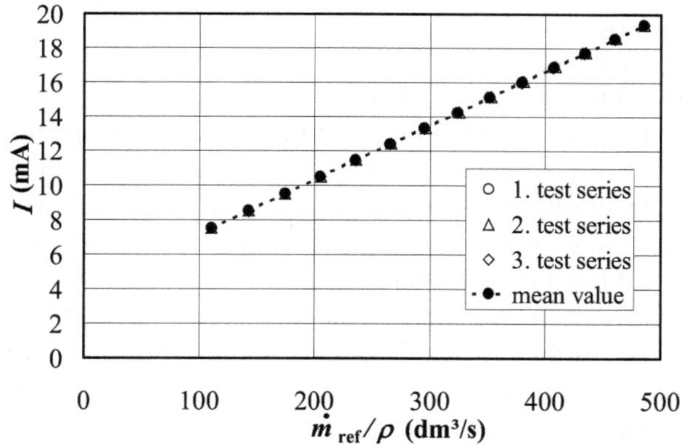

Fig. 15.3. Characteristic curve of the ultrasonic meter with probe flange for developed pipe flow.

15.2.4 Orientation

In all of the present experiments the probe flange and the ultrasonic meter were assembled as a fixed unit. Figure 15.4 gives the relative angular position of probes and meter and also the distance between the two which is $2D$. In this investigation the unit takes a number of different positions relative to the upstream installations in terms of distance and 12 different positions in terms of angle. (The manufacturer recommends a distance of $3D$ to the next upstream installation to keep accuracy specifications. He stipulates no angular orientation of the meter.) As shown in Fig. 15.1 the distance is measured from the meters center plane and the angle is given in terms of Θ_{UZ}. The latter measures from $\Theta = 0$ as defined by the section A-A in Fig. 15.1 up to the meters circumferential reference line pointed out in Fig. 15.4. According to the top view of Fig. 15.1 $\Theta = 0$ refers to the top position while 270° lies on the inside of the bend.

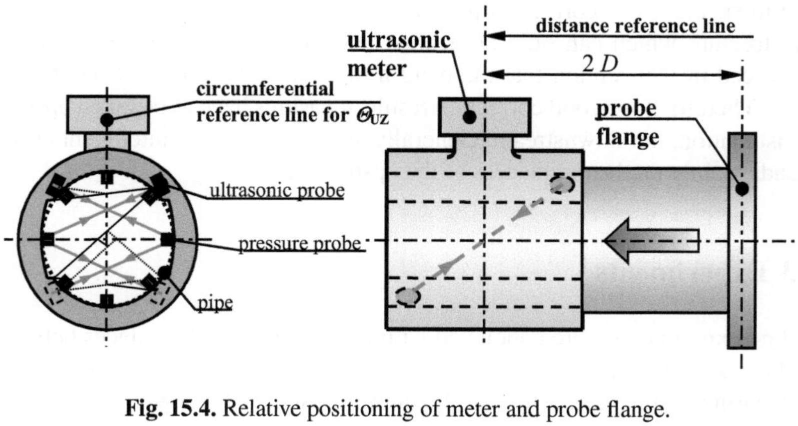

Fig. 15.4. Relative positioning of meter and probe flange.

15.2.5 Installations

The error shifts caused by two typical installations, the 90°-bend and the double bend, are studied. Their radius to diameter ratio is 1.5 following DIN 2605.

The general flow patterns induced by these bends is as follows (more details in Peters and Ruppel (2004)). When axial pipe flow (free of swirl) enters the 90°-bend a counterrotating vortex pair appears at the exit that was already described by Dean (1928). When the bend merges into a straight pipe this vortex pair looses strength along the pipe, yet it maintains its circumferential position. The entire flow disturbance induced by the bend vanishes after roughly $50D$ and developed pipe flow prevails. This was shown experimentally by Schlüter and Merzkirch (1996) and also by Kalkühler (1998).

When axial pipe flow (free of swirl) enters the double bend CFD calculations (Hilgenstock and Ernst, 1996) and measurements (Xiong et al., 2003) show an excentric main vortex with an embedded smaller one. The smaller one disappears rapidly along the pipe while the excentric pattern of the bigger one rotates about the pipe axis with the flow. At a fixed axial position in the pipe the pattern stays put. Decay of the main vortex is rather slow with developed pipe flow restored only after $100D$ (Xiong et al., 2003; Schlüter and Merzkirch, 1996).

In a real piping system the bend will be fed from an upstream pipe or just another bend or installation of some sort which means that the inlet conditions to the bend creating the vortex flow is a disturbed flow itself which is most likely not stable. Any numerical calculation on the bend flow or any experiment has, in principle, to take into account the individual inlet conditions which they normally don't. Because of lacking lab space we cannot extend the inlet pipe to a length that provides undisturbed developed flow. Yet, in order to secure stable inlet conditions, irrespective of the radial profile, we found the Akashi flow straigthener (Akashi et al., 1978) to work well (Fig. 15.1) in contrast to an inlet nozzle which suggests smooth flow but enhances swirl and instabilities.

This aspect of the upstream stability in context with meter correction has not been addressed before by other investigators although it is a very important prerequisite to success. The correction procedure relies on a detectable and reproducible flow structure which can only be achieved by stable upstream conditions. It is a widespread misconception that the bend flow itself is instable and one has to live with it. Therefore, for good correction results measures have to be taken upstream of the installation, not downstream. Generally, the quality of the intelligent correction depends mainly on the uniqueness of the upstream flow.

15.3 Experiments

The first experiments were conducted with the 90° bend at 7 distances between $4D$ and $20D$. The probe flange records the swirl angle θ versus the circumferential angle Θ appearing in Fig. 15.5a for the $4D$ and in Fig. 15.5b for the $12D$ distance. It is an eye-catching feature that the closer distance has the higher amplitudes with

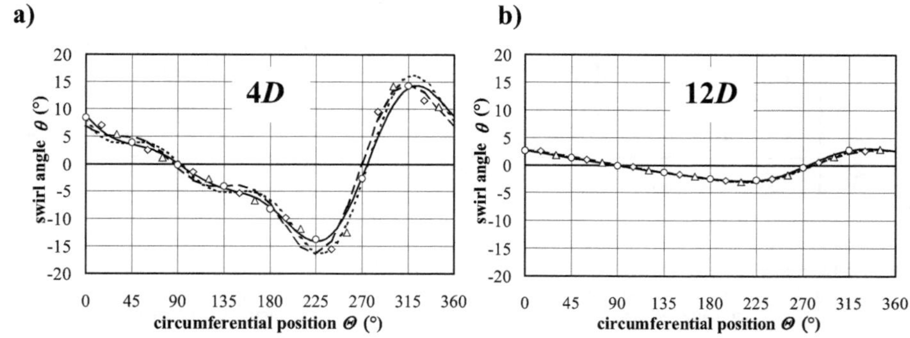

Fig. 15.5. Swirl angle vs circumferential angle. Symbols for measured points (○) $\Theta_{\mathrm{UZ}} = 0°$, (△) $\Theta_{\mathrm{UZ}} = 30°$, (◇) $\Theta_{\mathrm{UZ}} = 60°$ and lines for Fourier series according to (15.1) (—) $\Theta_{\mathrm{UZ}} = 0°$, (- - -) $\Theta_{\mathrm{UZ}} = 30°$, (· · ·) $\Theta_{\mathrm{UZ}} = 60°$. Installation: single bend; Re=200,000.

swirl angles up to $\pm 15°$. This reflects the amplitude decay along the pipe that has been observed before (Ruppel, 2004). Also notable is that the swirl angle profile stays put from distance to distance with zeros at 90° and 270° and extreme values symmetric about the second zero. This agrees perfectly with the vortex pair found behind the bend (see Chap. 14) with the zeros corresponding to opposite stagnation points where the vortices part at 90° and merge at 270°. In this regard an angular turn of the unit should make no difference. In order to see that 12 different positions were tested. Three of them are plotted by different symbols as identified in the legend. The symbols line up nicely on a smooth common curve which is not plotted. The curves that are plotted are approximations of the data to be explained next.

As pointed out earlier the first step of the intelligent correction procedure is the modeling of the upstream flow which is currently characterized by the swirl angle distribution $\theta(\Theta)$. Now, as this distribution oscillates about a mean value and as it returns in itself after 2π a Fourier series is appropriate for modeling (see Chap. 14).

$$F_\theta(\Theta) = \theta(\Theta) = a_0 + \sum_{i=1}^{3} a_i \cdot \sin(i\Theta) + b_i \cdot \cos(i\Theta) \tag{15.1}$$

The series contains 7 coefficients corresponding to the 8 data points provided by the probe flange. With the data already acquired by LabView a best fit calculation of the coefficients was done in one go. Despite some residual sinusoidal oscillations the resulting curves represent the data quite well (a better fit would ask for more coefficients and more probes). The Reynolds number effect is marginal. For the smaller amplitude at the larger distance the deviations shrink correspondingly and the curves seem to coincide.

Along with the swirl angle the error shift ΔF_{D}

$$\Delta F_{\mathrm{D}}\,[\%] = \frac{\dot{m}_{\mathrm{ref}} - \dot{m}_{\mathrm{UZ}}}{\dot{m}_{\mathrm{ref}}}\,100 \tag{15.2}$$

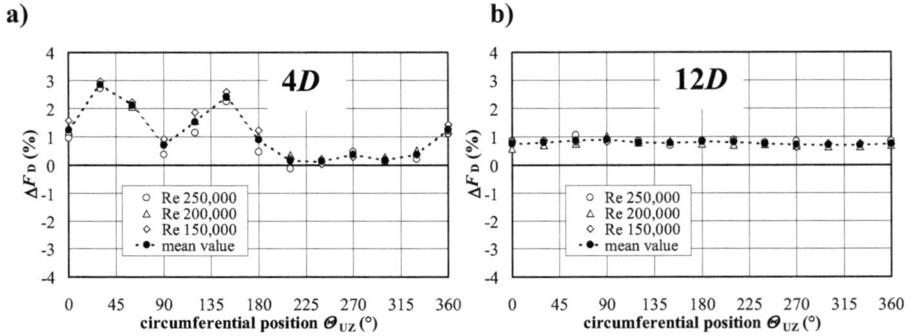

Fig. 15.6. Error shift of the meter vs circumferential angle. Installation: single bend.

of the ultrasonic meter was recorded. It is defined as the mass flux deviation of the meter relative to the reference in %. In Figs. 15.6a and 15.6b we find the results plotted versus the circumferential position of the unit in terms of Θ_{UZ} giving the turn of the meter on the Θ scale (see Fig. 15.1 section A-A). At $4D$ a pronounced error shift profile appears, independent of Reynolds number. Notable three percent are found at two distinct positions qualifying the meter as being generally sensitive to the identified flow structure. However, with the availability of this curve the meter may be set to specific positions around 270° where the error is avoided to the greatest possible extent.

Figures 15.7a and 15.7b display the equivalent results for the double bend installation. Strikingly, the $4D$ profile is similar to the previous one, it just appears shifted downwards to negative swirl angles and along the abscissa. Details of this flow have not yet been explored sufficiently. The general trend, however, can be explained by an additional rotationally symmetric vortex supposedly induced by the second bend. This vortex superposed to the vortex pair of the single bend obviously reduces all

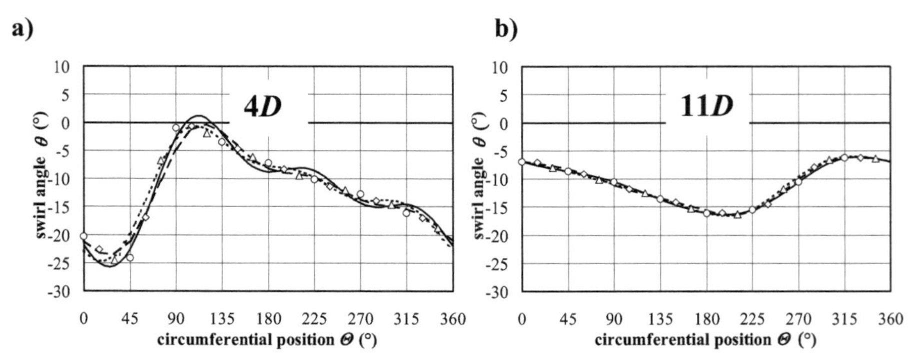

Fig. 15.7. Swirl angle vs circumferential angle. Symbols for measured points (○) $\Theta_{\mathrm{UZ}} = 0°$, (△) $\Theta_{\mathrm{UZ}} = 30°$, (◇) $\Theta_{\mathrm{UZ}} = 60°$ and lines for Fourier series according to (15.1) (—) $\Theta_{\mathrm{UZ}} = 0°$, (- - -) $\Theta_{\mathrm{UZ}} = 30°$, (· · ·) $\Theta_{\mathrm{UZ}} = 60°$. Installation: double bend; Re=200,000.

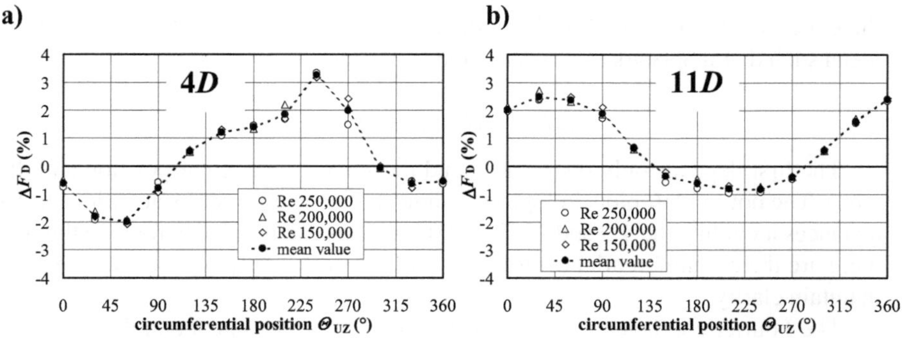

Fig. 15.8. Error shift of the meter vs circumferential angle. Installation: double bend.

swirl angles along the circumference likewise shifting the curve downwards (direction is due to sign convention of course). At the same time the entire vortex pattern starts to revolve about the pipes axis as the pattern moves along. This comes forward when comparing the positions of the $4D$ and the $11D$ profiles with respect to Θ. The amplitude decays with traveling distance but not as strong as in the 90° case, an observation already made by Kalkühler (1998) and Schlüter and Merzkirch (1996) (see also Chap. 4).

The error shifts come out accordingly. With the smaller amplitude decay a considerable sensitivity to angular position remains at $11D$ which had vanished in the single bend case.

15.4 Correction

Now, that we know how to identify the flow downstream of bend and double bend by Fourier coefficients and how to monitor the corresponding error shifts of the ultrasonic flow meter a mathematical link is required between the Fourier coefficients and the error shifts. This link is to be established on basis of the experimental data obtained for bend and double bend. In its final form it should be capable of predicting the measured error shifts from the measured Fourier coefficients. When this has been achieved the elimination of the error shifts (the correction) can be performed for any installation situation in terms of distance and circumferential angle that was not among the ones used for setting up the link. Eventually, it will not be necessary to know which bend of the two is actually installed. The Fourier coefficients provide all the input needed.

Detailing all the necessary equations would go beyond the scope of this paper. The reader is referred to the thesis of Ruppel (2004). Here we resort to explaining the general concept with the help of Fig. 15.9.

As already shown the error shift of the ultrasonic flow meter depends on circumferential position and distance while Reynolds number effects are marginal. It makes sense, therefore, to compose the error shift of a mean value ΔF_{mean} that averages the circumferential values and the circumferential value $\Delta F_{\text{circ.}}$ itself offset by the mean such that it appears as an oscillation about zero.

$$\Delta F = \Delta F_{\text{mean}} + \Delta F_{\text{circ.}} \tag{15.3}$$

The first depends only on distance and the second on angular position and distance. The bottom left plots in Fig. 15.9 show the mean error shifts at various tested distances including the $4D$ and $12D$ $(11D)$ positions exemplified before. To the right there are the circumferential contributions displayed for only two positions each to maintain clarity.

The Fourier coefficients a_0, a_i and b_i (i=1,2,3) of (15.1) contain the full information on the swirl angle distribution. Particular combinations of these coefficients provide measures for the amplitude and the phase of the signal (see Fourier analysis literature). The amplitude is associated with

$$c_0 = \sqrt{a_0^2}\,, \quad c_i = \sqrt{a_i^2 + b_i^2} \tag{15.4}$$

The evaluated c_0 and c_i are plotted vs distance in the top left graphs with the decay clearly visable. The points for each c_0 and c_i correspond to different angular positions of the unit and should lie on a common curve. The residual scatter is due to the limited number of probes (see Fig. 15.5, 15.7).

The coefficients d_i and e_i

$$d_i = \frac{a_i}{c_i}\,, \quad e_i = \frac{b_i}{c_i} \tag{15.5}$$

indicate the angular position of the signal (phase). The top right plots present in terms of these coefficients what was said before: in the 90° case the position of the swirl angle distribution stays put from one distance to the next and in the double bend case the whole pattern rotates about the pipes axis. For clarity only the e_1 and d_2 are plotted at two distances. In principle the 90° plot contains 36 curves and the double bend plot 42 curves.

We now link the upper and lower parts of Fig. 15.9. The modeling of ΔF_{mean} relies on the c_j $(j = 0, 1, 2, 3)$ as both of them are independent of angle. The ansatz for ΔF_{mean} is based on a series of power functions of the form

$$\text{factor}_j \, c_j{}^{\text{exponent}_j} \tag{15.6}$$

in which factor and exponent are determined by data fitting such that ΔF_{mean} in the bottom plots is predictable by the c_j of the upper plots.

The ansatz for $\Delta F_{\text{circ.}}$ is similar. The power functions are now extended by d_i and e_i to account for the angular dependence.

$$\text{factor}_{ji} \, c_j{}^{\text{exponent}_{ji}} \, d_i \qquad \text{and} \qquad \text{factor}_{ji} \, c_j{}^{\text{exponent}_{ji}} \, e_i \tag{15.7}$$

Again factors and exponents come from data fitting. When the two series of power functions are completed the error shift of the ultrasonic flow meter can be

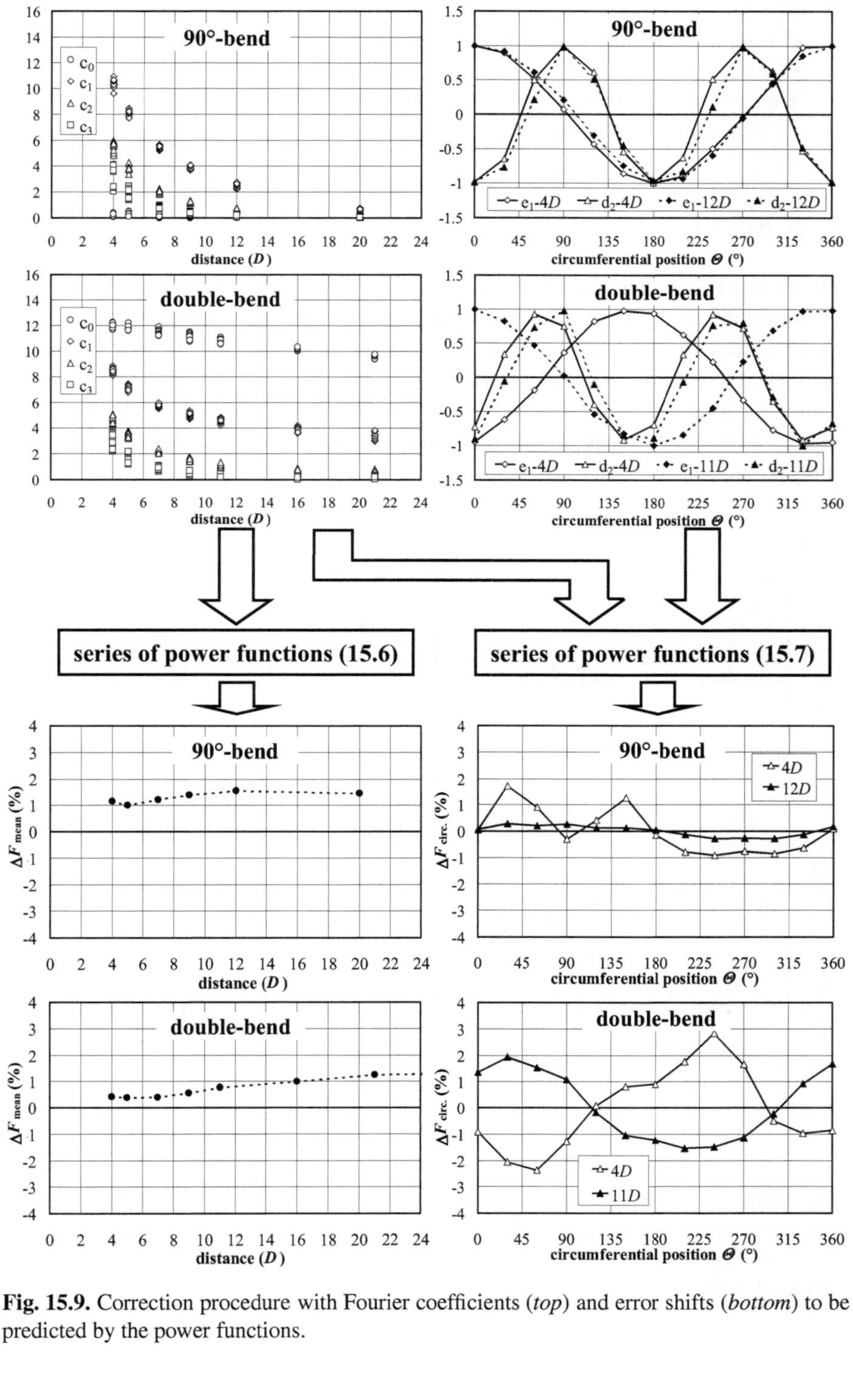

Fig. 15.9. Correction procedure with Fourier coefficients (*top*) and error shifts (*bottom*) to be predicted by the power functions.

predicted, which means it can be corrected. At any installation distance or angle just the Fourier coefficients have to be determined by the probe flange to calculate the correction.

The completed correction procedure is bound to the respective unit. An exchange or modification of the unit requires a new cycle through the whole procedure.

15.5 Results

Figure 15.10 shows the results of correction for the bend (top) and the double bend (bottom) obtained for a set of data measured at the Reynolds number $200,000$. Again the distances $4D$ and $12D$ ($11D$) are selected for demonstration. ΔF is the measured error shift (compare Figs. 15.6, 15.8) while K represents the predicted error shift. R stands for the residual error which is the deviation between the two. The agreement is not perfect but most importantly the residual error is much smaller than the initial one.

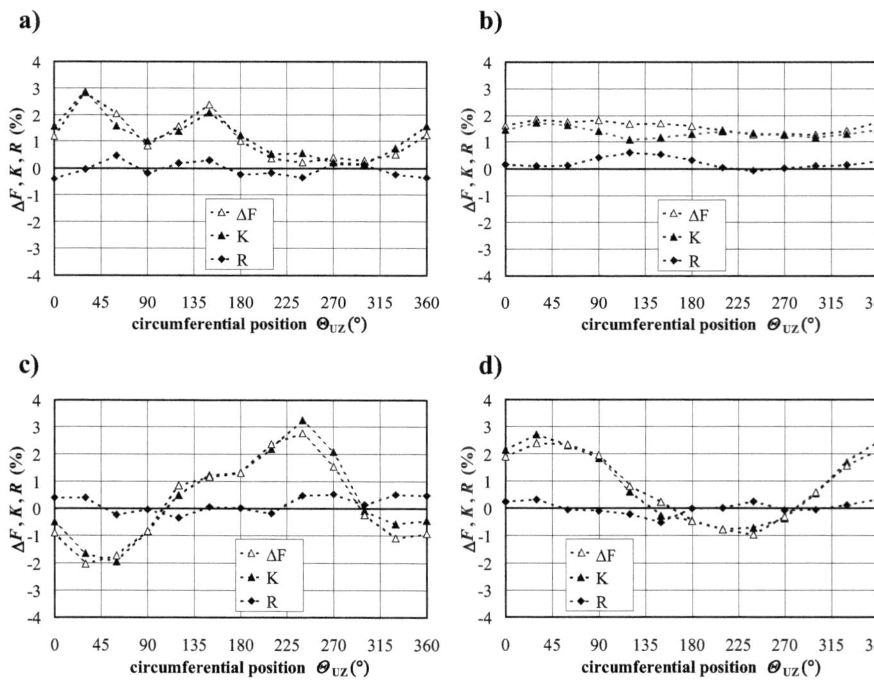

Fig. 15.10. Measured error shifts ΔF, predicted error shift K and residual error R vs circumferential position of the meter Θ_{UZ}. Mounting positions: (**a**) $4D$ downstream the single bend, (**b**) $12D$ downstream the single bend, (**c**) $4D$ downstream the double bend, (**d**) $11D$ downstream the double bend. Re=200,000.

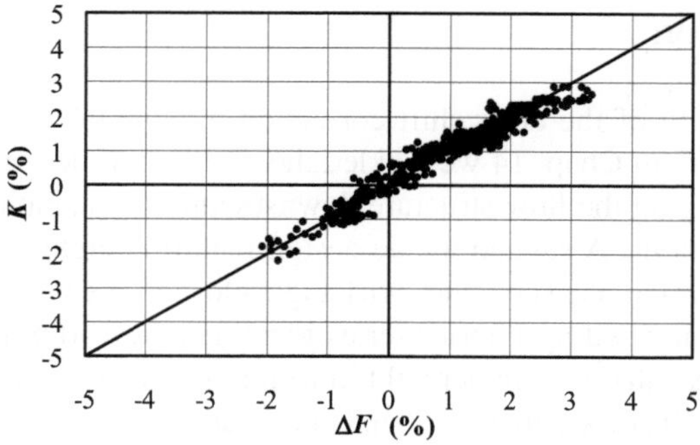

Fig. 15.11. Predicted error shift vs measured error shift. All investigated mounting positions.

Figure 15.11 plots K versus ΔF for all data that where taken at distances from $4D$ to $21D$ at angular positions between $0°$ and $360°$ and different Reynolds numbers from $150,000$ to $250,000$. Out of these data about a third was used to establish the mathematical link. If the correction was perfect all points would lie on the diagonal line. Therefore, the scatter about this line visualizes the achieved degree of correction. In order to quantify the distribution of points about the diagonal a probability density function (PDF) of R is provided in Fig. 15.12. It shows how many points in $\%$ on the ordinate fall into error classes on the abscissa. Noteable 31% of R fall into the $\pm 0.1\%$ class which means that almost a third of all measurements correspond to an error shift of less than $\pm 0.1\%$. The curve drops off rapidly within the 1% class.

The largest R appear between 0.75% and -0.79% at vanishing probabilities. For comparison the original distribution of ΔF data is given. They peak at 1.4% and lie between $+3.34\%$ and -2.09%. In order to characterize the correction by an overall benchmark one may take the average of all absolute ΔF which amounts to 1.26% and compare it with the average of all absolute R which is 0.21%. Then the error shift is reduced by 83%.

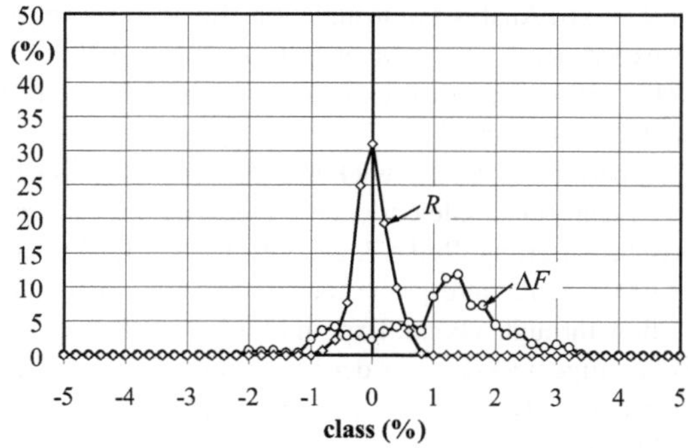

Fig. 15.12. Frequency of occurance vs class of error shift for residual error R and measured error shift ΔF.

15.6 Conclusion

After the feasibility of the error shift correction of flow meters had been demonstrated by the work in Chap. 14 we tackled the problem of designing an instrument capable of identifying the flow structure downstream of installations under realistic application conditions. A rugged flange design featuring eight differential pressure probes emerged which measures the swirl angle close to the pipe wall within $\pm 20°$. The swirl angle modeled by Fourier series identifies the vortex flow structure successfully. A representative commercial ultrasonic flow meter was chosen to be corrected for error shifts downstream of bend and double bend, respectively. The flange was combined with the meter to form a unit mountable at user-defined distances and circumferential positions. A straightforward mathematical link between flow structure and error shift was established with great success.

Thus, as a general result, we achieved to develope a correction procedure applicable to realistic conditions. As to the tested ultrasonic flow meter it turned out that the meter should be corrected not only for distance but also for circumferential position. By the correction performed the reading of the meter could be improved by 83%.

An important fact that we have adressed before and want to stress again is that the success of the correction depends heavily on the stability of the flow. In case of bend flow a simple flow straightener helps (upstream the bend!).

To conclude, we think that there is still potential for further improvements because information beyond the swirl, contained in the axial velocity profile, has not yet been exploited.

References

Akashi K, Watanabe H, Koga K (1978) Flow rate measurement in pipe line with many bends. Mitsubishi Heavy Ind 15:87–96

Baker RC (2000) Flow Measurement Handbook. Cambridge University Press

Dean WR (1928) The streamline motion of fluid in a curved pipe. Phil.Mag.Ser.7(5):673–695

Hilgenstock A, Ernst R (1996) Analysis of installation effects by means of computational fluid dynamics-CFD vs experiments?. Flow Meas.Instrum. 7:161–171

Holden JL, Peters RJW (1991) Practical experiences using ultrasonic flow meters on high pressure gas. Flow Meas. Instrum. 2:69–73

Kalkühler K (1998) Experimente zur Entwicklung der Geschwindigkeitsprofile und Turbulenzgrößen hinter verschiedenen Gleichrichtern. Dissertation, Universität Essen; also published as VDI Fortschritt-Bericht, Reihe 7, Nr. 339, 1998, VDI-Verlag, Düsseldorf

Peters F, Ruppel C (2004) A pressure probe for the detection of the flow direction close to walls. Case study: flow through a bend. Exp. in Fluids 36:813–818

Ruppel C (2004) Die intelligente Korrektur der Fehlerverschiebung eines Ultraschall-Zählers bei gestörter Zuströmung. Dissertation Universität Duisburg-Essen. See also: Shaker-Verlag Aachen ISBN 3-8322-2414-9

Ruppel C, Peters F (2004) Effects of upstream installations on the reading of an ultrasonic flowmeter. Flow Meas. Instrum. 15:167–177

Schlüter T, Merzkirch W (1996) PIV measurements of the time-averaged flow velocity downstream of flow conditioners in a pipeline. Flow Meas. Instrum. 7:173–179

Wildemann C, Merzkirch W, Gersten K (2002) A universal, nonintrusive method for correcting the reading of a flow meter in pipe flow disturbed by installation effects. J Fluids Eng 124:650–656

Xiong W, Kalkühler K, Merzkirch W (2003) Velocity and turbulence measurements downstream of flow conditioners. Flow Meas. Instrum. 14:249–260

Index

MIX
Papier aus verantwortungsvollen Quellen
Paper from responsible sources
FSC® C105338

If you have any concerns about our products,
you can contact us on
ProductSafety@springernature.com

In case Publisher is established outside the EU,
the EU authorized representative is:
Springer Nature Customer Service Center GmbH
Europaplatz 3, 69115 Heidelberg, Germany

Printed by Libri Plureos GmbH
in Hamburg, Germany